Thermal Energy Storage Technologies for Sustainability

Thermal Energy Storage Technologies for Sustainability
Systems Design, Assessment and Applications

by

S. Kalaiselvam
*Department of Mechanical Engineering, Anna University,
Chennai – 600 025, India*

R. Parameshwaran
*Department of Mechanical Engineering, Centre for Nanoscience and
Technology, Anna University, Chennai – 600 025, India*

AMSTERDAM · BOSTON · HEIDELBERG · LONDON
NEW YORK · OXFORD · PARIS · SAN DIEGO
SAN FRANCISCO · SINGAPORE · SYDNEY · TOKYO
Academic Press is an imprint of Elsevier

Academic Press is an imprint of Elsevier
32 Jamestown Road, London NW1 7BY, UK
525 B Street, Suite 1800, San Diego, CA 92101-4495, USA
225 Wyman Street, Waltham, MA 02451, USA
The Boulevard, Langford Lane, Kidlington, Oxford OX5 1GB, UK

Notice

No responsibility is assumed by the publisher for any injury and/or damage to persons or property as a matter of products liability, negligence or otherwise, or from any use or operation of any methods, products, instructions or ideas contained in the material herein. Because of rapid advances in the medical sciences, in particular, independent verification of diagnoses and drug dosages should be made.

Library of Congress Cataloging-in-Publication Data
Kalaiselvam, S.
   Thermal energy storage technologies for sustainability: systems design, assessment, and applications/by S. Kalaiselvam, R. Parameshwaran. – First edition.
      pages cm
   Includes index.
   ISBN 978-0-12-417291-3
1. Heat storage devices. 2. Heat storage. I. Parameshwaran, R. II. Title.
   TJ260.K255 2014
   621.402'8–dc23

                                                                                                    2014011943

British Library Cataloguing in Publication Data
A catalogue record for this book is available from the British Library

For information on all **Academic Press** publications
visit our web site at store.elsevier.com

Printed and bound in USA

ISBN: 978-0-12-417291-3

Contents

Acknowledgments

We wish to express our deep sense of gratitude to all who have helped us in successfully bringing out this edition of our book.

We are deeply indebted to Dr. S. Sivanesan, Professor and Dean, AC Tech Campus, Anna University, for his constant encouragement, timely help, and excellent support.

We express our immense thanks to Chelsea T. Johnston, Editorial Project Manager, Engineering, Elsevier, for the excellent support and encouragement in the development of this book from the inception through the review process to the final outcome. We would like to especially thank the reviewers and Elsevier for the encouragement and recognition given to us in representing a promising field of global interest through this edition of our book.

The support obtained from Anna University, Chennai, India, is gratefully acknowledged.

We are especially grateful to our colleagues and research fellows, D. Madhesh, J. Sandhya, K. R. Suresh Kumar, P. Subburam, and Subin David, for their time and efforts devoted toward the preparation and formatting of tables and references.

We are also deeply thankful to our family, A. Sivakumar and S. Jeyasheela, A. Ameelia Roseline, Harris and Deepti, N. Rajagopalan, and R. Banumathy, for their strong encouragement, love, and understanding throughout this project.

Preface

Energy is considered the lifeline of all human activities and, as such, it has to be conserved at every stage, starting at societal and going to national development. Energy management among the spectrum of sectors of a country can facilitate proper usage of energy based on actual demand. Proper guidance in handling and utilizing energy systems can further maximize energy conservation by the energy producer and the end user.

From this perspective, this book addresses the key energy challenges to be met through gaining knowledge on thermal energy storage (TES) technologies that can lead to a sustainable energy future. This book contains rich information on bridging energy gaps through the incorporation of TES technologies.

The publication of this book at this point in time suggests that the described ideologies and assessments can provide immediate solutions to the current energy market and strategic planning for combining such technologies with real-world engineering systems.

Our primary objective is to deliver a quality work demonstrating the concepts of TES that would be of significant interest to students, researchers, and academics, as well as industrialists, to whom this book can serve as a comprehensive tool for immediate reference while providing information pertaining to the multitude of aspects on TES technologies for sustainable development.

From this perspective, Chapter 1 explains the energy concepts, project energy demand/consumption, and possible energy management techniques that would be helpful for the development of a sustainable future.

In Chapter 2, the significance and functional aspects of a variety of energy storage technologies intended for meeting demand side energy requirements are demonstrated.

TES technologies, on the other hand, offer a wide range of opportunities and benefits to end-use energy and demand side management facilities, primarily in terms of cost effectiveness and energy savings, which are covered in Chapter 3.

Chapter 4 discusses the potential application of sensible TES technologies in residential buildings, in which the implementation of passive and active thermal storages can result in the enhancement of energy efficiency and thermal efficiency.

The nucleus of Chapter 5 is divided into two topics: (1) apposite latent thermal storage materials having excellent thermophysical properties and (2) the potential opportunity for such materials to be effectively integrated into real-time passive and active cooling applications in buildings.

The reversible chemical reactions occurring between working reactants or reactive components help to store and release the required heat energy. In this context, the concepts and inherent operational characteristics of various thermochemical energy storage systems are explained in Chapter 6.

In Chapter 7, the description and the operational characteristics of a variety of sessonal TES technologies are elaborately discussed.

Chapter 8 is exclusively dedicated to nanotechnology-based TES systems. This is an ever-growing and emerging field of interest to a variety of research communities, as well as to industrial professionals worldwide. It is suggested that heat storage materials embedded with nanomaterials exhibit improved TES properties, which enable them to be considered as suitable candidates for future TES applications.

Chapter 9 explains the energy efficiency and cost-energy savings potential of TES systems integrated with conventional and renewable energy systems that can collectively contribute to a reduction in green house gas (GHG) emissions and pave way for the development of a sustainable future.

To enable thermal storage systems to be fully functional, some crucial factors need to be considered during the design phase. From this perspective, Chapter 10 demonstrates the basic design of some sensible and latent thermal energy storage systems with example calculations.

Prior to the implementation of phase change materials into real-time building applications, their operational performance can be effectively analyzed by modeling and simulation methods. In this context, the major attributes of a variety of modeling and simulation approaches are reviewed and presented in Chapter 11.

Chapter 12 presents the exergetic assessment of thermal energy storage systems that reveal positive attributes on reducing the greenhouse gas emissions for the development of a sustainable future with adequate results on energy efficiency of the system.

The significance of providing proper control and optimization schemes in buildings integrated with TES systems to accomplish reduction in the operating cost without compromising energy efficiency are elaborately discussed in Chapter 13.

In Chapter 14, the economic and societal prospects of thermal energy storage technologies are presented. The application potential of thermal energy storage technologies and the scope for futuristic developments are included in Chapter 15.

The authors are grateful to the publisher for permitting the use of research publications cited in this book. The authors would like to thank other sources for the use of information cited in this book. Efforts have been made to cite sources at appropriate locations in the book and to obtain full reproduction rights. Apologies are offered for any discrepancy in identifying the copyright holder, and error correction is welcomed.

Energy and Energy Management

1.1 INTRODUCTION

Energy and energy management are two facets of a mature technology that would move the economic status of a country from normal to the height of societal development. A nation with a strong mission of ensuring energy efficiency at each step of its societal development can sustain higher economic growth on a long-term basis. The increasing concerns about climate change and environmental emissions have led to conserving energy through the development of several energy-efficient systems. The underlying concept behind this is the reduction of extensive utilization of fossil fuel or primary energy sources and their associated carbon emissions. From this perspective, the following sections are designed to explain energy concepts, project energy demand/consumption, and describe possible energy management techniques that would be helpful for the development of a sustainable future.

1.2 ENERGY RESOURCES, ENERGY SOURCES, AND ENERGY PRODUCTION

In the spectrum of energy and energy management, energy resources, energy sources, and energy production are extremely vital starting from their discovery, conversion, and production to end-use consumption. Although the terminologies related to energy resources and energy sources seem to be associated, a basic difference exists that helps the scientific community to move the task of energy production forward to meet energy demand.

Energy resource refers to a reserve of energy, which can be helpful to mankind and society in many ways. On the other hand, energy source also means the system that is devised for extracting energy from the energy resource. For example, the availability of fossil fuels under the earth in the form of coal can be categorized as an energy resource. The system or the technology that is incorporated to extract the energy available from the fossil fuel (coal) can be classified as the energy source.

Earth has large energy resources or basins including solar, hydro, wind, biomass, ocean, and geothermal. Through the application of the human ideologies and emerging technologies, tapping the energy from these reserves in an efficient manner has always been a paramount task. Earth's finite and renewable energy reserves along with recoverable energy from these resources are depicted in Fig. 1.1.

It is not only important that the energy be extracted from these reserves or reservoirs; the real success of the task depends on efficient transformation to the actual societal requirements. In simpler words, the extracted energy has to be generated or produced in a more usable form and has to be transported so that it caters to end-user energy demand. To sustain the living standards in developed nations as well as improve societal and economical status in developing countries, it is of great importance to balance the huge gap between energy generation and consumption.

Thermal Energy Storage Technologies for Sustainability

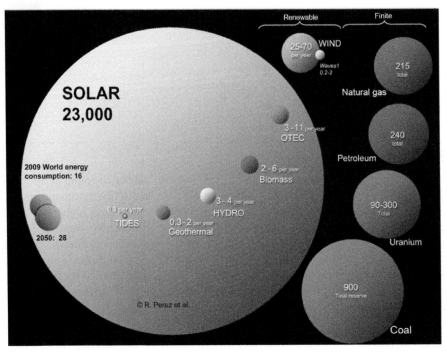

FIGURE 1.1

Finite and renewable planetary energy reserves (Terawatt-years, TWy). Total recoverable reserves are shown for the finite resources. Yearly potential is shown for the renewable [1].

The availability of reserves and the possible recovery of energy projected in Fig. 1.1 are more attractive and helpful. This is a basic step in the process of energy planning and energy management. It can be seen clearly from Fig. 1.1 that the total energy reserves available for the fossil fuel category account for nearly 2000 TW per year (TW—Terawatt). The reserves available for nuclear energy are comparatively less compared to fossil fuel reserves. The ratio of fossil fuel reserves to production globally at the end of 2012 is shown in Fig. 1.2.

The projected ratio of fossil fuel reserves to their production in Fig. 1.2 infers that the reserves for coal, oil, and natural gas in some parts of the world have increased over time. This could be attributed to emerging technological advancement in the search for new fossil fuel reserves or beds. It can also be seen clearly from Fig. 1.1 that energy recovery from nuclear energy can now help fulfill immediate energy needs. However, from the long-term energy perspective, dependence on nuclear fuels imposes certain environmental risk factors and unsafe conditions in terms of nuclear emissions and radioactive decay.

It is interesting to note that after the Industrial Revolution, human inventions (interventions) for using fossil fuels to satisfy the energy demand increasingly grew from region to region worldwide.

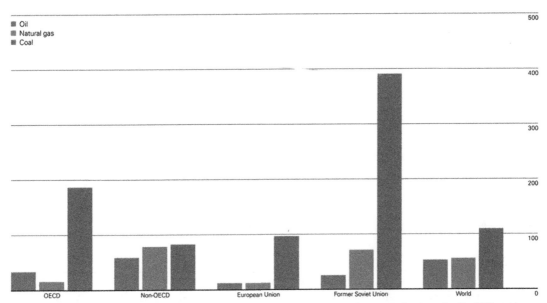

Coal remains the most abundant fossil fuel by global R/P ratio, although global oil and natural gas reserves have increased significantly over time. Non-OECD countries possess the majority of proved reserves for all fossil fuels and have a higher R/P ratio than the OECD countries for oil and natural gas.

FIGURE 1.2

Global projections on the ratio of the fossil fuel reserves to production at the end of 2012 [2].

The values projected in Fig. 1.3 infer continuous growth in fossil fuel-based primary energy sources in recent years as well as in the near future.

Tough competition exists between the world's nations in the search for new reserves of oil, natural gas, and coal. This process is even more encouraged in developed countries. This is in some ways advantageous, but the uncontrollable exploitation of such energy reserves leads to carbon emissions and other environmental risk factors.

The projections on the additions of world power generation capacity and retirements from 2013-2035 shown in Fig. 1.4 infer that the participation of developing nations including India and China is considerable. This means that developing countries are more interested in resolving issues related to energy usage per person, as compared to developed nations. Nearly 40% of the world's new power-generation capacities is being made by India and the China. At the same time, almost 60% of the power capacity additions have contributed for the replacements of retired plants in the Organization for Economic Co-operation and Development (OECD) countries.

On the other hand, developed nations are also equally interested in developing renewable energy sources-based systems for accomplishing demand-side energy management. However, in this type of task, aside from the cost implications involved, adding renewable energy as the source for power generation (electricity production) as depicted in Fig. 1.5 would facilitate maximum energy advantage with reduced or net zero emissions to the environment.

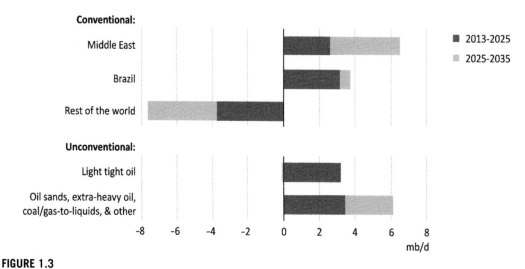

FIGURE 1.3

Projections on global oil production growth contributors [3].

World Energy Outlook 2013 Launch – a presentation by Maria van der Hoeven in London © OECD/IEA, 2013, page 8.

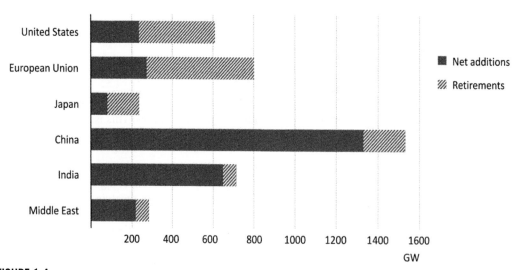

FIGURE 1.4

Projections on additions of power generation capacity and retirements from 2013–2035 [3].

World Energy Outlook 2013 Launch – a presentation by Maria van der Hoeven in London © OECD/IEA, 2013, page 10.

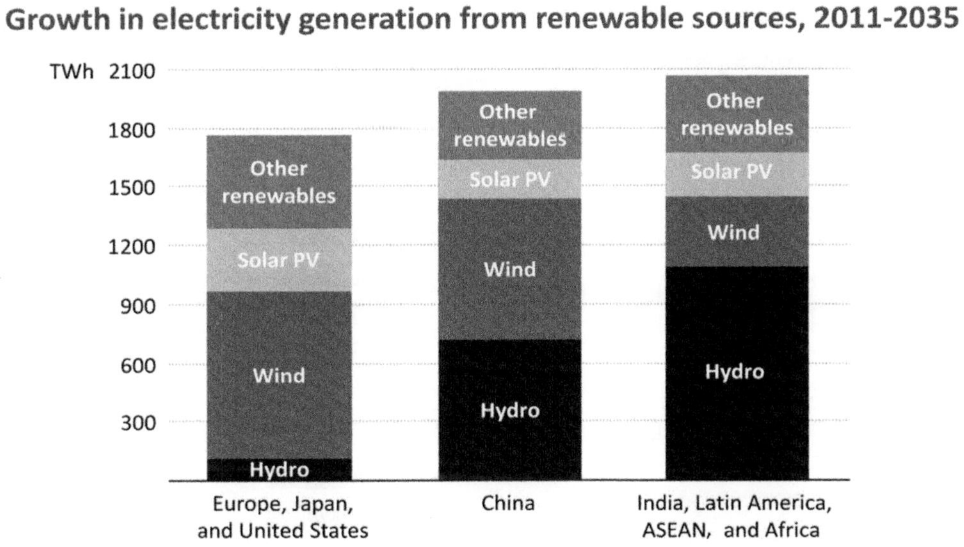

Growth in electricity generation from renewable sources, 2011-2035

FIGURE 1.5

Electricity generation growth using renewable energy sources from 2011–2035 [3].

World Energy Outlook 2013 Launch – a presentation by Maria van der Hoeven in London © OECD/IEA, 2013, pages 11.

1.3 GLOBAL ENERGY DEMAND AND CONSUMPTION

The statistical references from the BP Statistical Review of World Energy [2], International Energy Agency IEA [4] and International Energy Outlook EIA [6] indicate an increase in energy demand and world marketed energy consumption among world nations as shown in Fig. 1.6(a–e). The projections in Fig. 1.6(a–e) show that the energy demand arising from coal and oil has been reduced significantly for the OECD countries. This may be due to the fact that OECD nations have shown greater interest toward using renewable and other nonfossil fuel–based energy sources for balancing and satisfying their energy production and demand.

Fig. 1.6(d) shows that world market energy consumption (WMEC) has been consistently increasing by 1.4% every year since 2007. In total, the WMEC has increased up to 49%, indicating that the imbalance between energy production and consumption has reached its limit. Based on Fig. 1.6(e), the share of the world energy consumption for the United States and China tends to reduce and increase, respectively, in future years, whereas for India the share of energy consumption may rise only marginally. In many research contributions addressing the issues related to consumption of global and total energy, carbon dioxide (CO_2) emissions were reported, and possible ways to reduce their effects especially as applied to buildings have been suggested [6].

Energy production and consumption play a vital role in deciding energy conservation at every step of economic development worldwide. Global total energy consumption (GTEC) can be

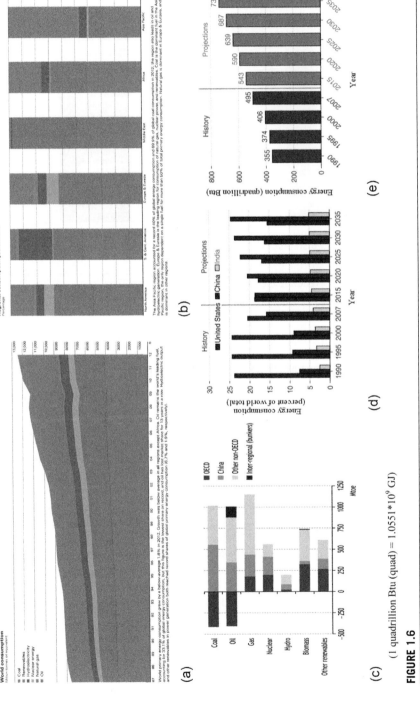

FIGURE 1.6

Energy demand and world marketed energy consumption [2,4,6].

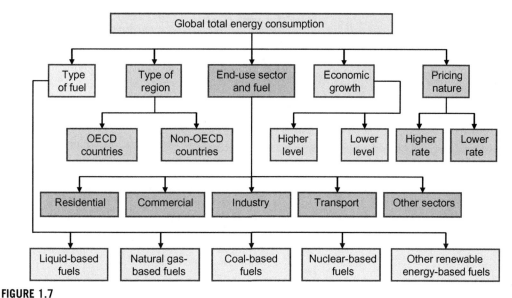

FIGURE 1.7

Classification of global total energy consumption [5].

broadly categorized as depicted in Fig. 1.7, and that has a direct influence on economical and environmental development. The classification of GTEC actually infers that energy consumption is an interconnected key element that has to be carefully dealt with while addressing an energy conservative and efficient design of systems and environment. Furthermore, the extent of energy consumed by the end-user is as important as other categories because it forms the baseline for generating, distributing, and conserving overall energy, which in turn would facilitate improving the economical status.

In addition to the predictions given for world energy consumption based on the statistical information provided by esteemed international organizations, a continuous impetus has been shown toward the prognostic approach in a different manner. One such type of energy prediction being made is presented in Fig. 1.8, which is primarily based on population projections [7,8]. Global energy demand/consumption has been formulated on the assumption that the average energy consumed per person would be 250 gigajoules (GJ) after the year 2100. At the same time, the energy demand/consumption has been categorized under the fossil, transition, and postfossil eras.

It is pertinent to note that world energy demand/consumption follows an increasing trend with respect to the world population over a period of years. As per the energy projections [7,8] in the fossil era, the dominion of the fossil fuels has shown a great impact on the economic status of the world's nations. This is why the fossil era has been designated for the years 1950–2010 [9]. It is framed so that a replacement for fossil fuels may be found in the forthcoming years through the development of alternative energy or fuels.

In the transition era, most of the reserves related to fossil fuel supplies and nuclear fuels may be fully exhausted at the end or beginning of the twenty-second century [10–12]. The status of fossil fuel

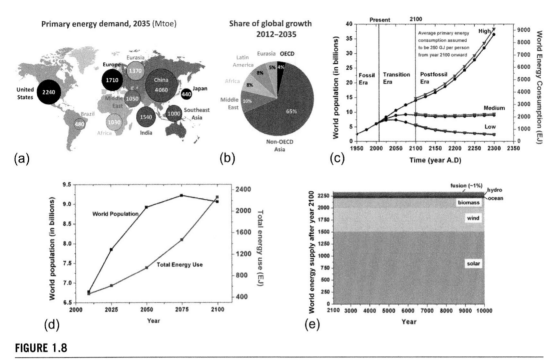

FIGURE 1.8

Global energy demand/consumption based on world population [3,9].

Parts (a) and (b), World Energy Outlook 2013 Launch – a presentation by Maria van der Hoeven in London © OECD/IEA, 2013, page 3.

reserves may be questionable during the transition era, but nuclear fission can be expected to sustain itself for quite some time after this era. This depends on nuclear breeder reactors safely handling the controlled fission processes in all aspects (technical, economical, and societal) without major catastrophe.

The postfossil era will provide an opportunity for the two major types of energy sources, which are sun-dependent and sun-independent energy sources. Energy resources such as solar, hydro, wind, ocean, and biomass represent the sun-independent type, whereas nuclear fusion and the geothermal resources are examples of the sun-independent type. Fig. 1.8 shows that the low and medium scenarios of world population tend to consume energy in a quite consistent manner. However, the high projection of population growth reaches the upper limits of energy demand/consumption after the year 2100.

The significant part of projections made for the postfossil era is that calculations have been made based on the medium population scenario, which is believed to range between 9 and 10 billion during this era. That is, with the advanced economic options available in this era, there would not be any problem meeting day-to-day energy demand/consumption requirements. Furthermore, the sun-dependent energy sources, especially solar energy, can be considered a viable solution that would contribute 70% of the world's total energy demand of ~2300 exajoules (EJ)/year beyond 2100.

As the second biggest energy source after solar energy, wind can satisfy energy demand requirements of around 640 EJ/year. Likewise, the technical potential of biomass, hydro, and ocean energy sources can account for 276, 50, and 74 EJ/year (with an assumption of 1% theoretical potential conservation of 7400 EJ/year), respectively [10]. The period during which these energy sources would prevail can be estimated to range from 5 to 10 billion years. The theoretical and technical potential, including the availability times for sun-dependent energy sources, are presented in Table 1.1.

Table 1.1 SDES—Theoretical and Technical Potential and Availability [9,10]

Energy Resource	Theoretical Potential (EJ/Year)	Technical Potential (EJ/Year)	Availability (Billion Years)
Hydro	146	50	5
Solar	3.9×10^6	1575	8–10
Wind	6000	640	5–7
Biomass	2900	276	5
Ocean	7400	74 (conservative 1% of theoretical potential)	5
Total	3.92×10^6	2615	5–10

Table 1.2 SIES—Theoretical and Technical Potential and Availability [13,14]

Energy Resource	Theoretical Potential (EJ/Year)	Technical Potential (EJ/Year)	Availability (Billion Years)
Geothermal	1.4×10^8 (from a depth of up to 3–10 km)	2.8×10^7	~12,000 years (with 100% share of total requirement of 2300 EJ/year
Fusion	1.564×10^{13}	1.564×10^{11} (conservative 1% of theoretical potential)	~68 million years (with 100% share of total requirement of 2300 EJ/year
		1.564×10^{12} (10% of theoretical potential)	~680 million years (with 100% share of total requirement of 2300 EJ/year
		7.82×10^{12} (50% of theoretical potential)	~3.4 billion years (with 100% share of total requirement of 2300 EJ/year

Likewise, the theoretical and the technical potential including the availability time for the sun-independent energy sources are summarized in Table 1.2. For the scenario of sun-independent energy sources, nuclear and geothermal reserves can have a future in the twenty-second century. Nuclear energy, including fission and fusion energy, can take part in solving the world's future energy demand/consumption. As pointed out earlier, the success rate of nuclear reactors (fission- and/or fusion based) is largely dependent on socioeconomic perspectives apart from the technical point of view. Even though the energy being extracted from nuclear fuel is technically feasible, the radioactive half-life and associated nuclear emissions cannot be completely removed.

The decommissioning of current nuclear plants in some developed countries can be considered a positive effort from the standpoint of environmentalists. This is because the availability of nuclear fuel (particularly uranium) could last to the end of the transition era (up to 80-90 years) based on the predictions given [9–12]. Thus, it is a smarter move to not depend on nuclear energy mostly, but to focus more on the development of new and renewable energy sources in the near future.

At the same time, the resource potential of geothermal energy can have a significant thrust on world energy demand/consumption criteria beyond 2100, as per predictions. The availability of geothermal energy sources can extend up to 12,000 years, provided the total energy requirement (100%) of the world (~2300 EJ/year) can be met using this energy source. On the other hand, fusion energy can also equally compete to counteract energy challenges in the near future. That is, the total potential of fusion energy is estimated to be 1.564×10^{13} EJ, theoretically [9].

Suppose the total energy requirement, as mentioned earlier, has to be met by fusion energy, the existence of which can be expected to be 68 million years (approx.) with an assumption of 1% technical potential of the theoretical potential. In case the technical potential is increased by 10% and 50% of the theoretical potential, then the expectancy of the fusion energy can be calculated approximately to ~680 million years and ~3.4 billion years, respectively.

From the projections given in Fig. 1.8 for the world energy supply after 2100 with respect to future years, it can be observed that the paramount of solar energy sources would be the scenario to meet world energy demand/consumption. Following this is wind energy and then biomass energy sources. Apart from these major energy sources, the contribution of ocean, hydro, and fusion energies can also be relevant from the perspective of energy efficiency. As mentioned earlier, the utilization of geothermal energy sources can have a great impact beyond 2100. However, the long-term energy storage (seasonal thermal energy storage) option using the geothermal energy source may be intermittent in nature.

The breakdown of primary energy demand and consumption as shown in Fig. 1.9(a) points out that the sectors of transport, industry, and residential (buildings) consume the majority of the energy that is generated worldwide. Of these sectors, residences or buildings have been indentified

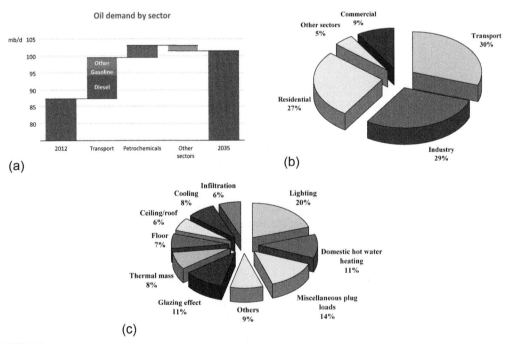

FIGURE 1.9

(a) Oil demand (b) the energy consumption breakdown by sector [3,5,15,16], and (c) relative average breakdown of end-energy usage and losses in buildings.

Part (a), World Energy Outlook 2013 Launch – a presentation by Maria van der Hoeven in London © OECD/IEA, 2013, page 6.

Table 1.3 Estimated Annual Primary Energy Consumption Worldwide by Buildings [17]	
Year	**Energy Consumption (Quads/Year)**
2004	72.2
2010	82.2
2015	90.7
2020	97.3
2025	103.3
2030	109.7

as the most important sector requiring incessant energy to function. For instance, a majority of people spend their time mostly in buildings (any type of work), which essentially consumes energy (minimum or maximum), depending on the utility services provided. The relative average breakdown of the end-energy usage and associated losses are graphically represented in Fig. 1.9(b)

Looking from the perspective of the economic status of a country, the value-added construction sector plays a vital role and becomes a decision-making arena for future market investments. This can be an acceptable factor for the economic growth on one side, but on the other side the primary energy consumption might be huge and challenging as well. The statistical report of the EIA (2007) suggests that buildings being constructed in developed countries would account for 40% of primary energy consumption, 70% of electrical energy utilization, and 40% of the greenhouse gas (GHG) emissions. Generically, buildings consume one-third to one-quarter of the overall energy being generated worldwide. The estimated annual primary energy consumption by buildings worldwide is presented in Table 1.3.

1.4 NEED FOR THE ENERGY EFFICIENCY, ENERGY CONSERVATION, AND MANAGEMENT

Energy efficiency and energy conservation are closely related, and their management can yield significant improvement in the performance of energy systems on a long-term basis. Basically, *energy efficiency* refers to the ability to achieve the required outcomes from the systems or modules with less energy expended. In other words, *energy efficiency* can be achieved in a process or a product through the reduced consumption of energy without influencing the specific output. Energy conservation is termed as the measure being taken to minimize energy consumption, which varies in magnitude depending on the type of applications or processes. Thus, by ensuring energy efficiency at every step of the process or product development and their utilization, energy conservation can be automatically achieved. The opportunities of energy conservation can be categorized into:

- Minor—Easy to implement; requires less investment and implementation time.
- Medium—Complex in nature; requires additional investment and moderate implementation time.

- Major—Yields energy savings substantially, but more intricate; requires more investment and longer implementation time.

In the circumstances of growing energy demand, the major energy challenges necessitating the implementation of energy efficiency measures can be listed as:

- Shortage of the primary energy resources
- Complexity involved in the sources enabled for the extraction of primary energy resources
- Atmospheric emissions, including GHG and carbon compounds
- Climate change, ozone depletion, and global warming potential on the environment
- Overall price hike in energy fuels

The cumulative energy-related CO_2 emissions and associated budget as projected in Fig. 1.10 reveal that the extensive consumption of primary energy may have a considerable impact on carbon emissions in the near future. This means that, due to the increasing levels of atmospheric emissions, the issues related to climate change and global warming potential are also increased.

Thus, to counteract energy challenges and energy issues, certain approaches can be useful in terms of enhancing energy efficiency, including the following:

- Develop innovative energy strategies and management techniques for reducing energy consumption and carbon footprints.
- Ensure the energy security through implementing energy conservative technologies.
- Optimize the operating characteristics of present or advanced materials to perform energy efficiently at all times irrespective of the erratic changes in the environmental conditions.

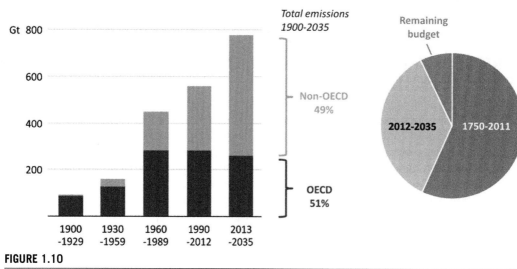

FIGURE 1.10

The cumulative energy-related CO_2 emissions and the carbon budget for 2°C [3].

World Energy Outlook 2013 Launch – a presentation by Maria van der Hoeven in London © OECD/IEA, 2013, page 5.

- Develop promotional activities pertaining to renewable energy integration with present energy systems to extend the existence period of primary energy reserves and reduce carbon footprints as well.
- Implement new or refined strategic energy policies and plan measures for addressing the foreseen future energy demand/consumption.

Among these approaches is a limelight for addressing the mitigation strategies for climate change through reducing carbon footprints (especially CO_2 emission) without affecting the stability of economic growth. To achieve this, value-added energy management techniques have to be introduced for the development of energy-efficient systems. In this context, carbon management concepts have become increasingly attractive in recent times, wherein the core theme is to make the energy-efficient systems design effective with cost optimization.

The management of the energy vectors (flow of a given quantity of energy in time and space from the originating point of the source to be used at the destination point in time and space) is rationally a viable approach for reducing carbon-based emissions. However, the suitability of the energy vectors depends on several factors, which include the following requirements [18]:

- Has high energy density by volume and weight
- Has ease of storage even at ambient temperature without the need for high pressure
- Is nontoxic, is reliable, and possesses handling safety
- Has limited risks during transit to place of utilization
- Amalgamates with existing infrastructure without necessitating new/supplemental dedicated equipment
- Is environmentally friendly during production and utilization

The major elements involved in energy-efficient systems design are presented in Table 1.4. The proper combination of energy resources locally available with the actual requirement can be very helpful in maintaining environmental sustainability. The characteristics of Earth's carbon reservoirs and the targeted carbon emissions for some major countries and regions are summarized in Tables 1.5 and 1.6, respectively. The target for the reduction of CO_2 emissions can be formulated as given by [18],

$$CO_2 \text{ emissions reduction target} = CO_2 \text{ emissions} - EACE \tag{1.1}$$

where, EACE refers to the ecologically allowable CO_2 emissions, and EACE can be represented by [18],

$$EACE = -CC - IC \tag{1.2}$$

where, CC means the common contribution, and IC means the individual contribution. For instance, the term CC is framed with a vision that CO_2 emissions can be reduced through a common approach (oceanic removals as common for all nations). Likewise, the IC can be thought of as reducing the CO_2 emissions by adopting the carbon management technologies by all the nations individually. In simple words, CC is independent of human activities, whereas IC may involve human involvement for carbon management.

It is interesting to note from Table 1.6 that CO_2 emissions have to be reduced by half the value (about 50%) of the present status of emission to stabilize the atmospheric concentration levels of carbon dioxide. Precisely, the nations that are interested in the targeted CO_2 emissions reduction scheme have to minimize the quantity of the CO_2 emitting out to the atmosphere, whereas their emission rate

Table 1.4 Energy System Design [18]

Resource: Examples	Energy Technology	Energy Vector	End Use
Fossil fuels		Electricity	Electric load
Coal/Lignite	PC/CFB	Heat	Heat load
Natural gas	CCGT	Carbon-based	Transport
Nuclear		Hydrogen	
Uranium/Thorium	Fission	Ammonia	
Deuterium	Fission		
Renewables			
Water	Hydro		
	Tides		
Solar	CSP		
	PV		
	Solar thermal		
	Thermolysis		
Wind	Wind turbine		
	(on/off-shore)		
	Kites, etc. (high altitude winds)		
Biomass	Anaerobic digestion		
	Combustion		
	Gasification		
	Fermentation		
	Esterification		
	Pyrolysis		
Geothermal	Geothermal		

Table 1.5 Characterisation of Earth's Carbon Reservoirs [18]

Carbon Reservoir	Mass (Pg C)	Share (%)
Carbon rocks (limestone, chalk, dolomite)	42,000,000	79.9150
Organic-rich rocks (coal, oil, natural gas)	10,500,000	19.9787
Hydrosphere	38,000	0.0723
Others (methane hydrates, marine sediments, etc.)	14,000	0.0266
Soil	2500	0.0048
Atmosphere	770	0.0015
Biosphere	560	0.0011
Total C	52,555,820	100.0000

Table 1.6 CO_2 Emissions Per Capita Per Year From Fuel Combustion as of 2008 [1], Ecologically Allowable CO_2 Emissions (EACE) and Calculated CO_2 Emissions Reduction Targets for Some Major Countries and Regions [18]

Country/Region	CO_2 Emissions[a] (Mg C Capita^{-1} Year $^{-1}$)	EACE[b] (Mg C Capita^{-1} Year $^{-1}$)	CO_2 Emissions Reduction Target (CO_2 Emissions[a] - EACE[b])	
			(Mg C Capita^{-1} Year $^{-1}$)	(% of CO_2 Emissions)
Australia	5.04	*0.60*	4.44	88
USA	5.01	*0.60*	4.41	88
Russian Federation	3.07	*0.60*	2.47	80
Netherlands	2.95	*0.60*	2.35	80
OECD	2.89	*0.60*	2.29	79
Germany	2.67	*0.60*	2.07	78
Denmark	2.40	*0.60*	1.80	75
United Kingdom	2.27	*0.60*	1.67	74
Norway	2.15	*0.60*	1.55	72
Poland	2.14	*0.60*	1.54	72
Middle East	2.05	*0.60*	1.45	70
France	1.57	*0.60*	0.97	62
Sweden	1.35	*0.60*	0.75	56
China	1.34	*0.60*	0.74	55
World	1.20	*0.60*	0.60	50
Asia (except China)	0.38	*0.60*	–	–
Africa	0.25	*0.60*	–	–

[a]*CO_2 emissions from fuel combustion only. Emissions are calculated using the IEA's energy balances and the Revised 1996 IPCC Guidelines and recalculated to C.*
[b]*The actual EACE can vary significantly among countries/regions. The EACE must be thus independently determined for all countries/regions. However, for simplicity, Table 1.6 uses only the globally averaged value of the EACE of 0.60 Mg C capita^{-1}Year $^{-1}$for all countries/regions. This fact is emphasized by providing these EACE estimates in italics.*

is higher. In turn, they can also motivate efforts related to increasing their EACE for accomplishing overall reduction in CO_2 emissions. Countries whose value is slightly over the targeted CO_2 emissions reduction can also counterbalance the effect by increasing their EACEs. The integration of the value-added carbon management with the conventional and the renewable energy-based technologies are presented in Tables 1.7–1.12.

A ray of hope would always mean that *whenever a problem is identified, a good solution exists nearby*. This is seen in the rapidly exhausting energy scenario that the development of alternative sources of energy can be considered a good solution for the energy issues the world's nations are facing today. Based on Fig. 1.1, the solution that is available for solving most of the current as well as the future energy demand can be found in solar energy. This is because the total energy reserve available from solar would account for 200 times more than any other energy resources (including nonrenewable and renewable energy resources). Thus, the development of new and energy-efficient systems based on renewable energy sources (solar energy) can help contribute to

Table 1.7 Value-Added Carbon Management in Fossil Fuel-Based Energy Technologies [18]

Energy Technology Integrated with Value-Added Carbon Management	Value-Added C-Rich Products	Additional C Management Effects
Enhanced oil, gas, and coal bed methane recovery	Oil, natural gas, methane	CO_2 sequestration in depleting wells
Methane hydrates	Methane	CO_2 sequestration as CO_2 hydrates
Underground coal gasification (UCG) followed by CO_2 storage in UCG voids	Syngas $(CO+H_2)$	CO_2 sequestration in UCG voids
Biogenic methane	Biomethane	CO_2 sequestration in biogenic methane-producing reservoirs

Table 1.8 Perspectives and Constraints of Fossil Fuel-Based Energy Technologies Integrated with Value-Added Carbon Management [18]

Energy Technology Integrated with Value-Added Carbon Management	Perspectives	Constraints
Enhanced oil, gas, and coal bed methane recovery	EOR is commercial and cost-effective. EOR, EGR, and ECBMR can be made cost-effective	EGR and ECBMR are at pilot phase and need further R&D
	The deployment of EOR and EGR significantly increased global accessible reserves of oil and natural gas	All technologies require supply of compressed CO_2
Methane hydrates. Underground coal gasification (UCG) followed by CO_2 storage in UCG voids	Large quantity potential of methane hydrates suitable for coal deposits for which conventional mining is not feasible	It is at research phase and needs significant R&D. Some unresolved problems exist regarding thermal and kinetic characteristics of UCG reactors
	CO_2 recycling improves coal gasification. Existing commercial interests can accelerate the deployment of UCG	UCG needs further R&D and on-site demonstration for technological validation
Biogenic methane	It has the potential of improving cost-effectiveness of conventional CCS technologies	Limited kinetics of microbial methane production
		Problems with appropriate site selection might appear

Table 1.9 Value-Added Carbon Management in Renewables-Based Energy Technologies [18]

Energy Technology Integrated with Value-Added Carbon Management	Value-Added C-Rich Products	Additional C Management Effects
Solar fuel synthesis	Syngas, synfuels	It can use captured CO_2
Catalytic hydrogenation of CO_2 CO, CH_4 and higher hydrocarbons It can use captured CO_2	CO, CH_4, and higher hydrocarbons	It can use captured CO_2
Electrochemical CO_2 conversion	HCOOH, CO, CH_3OH, CH_2CH_2, CH_4	It can use captured CO_2. The obtained chemicals can be used to synthesise stable carbon-rich materials (e.g., polymers) and thus store more carbon for longer time
Trireforming	Methanol, urea	It can use captured CO_2
Conventional value-added fertilizers, materials, and chemicals	Urea, isocyanates, polycarbonates, salicylic acid, carboxylates and lactones, carbonates, carbamates, polymers	They can use captured CO_2

Table 1.10 Perspectives and Constraints of Renewables-Based Energy Technologies Integrated with Value-Added Carbon Management [18]

Energy Technology Integrated with Value-Added Carbon Management	Perspectives	Constraints
Solar fuel synthesis	Solar energy has the largest available capacity among all renewable energy sources	Pilot demonstrations to validate the technology are needed. Several issues still need significant R&D efforts
	Novel materials such as cerium oxides increase the potential for technology deployment	Artificial photosynthesis is limited by effective catalysts, which are active in visible light
Catalytic hydrogenation of CO_2	It is one of the most promising routes to react CO_2	It requires renewable hydrogen and the development of effective catalysts
Electrochemical CO_2 conversion	It enables low-temperature reacting of CO_2	It needs renewable electricity Further R&D is still needed
	Formic acid is an attractive market product	
Trireforming	Mature technology that enables to react CO_2	Cost-effectiveness cannot be ensured today
Conventional value-added fertilizers, materials, and chemicals	There exists a well-defined market for such products	The capacity potential for these conventional CO_2 uses is relatively limited worldwide

Table 1.11 Value-added Carbon Management in Nonconversion Use of CO_2-Based Energy Technologies [18]

Energy Technology Integrated with Value-Added Carbon Management	Value-added C-rich products	Additional C management effects
Refrigerants	CO_2 as a heat transfer fluid	It can store CO_2
Geothermal heat transfer Fluids	CO_2 as a heat transfer fluid	It can store CO_2

Table 1.12 Perspectives and Constraints of Nonconversion Use of CO_2-Based Energy Technologies Integrated with Value-Added Carbon Management [18]

Energy Technology Integrated with Value-Added Carbon Management	Perspectives	Constraints
Refrigerants	The growing number of cars with air-conditioning increases the potential of cooling applications	Low overall CO_2 storage capacity
	It can minimize leakages of conventional nonenvironment-friendly refrigerants	Dedicated refrigeration applications must be developed
Geothermal heat transfer fluids	It can minimize leakages of conventional nonenvironment- friendly refrigerants	Low overall CO_2 storage capacity
	Promising for geothermal power plants and for heat pump applications	Dedicated energy applications must be developed for CO_2 cycles

reducing the extensive utilization of the depleting fossil fuels, carbon, and radioactive emissions substantially.

1.5 CONCISE REMARKS

Energy, the integral source for all human-based activities is vital in the sense that it has to be conserved and managed at every step of the development of society. The ever increasing concerns on energy and the environment have to be confronted with the most energy-efficient and cost-effective measures through proper energy management techniques. The adoption of several energy savings methodologies starting from the scheme inception of extraction of the energy sources to the end-energy utilization/demand requirements can be helpful in extending the availability of fossil fuel resources. Besides, the concepts of renewable energy integration with the conventional systems or newly developed systems can result in the augmentation of energy conservation potential and reduction of carbon emissions from the present status without sacrificing energy efficiency and environmental sustainability.

References

[1] Perez R, Zweibel K, Hoff TE. Solar power generation in the US: too expensive, or a bargain? Energ Policy 2011;39:7290–7.

[2] BP Statistical Review of World Energy; 2013. bp.com/statisticalreview [accessed on January 2014].

[3] World Energy Outlook 2013 Launch – a presentation by Maria van der Hoeven in London © OECD/IEA, 2013, pages 3, 5, 6, 8, 10 & 11.

[4] IEA. World energy outlook 2010. International Energy Agency; 2010.

[5] Parameshwaran R, Kalaiselvam S, Harikrishnan S, Elayaperumal A. Sustainable thermal energy storage technologies for buildings: a review. Renew Sustain Energy Rev 2012;16:2394–433.

[6] EIA. International energy outlook. Energy Information Administration; 2010.

[7] World population to 2300. United Nations, New York: Population Division; 2004. ST/ESA/SER.A/236.

[8] The world at six billion. UN Population Division; 1999. ESA/P/WP.154.

[9] Siraj MS. Energy resources—the ultimate solution. Renew Sustain Energy Rev 2012;16:1971–6.

[10] World Energy Assessment. Energy and the challenge of sustainability. United Nations Development Programme, http://www.undp.org/energy/activities/wea/drafts-frame.html; 2000. Online (28.04.2011).

[11] Survey of Energy Resources. World Energy Council; 2010.

[12] Global Uranium Resources to Meet Projected Demand. International Atomic Energy Agency, http://www.iaea.org/newscenter/news/2006/uranium_resources.html; 2011. Online (28.04.2011).

[13] MIT Report, The Future of Geothermal Energy, http://geothermal.inel.gov/publications/future_of_geothermal_energy.pdf; 2006. Online (28.04.2011).

[14] Hamacher T, Bradshaw AM. Fusion as a future power source: recent achievements and prospects. In: 18th world energy congress; 2001.

[15] DOE. Building energy data book 2006. Energy Efficiency and Renewable Energy; 2006.

[16] IPCC. Climate change 2001: the scientific basis. Cambridge: Cambridge University Press; 2001.

[17] Praditsmanont A, Chungpaibulpatana S. Performance analysis of the building envelope: a case study of the Main Hall, Shinawatra University. Energy Build 2008;40:1737–46.

[18] Budzianowski WM. Value-added carbon management technologies for low CO_2 intensive carbon-based energy vectors. Energy 2012;41:280–97.

Energy Storage

2.1 INTRODUCTION

In the global scenario of growing energy consumption and depletion of fossil fuel resources, the thirst for technologies that enable energy storage is indubitably currently increasing. Remember the law of conservation of energy, which means that energy can neither be created nor destroyed but can be transformed from one form to another. It also signifies that the total amount of energy in a system stays constant or is conserved. This is very true for energy storage in the sense that energy being stored at one point in time (off-peak condition) can be discharged at another time when it is required (on-peak condition).

For instance, electricity being generated through some energy conversion technique (energy transformation) can be stored in a system in the form of electric charges or charged particles at one instant of time (low demand period). During peak demand conditions, by discharging the stored electricity from the system, the required load level can be effectively achieved. The continuous consumption of energy by all means has shown terrible effects on centralized power distribution and network instability toward fluctuating load demand occasions. To ameliorate existing energy shortage episodes, the development of technologies related to energy storage is essential. Energy storage, in a broad sense, is thought of as a technique dedicated to storing and discharging electricity. Thus, the majority of energy storage systems have been devised to meet this purpose. Demand-side energy management can also be accomplished through the development of thermal energy storage technologies. Overall energy storage capacity and the operational performance of energy storage systems can be enhanced by renewable energy integration.

However, some constrictions may be involved in energizing the storage system configured with renewable energy sources compared to conventional energy sources. Precisely, energy storage technologies in any form can be a viable option for present as well as future energy needs. The significance and functional aspects of a variety of energy storage technologies intended for meeting demand-side energy requirements are demonstrated in the following sections.

2.2 SIGNIFICANCE OF ENERGY STORAGE

In the present scenario and future predictions about energy generation and energy consumption, it is important to confront the issues related to energy security and climate change. It is also uncertain whether the energy that is being extracted or generated is delivered to the end-energy user at all times. There are many opportunities for energy loss to take place, starting with the inception of the extraction to end-use consumption. To bridge the huge gap present between energy supply and energy demand, the concept of energy storage has been gaining momentum in recent years. The significance of energy storage is to [1]:

- Meet the short-term fluctuating energy demand requirements
- Supply energy during transitory power disturbances or surges
- Reduce the need for emergency power generators or part-load operated power plants that would otherwise consume primary energy sources
- Redistribute the energy required during on-peak demand conditions (daytime) through the energy produced during off-peak hours (nighttime)
- Make use of the energy (i.e., electricity) generated from the renewable energy sources for the fluctuating load demand
- Provide energy security with less environmental impact
- Improve operational performance of energy systems

2.3 TYPES OF ENERGY STORAGE

Energy storage is a widespread subject of interest to scientists, engineers, technologists, and industrial professionals worldwide. Numerous research works have been reported on energy storage and methodologies in past literature. A variety of suggestions have been provided for fostering energy storage technologies, either electrical energy or thermal energy storage technologies. The storage of electrical energy can be performed either directly or indirectly through electrical input and output. On the other hand, thermal energy storage uses the sensible heat or latent heat or thermochemical reaction capabilities of materials to store and retrieve heat energy on demand. The technologies for energy storage can be classified into four major types based on their applications, namely:

- Applications requiring low power in isolated regions for supplying energy to the transducers and terminals meant for urgent situations
- Applications requiring medium power in isolated regions, including individual electrical systems and town supply
- Applications requiring peak-load leveling (for example, network connections)
- Applications requiring power-quality control

The classification of energy storage systems is presented in Table 2.1. It is pertinent to note that the first two types are well suited for small-scale systems where energy storage can be made effective

Table 2.1 Classification of EES systems [2]

Electrical or Mechanical High Power EES Systems	Mechanical High Energy EES Systems	Electrochemical EES Systems (HP & HE Potential)[a]
Superconductive magnetic energy storage (SMES)	Compressed air energy storage (CAES)	Accumulators with internal storage (e.g., Pb/PbO2, NiCd, Li-ion, NiMH, NaNiCl, NaS)
Supercapacitors/electrochemical double-layer capacitors (EDLCs) Flywheels	Pumped hydro energy storage (PHS)	Accumulators with external storage (e.g., hydrogen storage system, flow batteries)

[a]Electrochemical EES systems can be used either in high power (HP) or high energy (HE) applications.

through kinetic energy, compressed air, hydrogen (fuel cells), chemical energy, supercapacitors, or superconducting magnetic storage. The last two types signify large-scale systems, where energy storage can be performed using hydraulic systems (gravitational energy), thermal energy storage, chemical energy (flow batteries and accumulators), and compressed air storage coupled with liquid or natural gas (NG). For brevity, only the major electrical energy storage (EES) technologies are discussed in the forthcoming sections, and the concepts related to the thermal energy storage technologies are largely covered in the forthcoming chapters.

2.4 ENERGY STORAGE BY MECHANICAL MEDIUM

The concepts of enabling kinetic and potential energies energy storage through a mechanical medium can be performed using flywheels, pumped hydroelectric, and compressed air energy storage (CAES) technologies. The significance of these technologies is discussed in the following sections.

2.4.1 Flywheels (kinetic energy storage)

As the name indicates, kinetic energy (or rotational energy) is the key source for flywheels to store energy mechanically. The integral parts of a flywheel energy storage system are shown in Fig. 2.1. They include the massive rotating cylinder component, which is supported over a stator through the magnetically levitated bearings. The flywheel is coupled with the electric generator/motor assembly for energy storage, and the entire flywheel system is placed in a low pressure (or vacuum) atmosphere to reduce shear disturbances/frictional losses due to wind or external forces.

The flywheel system operates on the principle of kinetic energy for storing and releasing energy depending on load demand. During the storage period, the flywheel system rotates at very high speed by means of electric motor activation. During the discharging cycle, the stored kinetic energy in the flywheel is utilized for regenerating the motor to function as an electric generator. By providing proper power controls and power converters, the required energy demand can be effectively met. Because the kinetic energy stored in the flywheel system is directly proportional to its mass and the square of the velocity (rotating speed), maximum energy storage density depends on the tensile strength of the

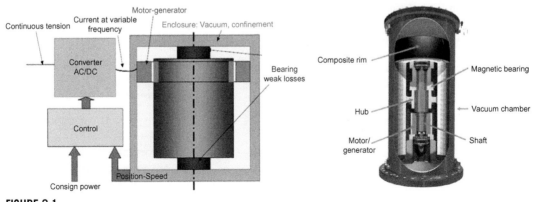

FIGURE 2.1

Flywheel energy storage system or accumulator [3–5].

flywheel material. Furthermore, the shape and the inertial effects of the rotating component also decide the quantity of energy being stored by the flywheel. The maximum specific energy of the flywheel system can be estimated from the relation given by

$$\text{Maximum specific energy} = \left(\text{Energy density/Density of the rotating disk material}\right) \quad (2.1)$$

Two types of flywheel are available in the energy market, the conventional steel rotor flywheel intended for low speed cycling (< 6000 rpm) and the advanced composite flywheel for high speed operations (10^4-10^5 rpm). Conventional low speed flywheels are suitable for the uninterruptable power supply. The low speed flywheels produced by the popularly known Piller's POWERBRIDGE can deliver 250-1300 kW. Similarly, the low speed (~3600 rpm) flywheel system can be expected to deliver 1.1 MW during 15 s of its functioning [6].

In recent years, the development of both long- and short-term flywheel energy storage systems have become increasingly popular. One such example can be quoted from the famous Beacon POWER flywheel products, which are dedicated to meeting the demand requirements of short- and long-term energy storage applications. The evolution of the company's product over years is presented in Fig. 2.2, from which the role of flywheel systems for the two kinds of applications can be easily observed. The flywheel farm constructed in late 2009 in Stephentown, New York, can deliver 20 MW for 15 min, which is a typical example of a high-power flywheel system for the power grid application [7].

The merits of the flywheel energy storage system are listed here:

- The reliability or cyclic ability very high (of the order of 105 cycles with deep discharge)
- Life span between 15 and 20 years

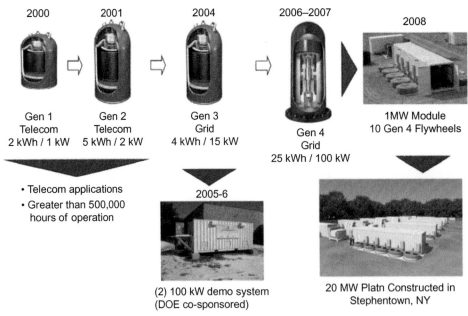

FIGURE 2.2

Evolution of Beacon POWER flywheel system [5,6].

- Swift charging and discharging capabilities (i.e., high power density)
- High performance in terms of cycle efficiency
- Low-cost maintenance

The demerits of the flywheel energy storage system include the following:

- High rate of self-discharging over time (approximately 20% per hour)
- Critical design aspects limits its usage for long-term energy storage applications
- High materials cost of production

Thus, flywheel systems can find suitability mostly to short-term energy storage applications requiring instant power delivery for meeting frequent load fluctuations as well as in power conditioning sectors.

2.4.2 Pumped hydroelectric storage (potential energy storage)

The concept of pumped hydro or hydroelectric storage is well known, and these storage systems have been established over years for their availability to satisfy the peaking load demand readily. The schematic representation of the pumped hydroelectric storage plant is shown in Fig. 2.3(a).

Basically, the potential energy of water stored in two reservoirs situated at different datum (height) levels is utilized for the generation and storage of electrical energy by means of turbines and pump facilities. During the low demand period, the water from the lower reservoir (or basin) is transferred to the higher reservoir by pumps. During the peaking or high demand periods, the water stored in the upper reservoir (or basin) is allowed to flow through the turbines to generate the required electricity. Depending on the equipment characteristics, the efficiency of the round-trip process performed using this system can be expected to vary from 60 to 80%. The life expectancy of this system is estimated to vary from 30 to 50 years.

It is noteworthy that the storage capacity of the pumped hydro system depends on two key parameters, which are the height from which the water falls and the volume of water required for the process. The graph representing the relationship between the height and the water volume is depicted

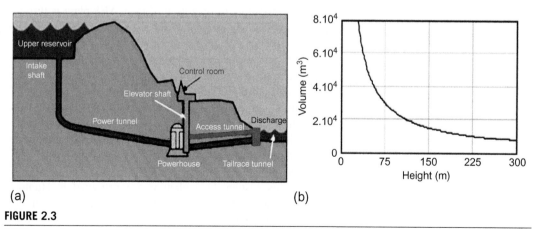

(a) (b)

FIGURE 2.3

(a) Schematic of pumped hydro storage plant, and (b) water volume needed at a given height to store 6 MWh [3,6,8,9].

in Fig. 2.3(b). For instance, a mass of 1 ton of water that is falling from a height of 100 m can generate about 0.272 kWh of energy [3]. The major limitation of this technology is the availability of a site with different elevations or nature of water bodies.

Pumped hydroelectric systems find their application in providing energy management, frequency control, reserve capacity, and power networks. This is the only system that has been established to achieve energy storage capacity from a few tens of GWh or 100 of MW). In the United States, pumped hydroelectric systems are installed in 39 sites, which have a capacity ranging from 50–2100 MW. These systems are highly capable of storing the excess power produced for more than 10 h and meet fluctuating demand requirements through load leveling. As expected, there are few limitations in the development of this system in terms of geographic site location, huge cost involved, and longer construction period for complete establishment of the plant.

2.4.3 Compressed air energy storage (potential energy storage)

CAES is considered to be the other commercially available technology, which has the ability to provide very large energy storage capacities next to the pumped hydroelectric storage plant. CAES essentially comprises five key components, which are (1) generator or motor assembly with special clutches for engaging and disengaging the compressor and turbine trains, (2) a two or more stages air compressor equipped with intercooler and aftercoolers for achieving compressor economics as well as to reduce moisture content present in the compressed air, (3) high and low pressure turbine trains, (4) an underground cavern or cavity for air storage, and (5) controls and auxiliaries of equipment including fuel storage and heat exchanger components. The configuration of the CAES system is schematically represented in Fig. 2.4.

The storage of air in the underground cavern or cavity can be formed by (1) excavating hard and impervious rocks, (2) dry-mining or solution of salt formation for creating salt caverns, and (3) caverns made up of porous media reservoirs including water-bearing aquifers or depleted gas or oil fields (for e.g., sandstone and fissured lime).

The prime theme behind CAES is to make use of the elastic potential energy of compressed air for acquiring the required energy storage. The basic principle of operation of the CAES plant is through the conventional gas turbine generation process, wherein compression and expansion cycles of the gas turbine are decoupled into two separate processes. Typically, during the low demand period, energy is being stored by compressing the air into an airtight space maintained at pressures ranging from 4 MPa to 8 MPa. The stored energy in the air can be extracted by subjecting the compressed air to elevated temperatures and expansion through a high pressure turbine.

By this, the turbine recovers some quantity of stored energy. With the exhaust being expanded from a low pressure turbine, the air is mixed with the fuel and the exhaust and combusted. The high pressure and low pressure turbines, which are coupled to the generator, produce the required electricity for offsetting the load demand. The waste heat of the exhaust gas is effectually retrieved by means of the recuperators before being vented out to the atmosphere.

Presently, only two plants worldwide use this system. One plant was constructed in Huntorf, Germany, and the other is located in McIntosh, Alabama. The CAES located in Germany has functioned since 1978 and has a cavern capacity of ~310,000 m³, which is converted from a solution minded salt dome beneath the ground at a depth of ~600 m. This system is integrated with a compressor capacity of 60 MW capable of providing pressure up to 10 MPa maximum. This plant is operated on a daily cycle

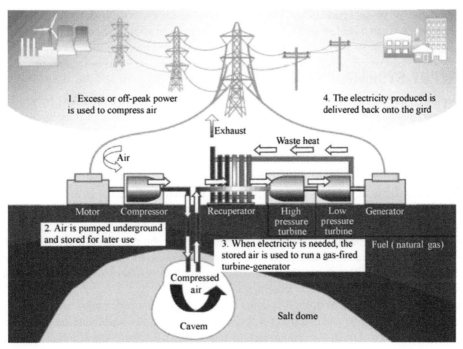

FIGURE 2.4

Schematic representation of CAES system [10,11].

basis of 8 h charging and 2 h of power generation up to 290 MW. This plant has been graded with 90% availability and 99% starting reliability.

The plant in the United States has been functioning since 1991 with the compressor unit capable of providing pressure ~7.5 MPa maximum. The cavern formed is the same as that of the German plant; however, the depth is about 450 m underground. The storage capacity of this plant is estimated to about 500,000 m³ with a generating capacity of 110 MW for a 26 h operating cycle. This plant utilizes the waste heat recovery option by means of recuperators, which adds value to this system by reducing the fuel consumption around 25% compared to the Huntorf CAES plant.

The limiting factors for widespread development of the CAES plant include the following:

- Geographic locations fulfilling the requirements for setting up aquifers, salt caverns, depleted gas fields, or rock mines
- Nonindependent system that requires gas turbine plant for effective functioning
- Nonsuitability for coupling with other major power plants such as thermal, nuclear, wind turbine, or solar photovoltaic power plants
- Combustion of fossil fuels or primary energy sources and their associated emissions to the environment causing serious concerns over GHG emissions and climate change

However, the average life span of the CAES system is estimated to be 40 years and is competent with the pumped hydroelectric storage systems in terms of long-term time scale energy storage option with underground air storage conditions. The self-discharging characteristics of the CAES system are very low, and so they can have low surface environment impact, in addition to cost effectiveness.

2.5 ENERGY STORAGE BY CHEMICAL MEDIUM

The concepts of enabling the storage of electrical energy through a chemical medium can be performed using flow batteries and thermochemical energy storage technologies. The significance of these technologies is discussed in the following sections:

2.5.1 Electrochemical energy storage

The storage of electricity in the form of chemical energy is usually referred to as electrochemical storage, and the famous example for this type of storage system is the battery. A battery essentially consists of two electrochemical cells each filled with liquid, paste, or solid electrolyte together and provided with positive (anode) and negative (cathode) electrodes. During the discharging cycle of the battery system, electrochemical reactions take place at the two electrodes, thereby generating electron flow through an external circuit. The electrochemical reactions can be reversed by applying an external voltage across the electrodes, which enables charging the battery.

Basically, the electrochemical storage systems are categorized into two major types, namely, integrated energy storage systems and external energy storage systems. Pb–acid batteries, NiCd batteries, NiMH batteries, Li-ion batteries, NaS batteries, and NaNiCl/ZEBRA batteries all fall under integrated energy storage systems. Likewise, V-redox, zinc bromine battery (ZnBr), Zn–air batteries, and hydrogen storage systems are grouped under external energy storage systems. External energy storage systems are also known as flow batteries.

In the integrated energy storage system, the charging and discharging electrochemical reactions take place within the active material of the battery system with no spatial separation. The energy storage capacity of integrated storage systems is relative to its charge/discharge power ability, and changes to the capacity specifications are generally not possible, which is not the case for flow battery systems.

In fact, battery energy storage systems can be effectively utilized for large-scale energy storage in stationary applications because they can serve to regulate frequency and manage fluctuating demand as well. The schematic diagram of the NaS battery system is shown in Fig. 2.5(a). The operating principle of this battery system is self-explanatory. NaS batteries exhibit almost four times the energy density compared to lead-acid batteries, and also cycling time is comparatively long (~2500 cycles on 90% discharge depth).

The swift response during the charging and discharging cycles makes them more attractive than other type of battery systems. The energy density and the energy efficiency of NaS batteries are estimated to be very high at about $151 kWh/m^3$ and 85%, respectively. Furthermore, the low maintenance, 99% of recyclability, and low self-discharge capabilities also add value to the NaS battery system. Recently, in China, the installation of the first industrial NaS battery station of 100/800 kWh capacity was announced at World Expo 2010 in Shangai. The pictorial view of the NaS battery station is shown in Fig. 2.5(b).

Another example of the NaS battery system installation is found in the island of Graciosa in the Canary Islands. A stand-alone renewable energy network is currently under development with this battery system. For achieving large-scale energy storage, two 500 kW capacity NaS battery systems have been installed in the island facility. In addition, Li-ion battery systems are also equally popular in recent times; a 12 MW/3 MWh Li-ion battery was installed in Chile in 2009.

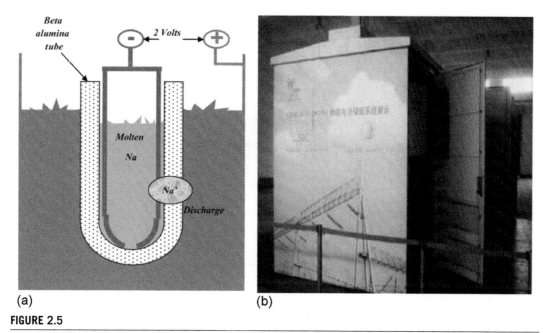

FIGURE 2.5

(a) Schematic diagram of a NaS battery system, and (b) the first NaS battery station in China [1,12].

On the other hand, flow batteries are well-known electrochemical energy storage technologies intended for storing energy for long durations. They are usually referred as redox flow batteries and have gained impetus in recent years. The flow battery works by principle of reversible chemical reactions between the two liquid electrolytes of the battery for enabling charging and discharging processes. The liquid electrolytes are contained in separate containers, which is not the case for conventional batteries. In a typical operation, the liquid electrolytes are pumped into an electrochemical reactor, and electricity is generated through a chemical redox reaction. By placing the liquid electrolytes separately, power and energy storage specifications can be made more easily.

The schematic diagram of the flow battery energy storage system is illustrated in Fig. 2.6. In addition, the design of the power cell can be optimized depending on the required demand rating by having flexibility in replacing or increasing the quantity of liquid electrolytes [14,15]. Flow batteries differ from fuel cells in the way that the electrolyte in the flow batteries can be recharged without replacing them entirely; also, the chemical reaction that occurs is reversible. Flow batteries can have a high discharging rate of up to 10h.

Flow batteries carry the following merits related to effective utilization:

- Energy storage capacity can be made flexible or scalable due to the volumetric storage capacity of the electrolyte stored separately.
- Installation costs can be minimumal if amalgamated with a larger system.
- Complete discharging takes place without any damage.
- There are relatively low self-discharging capabilities.
- They exhibit longer life and low maintenance over a longer period of energy storage.

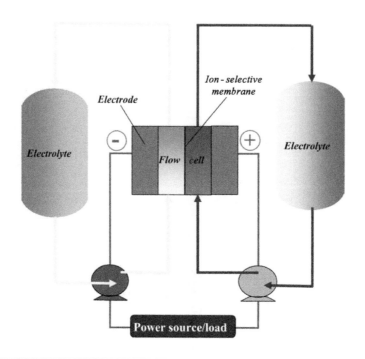

FIGURE 2.6

Schematic overview of a redox flow cell energy storage system [1,13].

Table 2.2 Comparison between different technologies of flow battery [1,10]			
Technology	**VRB**	**PSB**	**ZnBr**
Efficiency (%)	85	75	75
Cycle life charge/discharge	13,000(12,000+)	—	2500(2000+)
Capacity (MW)	0.5–100()	1–15	0.05–1(0.05–2)
Operation temp. (°C)	0–40()	50()	50()
Energy density (Wh/kg)	30(10–30)	—	50(20–50)
Self-discharge	Small	Small	Small

In the recent past, three major types of flow batteries have been developed for commercialization and demonstration purposes: vanadium redox battery (VRB), polysulphide bromide batteries (PSB), and ZnBr flow batteries [16–19]. The essential characteristics of these flow batteries are summarized in Table 2.2.

Based on the characteristic comparison of the flow battery technologies, VRB is considered to be more efficient than the other two types because they possess low cycling life and low capacities. The VRB flow batteries can take advantage of operating cost, life expectancy, safety, and maintenance provided the electrolyte used is of only vanadium rather than a blend of electrochemical elements. The comparison of different battery technologies is summarized in Table 2.3.

Table 2.3 Comparison of battery technologies [6]

Battery Type	Advantages	Disadvantages
Lead-acid	Low cost Low self-discharge(2–5% per month)	Short cycle life (1200–1800 cycles) Cycle life affected by depth of charge Low energy density (about 40 Wh/kg)
Nickel-based	Can be fully charged (3000 cycles) Higher energy density (50–80 Wh/Kg)	High cost, 10 times of lead acid battery High self-discharge (10% per month)
Lithium-ion	High energy density (80–190 Wh/Kg) Very high efficiency 90–100% Low self-discharge (1–3% per month)	Very high cost ($ 900–1300/KW h) Life cycle severely shortened by deep discharge Require special overcharge protection circuit
Sodium sulfur (NaS)	High efficiency 85–92% High energy density(100 Wh/kg) No degradation for deep charge No self-discharge	Be heated in stand-by mode at 325 °C
Flow battery	Independent energy and power ratings Long service life (10,000 cycles) No degradation for deep charge Negligible self-discharge	Medium energy density (40–70 Wh/kg)

2.6 ENERGY STORAGE BY ELECTRICAL MEDIUM

The concept of enabling the electric charge for the storage of energy through a dielectric medium can be performed using capacitor and supercapacitor energy storage technologies. The significance of these technologies is discussed in the following sections.

2.6.1 Electrostatic energy storage

In principle, energy can be stored as electric charges between two metal or conductive plates separated by a dielectric (insulating) medium by means of the applied voltage across the conductive plates. By supplying the direct current (DC) to one metallic plate, the charge of opposite nature would be induced on the other metallic plate. Thus, the storage capacitance mainly depends on the size of the metal plates, distance between the plates, and the material type of the dielectric medium used.

It can be noted that the energy being stored in a capacitor is directly proportional to the capacity and the square of the applied voltage across the terminals of the electrochemical cell. The capacity of the capacitor is directly proportional to the distance between the two terminals (or electrodes). The instantaneous recharge capability and long cycle lifetime can be regarded as the major advantages of capacitors, which are highly suitable for utility small-scale power control applications. However, the low energy density limits the capacitor's suitability for large-scale applications because it requires a larger area of dielectric medium, which is unrealistic and expensive. The schematic diagram of the capacitor storage system is shown in Fig. 2.7.

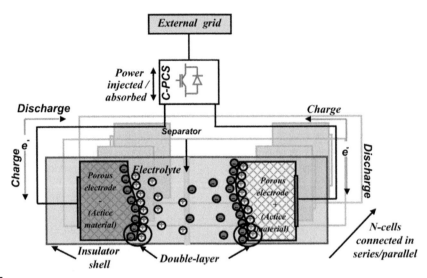

FIGURE 2.7

Schematic representation of the capacitor storage system [1].

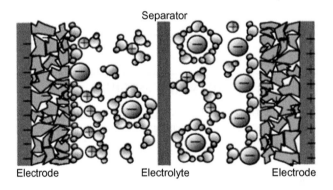

FIGURE 2.8

Schematic view of individual cell of a supercapacitor [6,20].

The operating principle of supercapacitors is very similar to that of capacitors, but with modification in the insulation medium. That is, in supercapacitors the insulating medium is replaced with an electrolyte ionic conductor. By this, the movement of ions along the conducting electrode can be achieved with a very large specific surface enabling higher energy density to the supercapacitor. The precision design of the electrodes and the selection of the electrolyte medium together would yield very high charge density on electrode surfaces. However, the voltage is limited to approximately 2.7 V per cell configuration. The schematic view of an individual cell of a supercapacitor is depicted in Fig. 2.8.

Although it has low voltage characteristics, by interconnecting the supercapacitors with each other, due to the high charge density, the energy storage capacity can be increased from 1 kWh to larger energy storage units. Supercapacitors also exhibit very high power output, which is most needed for scaling up the technology to reach approximately 50–100 kW. The cycling life of supercapacitors is estimated to be around 500,000 times with a life expectancy of 12 years. By ensuring proper controls over the operation of the supercapacitors, their power output can be tuned for extremely high content when they are connected in series configuration.

However, the major limitations of supercapacitor energy storage systems are their huge costs, which can be nearly five times that of conventional lead acid batteries, self-discharge rate, and very low energy density (5 Wh/kg). Supercapacitors find their potential utilization in hybrid electric vehicles for swift acceleration and electrical braking, doubly fed induction generator, and permanent magnet synchronous generator wind turbine applications.

2.7 ENERGY STORAGE BY MAGNETIC MEDIUM

The concept of enabling the magnetic field for storage of electrical energy can be performed using superconducting magnetic energy storage (SMES) technology. The significance of the technology is discussed in the following section.

2.7.1 Superconducting magnetic energy storage

In this type of storage system, electrical energy can be stored in a magnetic field without the need for a conversion to either mechanical or chemical forms. The SMES can be made possible by inducing the DC into superconducting coil cables, which has zero resistance to the flow of current. Usually, superconducting coils are made of niobiumtitane (NbTi) filaments, which is subjected to very low temperature ($-270\,°C$). The schematic representation of the SMES system is depicted in Fig. 2.9. Although there is a possibility for higher energy consumption in refrigeration and the associated resistive losses occurring in solid-state switches during the functioning of the system, the overall efficiency in commercial applications in terms of MW capacity is very high.

The swift charging/discharging and independent energy content for the fluctuating discharge rating are the meritorious characteristics of the SMES system. The energy content of the commercially available SMES system is extended up to 1 kWh. Response time is limited to only a few milliseconds by virtue of the energy release rate and the speed of the switching operation of the power electronics. SMES systems can have a higher life expectancy and cycling time compared to any other energy storage systems. However, the mechanical stress being introduced in the components may lead to material fatigue. Due to the high cost factors involved, the development of the SMES systems for utility small-scale or large-scale power/energy management and load leveling applications is limited.

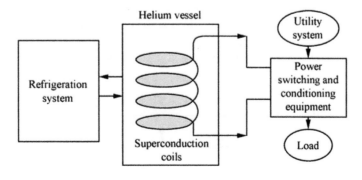

FIGURE 2.9

Schematic representation of the SMES system [10,21].

2.8 ENERGY STORAGE BY HYDROGEN MEDIUM

The concepts of enabling the storage of energy through hydrogen medium can be performed using hydrogen-based fuel cells and solar hydrogen production technologies. The significance of these technologies is discussed in the following sections.

2.8.1 Hydrogen-based fuel cells

Fuel cells are basically meant for restoring the energy that is spent to produce hydrogen (H_2) through the electrolysis of water. The reason for giving main focus to hydrogen is that it is considered to be the cleanest, lightest, and most efficient fuel. It does not occur naturally, but has to be derived from primary energy sources [22,23]. Hydrogen may be one of the alternative fuels for future energy systems that can be effectively stored and reused. The electricity produced from hydrogen can be used as such, but it has to be converted to another energy form for storage.

The storage system primarily consists of three main components, namely, electrolysis, fuel cell, and hydrogen buffer tank. In electrolysis, off-peak electricity is consumed for producing hydrogen. The fuel cell utilizes the produced hydrogen and oxygen from the atmospheric air to generate electricity for offsetting on-peak electricity demand. The buffer tank makes sure of the sufficient quantity of the resource's availability during demand periods.

The schematic of the hydrogen-based fuel cell is shown in Fig. 2.10. The hydrogen fuel cell operates by the principle of oxidation-reduction between the hydrogen and oxygen within the electrochemical cell comprised of two electrodes separated by an electrolyte medium. The transfer of ions or charged particles is enabled by means of the electrolyte medium. The fuel cell can also be operated reversible by supplying external electric current to produce or split hydrogen and oxygen in the presence of water.

There are a variety of hydrogen-based fuel cells including proton exchange membrane fuel cells, phosphoric acid fuel cells, regenerative fuel cells, and alkaline fuel cells. Hydrogen-based fuel cells

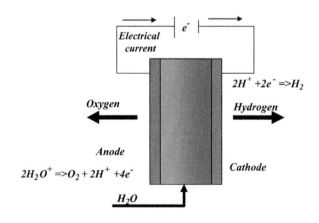

FIGURE 2.10

Schematic representation of the hydrogen fuel cell [1].

possess high energy density, suitable for small- and large-scale energy storage applications because of their simple modular usage, high charging/discharging rate on the order of 20,000 cycles, and a 15-year life span. They are highly preferred for applications requiring demand-side management, that is, shifting of the peak load demand. To prove their capability, the hydrogen fuel cells are integrated to the wind turbines, solar photovoltaic, and other renewable energy systems in recent years. The most concerning issues related to the development of hydrogen fuel cells are their economics ($6-$20/kWh), relatively low round-trip efficiency about 20 to 50%, and limited life expectancy.

2.8.2 Solar hydrogen production

The concept of solar hydrogen production originated from the basic principle of utilizing solar heat energy at high temperatures for performing an endothermic chemical transformation or reaction to produce a fuel that can be stored and/or transported to the end user. The principle of converting solar energy into solar fuel and the available thermochemical routes for solar hydrogen production are illustrated in Fig. 2.11(a,b).

Heat energy in the form of solar radiation being captured using the parabolic collectors can be processed further to obtain the same at a much higher temperature required for performing the endothermic chemical reaction. By selecting the proper reactants, the required fuels can be obtained as the outcome of chemical reaction, which can then be stored and transported to the end-user side for generating electricity.

The solar hydrogen production routes shown in Fig. 2.11(b) infer that the chemical sources required for producing hydrogen include (1) water for carrying out the solar thermolysis and solar thermochemical cycles; (2) fossil fuels such as NG, oil, coal for performing solar cracking process; and (3) the combination of the fossil fuels and water for enabling the solar reforming and solar gasification processes to occur. It is noteworthy that solar energy stands as the prime energy source for all these high temperature processes to be performed.

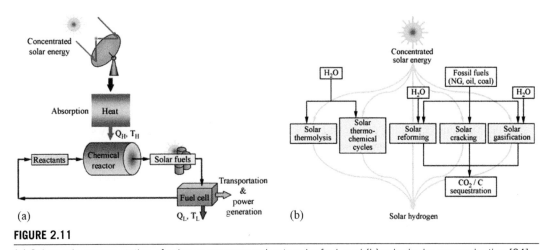

FIGURE 2.11

(a) Schematic representation of solar energy conversion to solar fuel, and (b) solar hydrogen production [24].

2.9 ENERGY STORAGE BY BIOLOGICAL MEDIUM

Energy obtained through the breakdown of glucose by means of enzymes can be effectively stored using a bio-battery storage system [25]. Glucose serves as the potential source for energy storage. The configuration of a bio-battery is very similar to that of the conventional battery, which consists of an anode, cathode, electrolyte, and separator. These components are layered one over the other, and the anode is located at the top of the battery, whereas the cathode terminal is provided at the bottom.

The energy storage capacity of the bio-battery is largely based on the quantity of glucose material. The materials that are rich in providing glucose for the purpose include old papers, soda, and glucose available in living organisms or natural sugar. The fundamental principle behind energy storage is the enzymatic hydrolysis process. This is a vital process in which the material (cellulose) can be converted into glucose under the presence of enzymes (like a catalyst). After glucose is obtained, oxygen and other enzymes can take advantage of it and produce protons and electrons further.

The process of producing the protons and electrons in a bio-battery can be represented through the following chemical reactions:

At anode:

$$Glucose \rightarrow Gluconolactone + 2H^+ + 2e^- \tag{2.2}$$

Equation (2.2) shows that the production of protons (H^+) and electrons (e^-) can be achieved by breaking down the glucose. The protons and the electrons thus produced are influenced by the separator, where the former is guided to the cathode side of the battery and the latter is redirected to the cathode. In the cathode, the well-known oxidation-reduction reaction occurs, in which the protons and electrons combine with oxygen to form water (or water vapor). The oxidation-reduction reaction at the cathode can be presented by:

$$O_2 + 4H^+ + 4e^- \rightarrow 2H_2O \tag{2.3}$$

Thus, the flow of electrons and protons across the electrodes plays a major role in generating the required electricity in a bio-battery system.

The fascinating features of the bio-battery include the following:

- Instant recharging capability
- Alternative renewable power source
- Environmentally safe and viable
- Nontoxic and nonflammable

The main parameter limiting the use of bio-batteries for long- and short-term energy storage applications is their self-discharge property (low retention rate of energy). However, in recent times, an increasing interest has been shown by researchers and scientists worldwide in producing bio-batteries with excellent energy storage properties, which can fulfill the requirements for the development of a sustainable future.

2.10 THERMAL ENERGY STORAGE

The concepts of storing thermal energy in the form of sensible heat, latent heat, and reversible thermochemical reactions have been put into practice over years for achieving energy redistribution and energy efficiency on short-term or diurnal and long-term basis. The definition and functional aspects of these three major types of thermal energy storage systems are largely dealt with in later chapters.

2.10.1 Low temperature thermal storage

In low temperature thermal energy storage, the heat energy can be stored and retrieved using a heat storage material, the operating temperature of which is quite comparable with that of the spatial temperature of the cooling/heating application. For instance, the peak load shaving of cooling demand during on-peak periods to off-peak hours using chilled water, phase change materials, or ice-thermal storage options can be grouped under this category. On the other hand, cryogenic energy storage (CES) is a type of storage principle in which the cryogen (e.g., liquid air or liquid nitrogen) is produced during off-peak power demand periods using renewable-based power sources or by mechanical work obtained from the hydro or wind turbines. During on-peak load demand periods, the cryogen being subjected to heat from the surrounding environment is boiled, and the heated cryogen is then utilized for generating electricity through the cryogenic heat engine arrangement. The schematic diagram of the CES system is depicted in Fig. 2.12.

It is pertinent to note that the energy density of CES system is relatively higher (100–200 Wh/kg) with low capital cost per unit of energy. The longer storage duration and environmental friendliness makes the CES system viable for commercialization. However, the energy efficiency of the CES system when compared to the energy consumption is relatively low (about 40–50%).

2.10.2 Medium and high temperature thermal storage

The medium temperature thermal storage system utilizes heat storage materials whose operating temperature is generally higher than that of the human comfort range. However, for some heating applications in buildings, the operating temperature of the heat storage materials are acceptable, which allow them to be incorporated into the fabric components of the building structure. In the case of solar water heating or air heating applications, the heat storage materials would function at an elevated temperature for storing the requisite heat energy. On the other hand, the high temperature thermal storage system

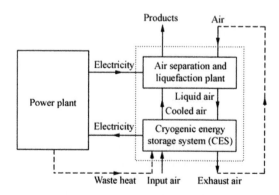

FIGURE 2.12

Schematic diagram of CES system [26].

utilizes molten salt materials for storing and releasing heat energy at very high operating temperature (> 300 °C). High temperature heat storage systems often find their potential application in concentrated solar power plants.

2.11 TECHNICAL EVALUATION AND COMPARISON OF ENERGY STORAGE TECHNOLOGIES

Energy storage technologies have unique characteristics, which may differ with respect to their application domain and energy demand. By performing an evaluation based on the technical as well as economical aspects of the energy storage systems, the selection of the proper storage system can be made feasible for the energy applications. The set of evaluation criteria set forth for energy storage systems are discussed next:

- Storage capacity
  - It is dependent on the quantity of available energy in the storage system after it is charged. Usually, as discharging is not complete, the storage capacity often signifies the measure of total energy stored in the system. The total energy being stored would be superior compared to what is being retrieved. The minimum-charge state of the storage system is represented by the depth of discharge, which influences the usable energy. During frequent charging and discharging, the amount of energy retrieved is lower than storage capacity due to the weakening of efficiency of the storage system.
- Energy and power density
  - Energy density is computed based on the energy being stored to that of the volume. Likewise, power density (W/kg or W/liter) is calculated as the ratio of rated power output to the volume of the storage device. Volume of the storage device is estimated as the volume of the whole/ total energy storage system. The comparison of energy density with respect to the power density of different energy storage systems is depicted in Fig. 2.13
- Response time
  - Response time is the rate or speed of discharge or storage of energy for offsetting load demand requirements considerably. The response time of the energy storage system may vary in terms of a few milliseconds to hours depending on the applications. Hence, the suitability of a particular energy storage system for a specific application depends largely on the response time to the energy and power density and storage capacity.
- Self-discharge
  - This is a property of the energy storage system in which the rate of retention of the stored energy is decreased through the energy being discharged during nonuse periods.
- Efficiency
  - Energy storage systems or devices functioning on a continuous or partial basis have inherent process energy losses or from the components associated to the system. The efficiency of the storage system can be enhanced by regulating the power-transfer chain losses during charging and discharging processes. Instantaneous power can be a decisive factor determining the efficiency of the energy storage system. The graphical form of the existence of an optimum time of discharge and maximum efficiency is depicted in Fig. 2.14(a). For an

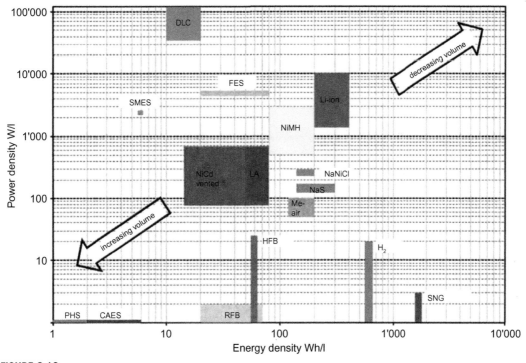

FIGURE 2.13

Comparison of power density (Watts/liter of storage facility) and energy density (Watt-hours/liter of storage facility) of EES technologies [27,28].

actual case, the results projected may vary due to the intricate nature of the components performing with respect to the operating point and the state of charge as presented in Fig. 2.14(b).

- Control and monitoring
  - The process of control and monitoring of the energy storage systems are regarded to be extremely important to ensure the operational performance and the safety aspects of the system/device for storing energy.
- Life expectancy
  - The estimation of the lifetime of the energy storage system is equally important for ensuring its reliability on the long run. The accuracy with which the estimation is carried out by including the original investment costs determines the overall energy storage cost and the projected life expectancy of the system.
- Economics of operation
  - The cost involved in the energy storage process can be considered an important factor for evaluating the economic prospects of the storage system. In some energy storage systems, the cost associated with the incorporation of auxiliary components in the system is determinant of the total cost of the system. For such reason, the economic viability of some energy storage systems

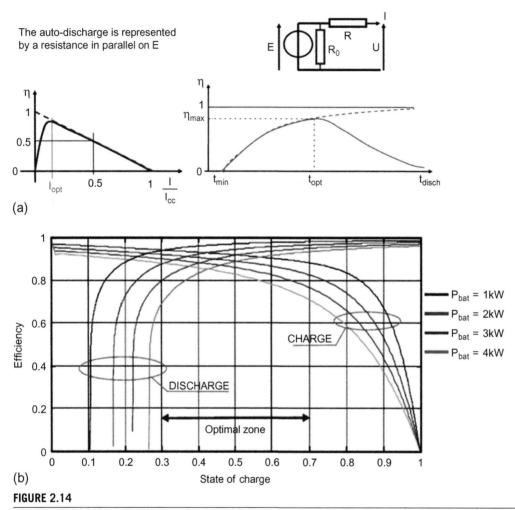

FIGURE 2.14

(a) Graphs representing the effect of current or time discharge, as well as the self-discharge effect on efficiency of electrochemical accumulator. The dotted lines correspond to a model with no self-discharge resistance (I: source of current, I CC: short-circuit current), and (b) power efficiency of a 48 V-310 Ah (15 kWh/10 h discharge) lead battery [29].

is limited by a minimum energy storage capacity and power output. The comparison of costs of different energy storage technologies based on the data from [Raster] including the required power conditioning equipment for relative energy storage system/device projected in Fig. 2.15.

- Self-sufficiency
    - It is the state or the maximum amount of time the energy storage system can release energy continuously. It can also be defined as the ratio of energy capacity to the discharge power. It is largely dependent on the type of application and the type of storage.

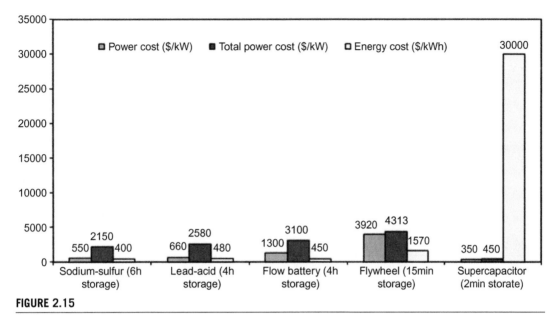

FIGURE 2.15

Costs for energy storage systems [6].

- Durability
  - This refers to the total number of times the energy storage system can perform by releasing energy for which it was designed after every recharging cycle. The more resilient the system is, the higher its performance over a period of time. The durability of the energy storage system strongly depends on the state of the charge with respect to the number of operating cycles as shown in Fig. 2.16. In most practical situations, the durability of the energy storage systems as predicted would vary dramatically, which indicates the elusive nature of quantifying the number of cycles based on the orders of magnitude.
- Operational constraints
  - The factors related to safety in terms of toxicity, flammability, durability, and other factors such as temperature and pressure have a direct effect on the selection of a particular energy storage system by virtue of satisfying energy requirements.

The comparison between the rated power, energy content, and time of discharge for a variety of energy storage technologies is presented in Fig. 2.17(a). The global installation capacity of different energy storage systems typically utilized in electricity grids is shown in Fig. 2.17(b). The feasibility chart of the energy storage technologies in the present as well as in the future representing the gap present in their development and the scope for further research as well is illustrated in Fig. 2.17(c). Information on the world's metal reserves for the EES, the operation and maintenance costs of EES technologies, and the major and the essential characteristics of EES technologies has been reviewed and presented in Tables 2.4–2.6, respectively.

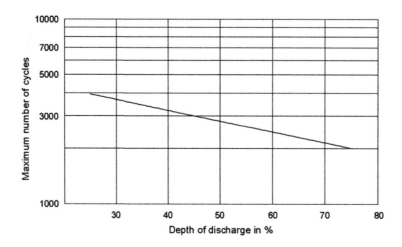

FIGURE 2.16

Evolution of cycling capacity as a function of depth of discharge for a lead–acid battery [30].

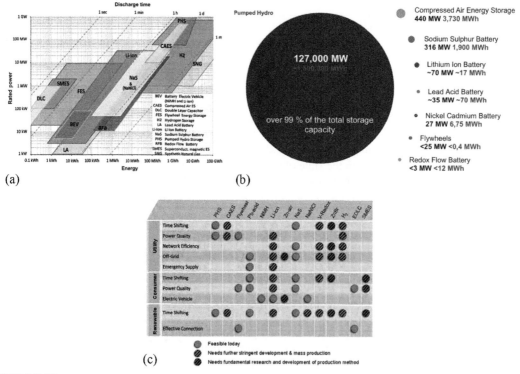

(a)   (b)   (c)

FIGURE 2.17

(a) Comparison of rated power, energy content, and discharge time of different EES technologies; (b) worldwide installed storage capacity for electrical energy; and (c) electrical energy storage present feasibility, future potential, need for further research, and development [27,28,31].

Table 2.4 Worlds metal reserves for electrical energy storage (EES) [32]

Metal	EES Usage	Reserves (ktons)	Reserves (ktons)	Years Left (Current Cons. Rate)
Antimony	Mg/Sb	1800	169	10.7
Bismuth	SMES	320	8.5	37.7
Barium	SMES	240,000	7800	30.8
Copper	SMES	690,000	16,100	42.9
Helium	SMES	31,300[a]	180[a]	173.9
Lead	Pb–acid, SMES	85,000	4500	18.9
Lithium	Li-ion	13,000	34	382.4
Magnesium	SMES, fuel cell, Mg/Sb	N/A	0.78	>1000
Nickel	NiCd, NaNiCl, fuel cell	80,000	1800	44.5
Palladium/Platinum	Fuel cell	66	0.40	165
Sodium	NaS	3,300,000	6000	550
Strontium	SMES	6800	380	17.9
Titanium	Flywheel, fuel cell	6550	283	23.1
Vanadium	V-redox	14,000	60	233.3
Yttrium	SMES	540	8.9	60.7
Zinc	ZnBr, Zn–Air	250,000	12,400	20.2
Zirconium	Fuel cell	52,000	1410	36.9

[a] In million m^3.

Table 2.5 Operation and maintenance costs of EES technologies [33]

	O&M Costs ($/kW-y)		
	Bulk Energy Storage	Distributed Generation	Power Quality
PHS	2.5	N/A	N/A
CAES	2.5	10	N/A
Hydrogen	N/A	3.8 (fuel cell) 2.5 (gas turbine) N/A (electrolyser)	N/A
Flywheel	N/A	$1000/year[a]	5[b]
SMES	N/A	N/A	10
EDLC	N/A	N/A	5
Conventional Batteries			
Pb–acid	5	15	10
NiCd	5	25	N/A
NiMH	N/A	N/A	N/A
Advanced Batteries			
Li-ion	N/A	25	10
NaS	20	N/A	N/A
NaNiCl	N/A	N/A	N/A
Flow Batteries			
V-Redox	N/A	20	N/A
ZnBr	N/A	20	N/A
Zn–air	N/A	N/A	N/A

[a] The costs are specific to a 18 kW/37 kWh flywheel system and are not generic.
[b] The costs refer to either high or low speed flywheel system.

Table 2.6 Characteristics of EES technologies [2]

	Power Rating (MW)	Energy Rating (kWh)	Specific Power (W/kg)	Specific Energy (Wh/kg)	Power Density (kW/m³)	Energy Density (kWh/m³)	Round-Trip eff. (%)	Critical Voltage (V)
PHS	100-5000	2×10^5-5×10^6	Not appl.	0.5-1.5	0.1-0.2	0.2-2	75-85	Not appl.
CAES	100-300	2×10^5-10^6	Not appl.	30-60	0.2-0.6	12	≤5	Not appl.
Hydrogen	<50	10^5	>500	33,330	0.2-20	600[b]	29-49	Not appl.
Flywheel	<20	$10^{-5} \times 10^3$	400-1600	5-130	5000	20-80	85-95	Not appl.
SMES	0.01-10	10^{-1}-10^2	500-2000	0.5-5	2600	6		Not appl.
EDLC	0.01-1	10^{-3}-10	0.1-10	0.1-15	40,000-120,000	10-20		0.5
Conventional Batteries								
Pb-acid	<70	10^2-10^5	75-300	30-50	90-700	75	80-90	1.75
NiCd	<40	10^{-2}-1.5×10^3	150-300	45-80	75-700	<200	70-75	1.0
NiMH	10^{-6}-0.2	10^{-2}-500	70-756	60-120	500-3000	<350	70-75	1.0
Advanced Batteries								
Li-ion	0.1-5	10^2-10^5	230-340	100-250	1300-10,000	250-620	90-98	3.0
NaS	0.5-50	6×10^3-6×10^5	90-230	150-240	120-160	<400	85-90	1.75-1.9
NaNiCl	<1	120-5×10^3	130-160	125	250-270	150-200	90	1.8-2.5
Flow Batteries								
V-Redox	0.03-7	10-104	N/A	75	0.5-2	20-35	75	0.7-0.8
ZnBr	0.05-2	50-4×103	50-150	60-80	1-25	20-35	70-75	0.17-0.3
Zn-air[c]	Several	$\times 103$	1350	400	50-100	800	60	0.9

	Discharge Time	Response Time	Lifetime (Years)	Lifetime (Cycles)	Operating Temp. (1C)	Self-Discharge %/day	Spatial Requirement (m²/kWh)†	Recharge Time[i]
PHS	h-days	s-min	50-100	$>5 \times 10^2$	Ambient	0	0.02	min-h
CAES	h-days	1-15 min	25-40	No limit	Ambient	0	0.10-0.28[g]	min-h
Hydrogen	s-days	ms-min	5-15	$>10^3$	−80 to +100[a]	0.5-2	0.005-0.06[g]	Instant.
Flywheel	15 s-15 min	ms-s	≥20	10^5-10^7	20 to +40	20-100	0.28-0.61[g]	<15 min
SMES	ms-5 min	ms	≥20	10^4	−270 to −140	10-15	0.93-26[g]	min
EDLC	ms-1 h	ms	≥20	$>5 \times 10^5$	−40 to +85	2-40	0.43	s-min
Conventional Batteries								
Pb-acid	s-3h	ms	3-15	2×10^3	+25	0.1-0.3	0.06	8-16h
NiCd	s-h	ms	15-20	1.5×10^3	−40 to +45	0.2-0.6	0.03	1h
NiMH	h	ms	5-10	3×10^2-5×10^2	−20 to +45	0.4-1.2	0.02?	2-4h
Advanced Batteries								
Li-ion	min-h	ms-s	8-15	$>4 \times 10^3$	−10 to +50	0.1-0.3	0.01?[h]	min-h
NaS	s-h	ms	12-20	2×10^3-4.5×10^3	+300	20	0.019	9h
NaNiCl	min-h	ms	12-20	10^3-2.5×10^3	+270 to +350	15	0.03?	6-8h
Flow Batteries								
V-Redox	s-10h	<1 ms	10-20	$>13 \times 103$	+0 to +40	0-10	0.04	min
ZnBr	s-10h	<1 ms	5-10	$>2 \times 103$	+20 to +50	0-1	0.02	3-4h
Zn-air[c]	6h	ms	30	$>2 \times 103[>104]$	0 to +50	N/A	<0.005?	N/A

Table 2.6 Characteristics of EES technologies—cont'd

	Investment Power Cost	Investment Energy Cost	Commercial Use Since	Technical Maturity	Environmental Impact[e]
PHS	500–3600	60–150	1929	mature	1
CAES	400–1150	10–40	N/A	medium[i]	1
Hydrogen	550–1,600[d]	1–15	2010	early	3
Flywheel	100–300	1000–3500	2008	mature	5
SMES	100–400	700–7000	2000's	early	3
EDLC	100–400	300–4000	1980s	medium	3
Conventional Batteries					
Pb-acid	200–650	50–300	1870	mature	2
NiCd	350–1000	200–1000	1915	mature [portable]	2
NiMH	120%×NiCd	120%×NiCd	1995	mature [mobile]	3
Advanced Batteries					
Li-ion	700–3000	200–1800	1991	mature [mobile]	4
NaS	700–2000	200–900	1998	Medium	4
NaNiCl	100–200	70–150	1995	mature [mobile]	N/A
0.5					
Flow Batteries					
V-Redox	2500	100–1000	1998	medium	3
ZnBr	500-1800	100–700	2009	medium	3
Zn-air	785	126N/A	2013/14	early	3

	Recyclability[e,f]	Maintenance[e]	Memory Effect	Transportability	Cumulative en.demand MJ/KWH[k]
PHS	Not appl.	3	N/A	no	N/A
CAES	Not appl.	3	N/A	no	N/A
Hydrogen	2–3?	1	N/A	yes	5501
Flywheel	4	3	no	yes	30,449
SMES	N/A	2	N/A	yes	N/A
EDLC	4	4	no	yes	N/A
Conventional Batteries					
Pb-acid	5	3	no	yes	652
NiCd	4–5	1	yes	yes	1372
NiMH	4-5?	1	yes	yes	N/A
Advanced Batteries					
Li-ion	4	5	no	yes	1156
NaS	5	3	no	yes	N/A
NaNiCl	5	5	no	yes	N/A
0.5					
Flow Batteries					
V-Redox	5	3	no	no	774
ZnBr	5	1	no	yes	N/A
Zn-air	3	3	no	yes	710

Table 2.6 Characteristics of EES technologies—cont'd

	Power Rating (MW)	Energy Rating kWh	Specific Power W/kg	Specific Energy Wh/kg	Power Density kW/m3	Energy Density kWh/m3
			References for Table 2.6			
PHS	[34]	[35]	Not appl.	[34,35]	[28]	[28]
CAES	[34]	[35]	Not appl.	[34]	[28]	[35,38]
Hydrogen	[34]	[28]	[34,38]	[28]	[28]	[28]
Flywheel	[34,35]	[35]	[34,35]	[34]	[28]	[28]
SMES	[34,35]	[35]	[34]	[34]	[28]	[28]
EDLC	[34]	[35]	[34,35]	[34]	[28]	[28]
Conventional Batteries						
Pb-acid	[36]	[35]	[34]	[34,35]	[28]	[36]
NiCd	[34]	[35]	[34]	[35]	[28]	[37]
NiMH	[35]	[35]	[28,50]	[35]	[28]	[37]
Advanced Batteries						
Li-ion	[37]	[35]	[37]	[35,37]	[28]	[37]
NaS	[34]	[35]	[34]	[34]	[28]	[28,55]
NaNiCl	[34]	[35]	[34]	[34]	[28]	[35]
Flow Batteries						
V-Redox	[34]	[31,60]	N/A	[34,36]	[28]	[35]
ZnBr	[34]	[35]	[34,35]	[35]	[28]	[35]
Zn-air	[64]	[64]	[28]	[65]	[28]	[64]

	Round-tripeff. %	Critical Voltage V	Discharge Time	Response Time	Life Time [years]	Life Time [cycles]
			References for Table 2.6—cont'd			
PHS	[28,36]	Not appl.	[28,34]	[34,35,37]	[34–36]	[37]
CAES	[36]	Not appl.	[34,39]	[34,35]	[34,35]	[35]
Hydrogen	[36]	Not appl.	[34]	[40]	[34]	[34]
Flywheel	[34]	Not appl.	[34]	[34,37,38,31]	[34,35,41]	[34,35]
SMES	[34,38,42,43]	Not appl.	[34]	[34,37,38,40]	[34,35]	[34,35]
EDLC	[34,44]	[45]	[34]	[34,38,40]	[34]	[36]
Conventional Batteries						
Pb-acid	[28,36]	[46,47]	[34]	[28,38,40]	[34]	[36]
NiCd	[36]	[47,48]	[34]	[28,38,49]	[34]	[35]
NiMH	[36]	[47,48]	[28]	[28,38,39]	[28]	[35]
Advanced Batteries						
Li-ion	[36,31,44]	[46,47,51,52]	[34,37]	[28,37,38]	[34,53,54]	[35,31]
NaS	[34]	[56,57]	[34,40]	[28,38,49]	[36]	[34]
NaNiCl	[34,58]	[59]	[34]	[28,38,39]	[36]	[35]
Flow Batteries						
V-Redox	[36,53]	[61,62]	[34]	[35]	[35]	[36]
ZnBr	[34,44]	[63]	[34]	[49]	[34,36]	[34]
Zn-air	[66]	[67,68]	[64]	[28,38,64]	[64]	[64]

Table 2.6 Characteristics of EES technologies—cont'd

	Operating Temp. °C	Self-Discharge%/ Day	Spatial Requirement m2/kWh	Recharge Time	Inv. Power Cost €/kW	Inv. Energy Cost €/kWh
References for Table 2.6						
PHS	assumption	[34,35]	[40]	[36]	[34,35]	[34]
CAES	assumption	[34]	[40]	[72]	[34,35]	[35]
Hydrogen	[36]	[34]	[40]	[40]	[34]	[34]
Flywheel	[34,35]	[34,36]	[40]	[35]	[34]	[34]
SMES	[35,42]	[34]	[40]	[78]	[34]	[34]
EDLC	[34,35]	[34]	[65]	[35]	[34]	[34]
Conventional Batteries						
Pb-acid	[79]	[34,77]	[40]	[35]	[34]	[34,36]
NiCd	[36]	[34]	[40]	[35]	[34]	[34]
NiMH	[35,36]	[35]	assumption	[35]	[35]	[35]
Advanced Batteries						
Li-ion	[35]	[34,77]	assumption	[37]	[34]	[34]
NaS	[37]	[34]	[40]	[80]	[34]	[34]
NaNiCl	[35,36]	[34]	assumption	[55]	[34]	[34]
Flow Batteries						
V-Redox	[34,35]	[34]	[49]	[28]	[34,35]	[34]
ZnBr	[41,82,83]	[34,35]	[40]	[49]	[34]	[34]
Zn-air	[84]	N/A	assumption	N/A	[64]	N/A

Font coding (related to the characteristic described in each column): bold roman = most favourable(s), bold italic = second most favourable(s), italic = least favourable(s).

The en-dash in some inputs (e.g., in column energy rating) is indicative of "to." It is used as a sign to indicate the range.

N/A = Not available in the literature, Not appl. = Not applicable.

[a] Hydrogen's operating temperature of 80-100 1C relates to polymer electrolyte fuel cells. The operating temperature of solid electrolyte fuel cells is 1000 °C [36].

[b] Hydrogen's energy density of 600 kWh/m³ is for a pressure of 200 bar.

[c] Zn–air energy storage system refers to rechargeable flow battery technology; it is an emerging technology that has been developed very recently.

[d] The power price reported for hydrogen relates to gas turbine based generator. The power price for fuel cells is in range of 2000–6600 €/kW [34].

[e] Environmental impact, recyclability, and maintenance are classified on a 5-point scale (1 to 5). For environmental impact and maintenance 1 = high, 2 = medium, 3 = low, 4 = very low, 5 = no. For recyclability 1 = poor, 5 = excellent.

[f] Spatial requirement and recyclability ratings are based either on literature or own judgement based on power and density figures (indicated with a question mark in this case).

[g] Spatial requirement: CAES, lower value for CAES storage in aquifer and higher value for CAES in vessels. Hydrogen, lower value for H₂ engine and higher value for H₂ fuel cell. SMES, lower value accounts for large SMES and higher value accounts for micro-SMES. Flywheels, low value for high-speed flywheels and higher value for low-speed flywheels.

[h] The spatial requirement for Li-ion electricity storage is found in the literature 0.03 m²/kWh [49]; though this figure seems to be quite large considering the energy density of Li-ion batteries currently found in the literature [37], so an adjustment on this figure is made (0.01 m²/kWh).

[i] The recharge time for each technology is proportionate to the size of the system.

[j] Technical maturity: diabatic CAES storage according to [28] is a mature technology, while adiabatic CAES is at an early development stage.

[k] Cumulative energy demand: only the energy input for the material supply is included.

References for Table 2.6—cont'd								
	Commercial Use Since	**Technical Maturity**	**Environ. Impact**	**Recycl- ability**	**Main- tenance**	**Memory Effect**	**Transpor- tability**	**Cumulativeen. demandMJ/ kWh**
PHS	[31,69]	[28,40,70]	[35]	Not appl.	[32,35]	N/A	[71]	N/A
CAES	N/A	[40,49]	[40]	Not appl.	[87]	N/A	[71]	N/A
Hydrogen	[73]	[28,40,49]	[38,40]	assumption	[58]	N/A	[40]	[74]
Flywheel	[75]	[28,49]	[35,40,41]	[76]	[40,77]	[87]	[86]	[74]
SMES	[35]	[28,49]	[40]	N/A	[35,38]	N/A	[79]	N/A
EDLC	[58]	[49]	[58]	[28]	[34,35]	[41]	[34]	N/A
Conventional Batteries								
Pb-acid	[37]	[28,49]	[35]	[36,31,76]	[32,77]	[74]	[34]	[74]
NiCd	[28]	[28]	[35]	[35,76]	[35]	[34,38,49,58,74]	[89]	[74]
NiMH	[54]	[28]	[35]	assumption	[35]	[34,74]	[71]	N/A
Advanced Batteries								
Li-ion	[35,31,74]	[36,49]	[58,76]	[35,58,76]	[35]	[35,37,74]	[31]	[74]
NaS	[49]	[49]	[35,49]	[41,49]	[32,77]	[88]	[31]	N/A
NaNiCl	[28]	[81]	N/A	[35,55]	[35]	[55]	[31]	N/A
Flow Batteries								
V-Redox	[35]	[34,49]	[35]	[35]	[49,53]	[60]	[71]	[74]
ZnBr	[49]	[34,49]	[35]	[35]	[32]	[41]	[31,71]	N/A
Zn-air	[64]	[28]	[38,64,65,71]	[58]	[64]	[85]	[31,64]	[74]

2.12 CONCISE REMARKS

Energy storage technologies basically facilitate achieving demand-side energy management, bridging the gap present between the power demand and the quality of power supplied and reliability on long-term basis. Through the amalgamation of energy storage systems, the power and the energy challenges faced by conventional systems can be effectively confronted. The attractive perspective of energy storage technologies is that they have numerous applications ranging from large-scale generation and transmission-based systems to network distribution systems. Despite having a variety of energy storage technologies, it can be seen that each storage system performs in a different manner with its inherent storage characteristics.

The interests shown toward the development of energy storage technologies are currently gaining impetus. It is foreseen that the level of storage capacity can be increased by 15–25% in the imminent future in developed countries, and this value may increase in developing nations. By this, the value chain in the electricity industry can be improved to a greater extent.

In the present scenario, pumped hydroelectric storage can be the leading technology compared to other storage systems falling under the category of large-scale energy storage. The CAES systems can compete with the pumped hydroelectric system; however, their development may be seen as growing only in nations with good and favorable geological conditions. The batteries technology, on the other hand, can gain momentum in the energy market for storing and discharging electricity on demand.

Likewise, the SMES, flywheel, flow batteries, fuel cells, thermal energy storage, and so on are also equally important in the energy sector in recent years. The swift response exhibited during the charging and discharging processes allow them to be utilized for many short- and long-scale applications, especially in the power quality sector. In short, the high energy density and energy capacity, increased storage benefits, durability, reliability, energy conservation, and environmental safety prospects of the energy storage technologies enable them to be preferred perpetually toward growing energy requirements.

References

[1] Kousksou T, Bruel P, Jamil A, El Rhafiki T, Zeraouli Y. Energy storage: applications and challenges. Sol Energy Mat Sol Cells 2014;120:59–80.

[2] Chatzivasileiadi A, Ampatzi E, Knight I. Characteristics of electrical energy storage technologies and their applications in buildings. Renew Sustain Energy Rev 2013;25:814–30.

[3] Ibrahim H, Ilinca A, Perron J. Energy storage systems—characteristics and comparisons. Renew Sustain Energy Rev 2008;12:1221–50.

[4] Multon B, Bernard N, Kerzrého C, Ben Ahmed H, Cognar JY, Dclamare J, Faure F. Stockage électromécanique d'énergie. Club CRIN Énergie Alternatives—Groupe Stockage d'énergie. Présentation du 23 Mai 2002.

[5] <http://www.beaconpower.com/products/presentations-reports.asp> (last accessed October 2012).

[6] Zhou Zhibin, Benbouzid Mohamed, Charpentier Jean Frédéric, Scuiller Franck, Tang Tianhao. A review of energy storage technologies for marine current energy systems. Renew Sust Energ Rev 2013;18:390–400.

[7] Lazarewicz ML, Rojas A. Grid frequency regulation by recycling electrical energy in flywheels. Proceedings of the IEEE Power Engineering Society General Meeting 2004;2:2038–42.

[8] Hessami M, Bowly DR. Economic feasibility and optimisation of an energy storage system for Portland Wind Farm (Victoria, Australia). Appl Energy 2011;88:2755–63.

[9] Faure F. Suspension magnétique pour volant d'inertie. Thèse de doctorat, France: Institut National Polytechnique de Grenoble; 2003, Juin.

[10] Chen H, Cong TN, Yang W, Tan C, Li Y, Ding Y. Progress in electrical energy storage system: a critical review. Proc Natl Acad Sci USA 2009;19:291–312.

[11] McDowall JA. High power batteries for utilities—the world's most powerful battery and other developments. In: Power engineering society general meeting, 2004; Denver, USA: IEEE; 2004. p. 2034–7.

[12] Tan X, Li Q, Wanga H. Advances and trends of energy storage technology in Microgrid. Elec Pow Energy Sys 2013;44:179–91.

[13] 〈http://www.leonardo-energy.org〉.

[14] Scamman DP, Gavin WR, Roberts EPL. Numerical modelling of a bromide polysulphide redox fl ow battery. Part 1: modelling approach and validation for a pilot-scale system. J Pow Sourc 2009;189:1220–30.

[15] Hall PJ, Bain EJ. Energy storage technologies and electricity generation. Energy Pol 2008;36:4352–5.

[16] Nguyen T, Savinell RF. Flow batteries. Electrochem Soc Inter 2010;19:54–6.

[17] Kazempour SJ, Moghaddam MP, Haghifam MR, Youse fi GR. Electric energy storage systems in a market-based economy: comparison of emerging and traditional technologies. Renew Energy 2009;34: 2630–9.

[18] Yamamura TT, Wu X, Ohta S, Shirasaki K, Sakuraba H, Satoh I. Vanadium solid-salt battery: solid state with two redox couples. J Pow Sourc 2011;196:4003–73.

[19] Price A. Technologies for energy storage—present and future: flow batteries. In: IEEE power engineering society summer meeting; 2000. p. 1541–5.

[20] Chen W, Ådnanses AK, Hansen JF, Lindtjørn JO, Tang T. Super-capacitors based hybrid converter in marine electric propulsion system. In: Proceedings of the IEEE XIX international electronics; 2010, 1–6 Mach. Conf Rome.

[21] http://www.beaconpower.com/products/EnergyStorageSystems/DocsPresentations.htm [2007-03-20].

[22] Christodoulou C, Karagiorgis G, Poullikkas A, Karagiorgis N, Hadjiargyriou N. Green electricity production by a grid connected H_2/fuel cell in Cyprus. In: Proceedings of the renewable energy sources and energy efficiency; 2007.

[23] Winter CJ. Hydrogen energy—abundant, efficient, clean: a debate over the energy system of change. Int J Hydro Energy 2009;34:1–52.

[24] Steinfeld A, Meier A. Solar fuels and materials. In: Encycl energy, vol. 5. Elsevier Inc; 2004. p. 623–37.

[25] http://en.wikipedia.org/wiki/Biobattery (accessed January 2014).

[26] Chen HS, Ding YL. A cryogenic energy system using liquid/slush air as the energy carrier and waste heat and waste cold to maximize efficiency, specifically it does not use combustion in the expansion process. UK Patent G042226PT, 2006–02–27.

[27] Schwunk S. Battery systems for storing renewable energy. Germany: Fraunhofer-Institut für Solare Energie; 2011.

[28] International Electrotechnical Commission. Electrical energy storage: white paper. Geneva, Switzerland: International Electrotechnical Commission; 2011.

[29] Gergaud O. Modélisation énergétique et optimisation économique d'un système de production éolien etphotovoltaïque couplé au réseau et associé à un accumulateur. Thèse de l'ENS de Cachan;décembre 2002.

[30] Messenger R, Ventre J. Photovoltaic systems engineering. Boca Raton, FL: CRC Press; 1999.

[31] Rastler D. Electric energy storage technology options: a white paper primer on applications, costs, and benefits. Palo Alto, California: Electric Power Research Institute (EPRI); 2010.

[32] U.S. Department of the Interior. Mineral commodity summaries 2012. United States: U.S. Department of the Interior, 2012.

[33] Schoenung S, Hassenzahl W. Long- vs. short-term energy storage technologies analysis: a life-cycle cost study. Sandia National Laboratories: California, USA; 2003.

[34] Sandu-Loisel R, Mercier A. Technology map of the european strategic energy technolgy plan. Luxembourg: Publications Office of the European Union; 2011.

[35] Alanen R, Appetecchi G, Conte M, De Jaeger E, Graditi G, Jahren S, et al. Electrical energy storage technology review. European Energy Research Alliance 2012

[36] Droege P. Urban energy transition: from fossil fuel to renewable power. 1st ed. Hungary: Elsevier; 2008, 664.

[37] Simbolotti G, Kempener R. Electricity storage: technology brief. IEA-ETSAP and IRENA 2012.

[38] Chen H, Cong TN, Yang W, Tan C, Li Y, Ding Y. Progress in electrical energy storage system: a critical review. Prog Nat Sci 2009;19(3):291–312, PubMed PMID: WOS:000263557200003.

[39] Hawaiian Electric Company. Energy storage 2012 [cited 2012 20 August]. Available from: ⟨http://www.heco.com/portal/site/heco/menuitem.508576f78baa14340b4c0610c510b1ca/?vgnextoid=94600420af0db110VgnVCM1000005c011bacRCRD&vgnextchannel=ab020420af0db110Vgn VCM1000005c011bacRCRD&vgnextfmt=default&vgnextrefresh=1&level=0&ct=article⟩.

[40] Schoenung S. Characteristics and technologies for long- vs. short-term energy storage: a study by the DOE Energy Storage Systems Program. California, USA: Sandia National Laboratories; 2001.

[41] Connolly D. A review of energy storage technologies for the integration of fluctuating renewable energy: Limerick; 2010

[42] Hassenzahl W, Hazelton D, Johnson B, Komarek P, Noe M, Reis C. Electric power applications of superconductivity. In: Proceedings of the IEEE; 2004. p. 1655–74.

[43] European Commission. DG ENERWorking Paper: The future role and challenges of energy storage. 2013 [cited 2013 10 April]. Available from: http://ec.europa.eu/energy/infrastructure/doc/energy-storage/2013/energy_storage.pdf/

[44] Teller O, Nicolai JP, Lafoz M, Laing D, Tamme R, Pedersen AS, et al. Joint EASE/EERA recommendations for a European energy storage technology development roadmap towards 2030. European Association for Storage of Energy and European Energy Research Alliance 2013.

[45] Willer B. Investigation on storage technologies for intermittent renewable energies: evaluation and recommended R&D strategy. WPST3-Supercaps. INVESTIRE-Network, 2003.

[46] Garche J. Encyclopedia of electrochemical power sources. The Netherlands: Elsevier; 2009.

[47] Buchmann I. Basics about discharging 2013 [cited 2013 10 April]. Availablefrom:/>http://batteryuniversity.com/learn/article/discharge_methods/>.

[48] Randolph C. Glossary of battery terms 2013 [cited 2013 10 April]. Available from: ⟨http://www.greenbatteries.com/batteryterms.html⟩.

[49] Beaudin M, Zareipour H, Schellenberglabe A, Rosehart W. Energy storage for mitigating the variability of renewable electricity sources: an updated review. Energy for Sustainable Development 2010;14(4):302–14.

[50] Evans B. EE382C embedded software systems—battery technology 2000 [cited 2012 21 August]. Available from: <http://users.ece.utexas.edu/_bevans/courses/ee382c/lectures/batteries.html/>.

[51] Large Battery. Lithium battery cutoff voltage 2011 [cited 2013 10 April]. Available from: ⟨http://www.large-battery.com/lithium-battery-discharge-cutoff-voltage.html⟩.

[52] Buchmann I. Premature voltage cut-off 2013 [cited 2013 10 April]. Availablefrom:/>http://batteryuniversity.com/learn/article/premature_voltage_cut_off//&gt.

[53] Inage S. Prospective on the decarbonised power grid. IEC/MSB/EES Workshop; 2011 31 May–1 June 2011; Germany.

[54] Rastler D. Overview of energy storage options for the electric enterprise. Palo Alto, California: Electric Power Research Institute (EPRI); 2009.

[55] Palmer L. Zebra battery 2008 [cited 2012 22 August]. Available from: <http://www.solartaxi.com/technology/zebra-battery/>.

[56] Crompton T. Battery reference book. 3rd ed. Oxford, UK: Reed Educational and Professional Publishing; 2000.

[57] Braithwaite J, Auxer W. Sodium Beta batteries. Sandia National Laboratories; 1993. [cited 2013 10 April]. Available from: <http://www.sandia.gov/ess/publications/SAND1993-0047j.pdf/>.

[58] Naish C, McCubbin I, Edberg O, Harfoot M. Outlook of energy storage technologies. European Parliament, Policy Department, Economic and Scientific Policy; 2008.

[59] Kim J, Li G, Lu X, Lemmon J, Sprenkle V, Cui J, et al. Intermediate temperature:planar Na–Metal Halide batteries. Pacific Northwest National Laboratory. DOEEES Program Peer Review and Update Meeting; 26-28 Sept; Washington DC 2012. [cited 2013 10 April]. Available from: <http://energy.gov/sites/prod/files/ESS%202012%20Peer%20Review%20-%20Intermediate%20Temperature%20Planar%20Na-Metal%20Halide%20Batteries%20-%20Jin%20Kim,%20PNNL.Pdf>.

[60] Tassin N. Investigation on storage technologies for intermittent renewable energies: evaluation and recommended R&D strategy. INVESTIRE-Network, 2006.

[61] Yamamura T, Wu X, Sato I, Sakuraba H, Shirasaki K, Ohta S. Vanadium battery 2013 [cited 2013 10 April]. Available from: <http://ip.com/patfam/en/43900326/>.

[62] Unknown author. UNIKEN the University of New South Wales. Energy Focus, August 1995; Magazine of the NSW Department of Energy, Australia. 1995. [cited 2013 10 April]. Available from: <http://www.oocities.org/infotaxi/VANADI4.HTM#Swe/>.

[63] Norris BJ, Lex P, Ball GJ, Scaini V. Grid-connected solar energy storage using the zinc–bromine flow battery, 2002. Available from: <www.zbbenergy.com/index.php/download_file/view/20/znbr/>.

[64] EOS Energy Storage. Energy storage. 2012.

[65] Mancey H. Rechargeable metal–air battery system. Next Generation Batteries, 2012. 2012, 19–20 July; Boston, USA.

[66] Lieurance D, Kimball F, Rix C. Modular transportable superconducting magnetic energy systems 1995 [cited 2012 19 August]. Available from: http://ntrs.nasa.gov/archive/nasa/casi.ntrs.nasa.gov/19960000275_1996900275.pdf

[67] Energizer. Energizer Zinc Air prismatic handbook 2009. Version 1.2. [cited 2013 10 April]. Available from: <http://data.energizer.com/pdfs/zincairprismatichandbook.pdf/>.

[68] Duracell. Technical bulletin 2004 [cited 2013 10 April]. Available from: http://www.duracell.com/media/en-US/pdf/gtcl/Technical_Bulletins/Zinc%20Air%20Tech%20Bulletin.pdf/.

[69] Baker J, Collinson A. Electrical energy storage at the turn of the millenium. Power Engineering Journal 1999;6:107–12.

[70] APS panel on public affairs. Challenges of electricity storage technologies: A report from the APS panel on public affairs committee on energy and environment. 2007. Available from: ⟨www.aps.org/policy/reports/popa-reports/upload/Energy_2007_Report_ElectricityStorageReport.pdf⟩.

[71] EOS Energy Storage. Energy storage: Utility opportunity summary. 2011.

[72] Das T, McCalley J. Educational chapter: compressed energy storage 2012 [cited 2012 21 August]. Available from: http://home.eng.iastate.edu.

[73] Murray J. French firm unveils "game-changing" hydrogen storage system 2010 [cited 2012 21 August]. Available from: ⟨http://www.businessgreen.com/bg/news/1805351/french-firm-unveils-game-changing-hydrogen-storage⟩.

[74] Buchmann I. Is Lithium-ion the Ideal Battery? 2012 [cited 2012 21 June]. Available from: <http://batteryuniversity.com/learn/article/is_lithium_ion_the_ideal_battery/>.

[75] Lazarewicz M, Judson J. Performance of first 20 MW commercial flywheel frequency regulation plant 2011 [cited 2012 21 August]. Available from: ⟨http://www.beaconpower.com/files/Beacon_Power_presentation_ESA6_7_11_FINAL.pdf⟩.

[76] Alsema E, Patyk A. Investigation on storage technologies for intermittent renewable energies: evaluation and recommended R&D strategy. WP5 Final report-environmental issues. INVESTIRE-Network, 2003.

[77] Kaldellis JK, Zafirakis D, Kavadias K. Techno-economic comparison of energy storage systems for island autonomous electrical networks. Renew Sustain Energy Rev 2009;13(2):378–92, PubMed PMID: WOS:000262905500005.

[78] http://www.azom.com. Superconducting Magnetic Energy Storage (SMES)Systems 2001 [cited 2012 21 August]. Available from: http://www.azom.com/article.aspx?ArticleID=1123-_How_fast_can.

[79] Buchmann I. Can the Pb-acid battery compete in modern times? 2012 [25 September 2012]. Available from:<http://batteryuniversity.com/learn/article/can_the_lead_acid_battery_compete_in_modern_times/>.

[80] Norris B, Newmiller J, Peek G. NaS battery demonstration at American electric power. California, USA: Sandia National Laboratories; 2007.

[81] Sauer D, Kleimaier M, Glaunsinger W. Relevance of energy storage in future distribution networks with high penetration of renewable energy sources. In: 20th International conference on electricity distribution; 8—11 June; 2009, Prague.

[82] Tong W. Wind power generation and wind turbine design. Southampton, UK: WIT Press; 2010, 725.

[83] McKenna P. Startup promises a revolutionary grid battery 2012 [cited 2012 20 August]. Available from: http://www.technologyreview.com/news/426535/startup-promises-a-revolutionary-grid-battery/.

[84] Steve's Digicams. Zinc-air battery packs 2012 [cited 2012 23 July]. Available from: <http://www.steves-digicams.com/accessories/batteries/other-chargers/zinc-air-battery-packs.html-b>.

[85] Worth B, Perujo A, Douglas K, Tassin N, Brüsewitz M. Investigation on storage technologies for intermittent renewable energies: evaluation and recommended R&D strategy.WPST9-Metal-air systems. INVESTIRE-Network, 2008.

[86] Van der Burgt J. Transportable, flexible, grid-connected storage 2012 [cited 2012 21 August]. Available from: http://smartgridsherpa.com/blog/transportable-flexible-grid-connected-storage.

[87] Reed C. Flywheel energy storage: energy storage via rotational inertia 2010 [cited 2012 21 August]. Available from: ⟨http://mragheb.com/NPRE%20498ES%20Energy%20Storage%20Systems/Christopher%20Reed%20Flywheel%20Energy%20Storage.pdf⟩.

[88] Okimoto A., editor NaS battery application. DERBI (Development of Renewable Energy Resources in Building and Industry) 2009 International Conference; 2009 June 11-13; Perpignan, France.

[89] Saft. Nickel–Cadmium (Ni–Cd)—SBLE, SBM, SBH - Block battery construction 2007 [cited 2012 22 August]. Available from: ⟨http://www.saftbatteries.com/Produit_SBLE__SBM__SBH___Block_battery_construction_293_38/Language/en-US/Default.aspx⟩.

Thermal Energy Storage Technologies

3.1 INTRODUCTION

Growing concerns over increased greenhouse gas (GHG) emissions and climate change have triggered attempts to conserve energy at every step of global technological and economic development. The increasing concerns related to energy security combined with excessive waste of useful energy have resulted in the development of a variety of energy-efficient technologies to meet the energy demand in the major energy sectors including industry, transport, construction, and buildings. As discussed in Chapter 1, energy efficiency and energy conservation are interrelated, and any effort to integrate breakthrough research concerning the technological aspect combined with economical viability can produce an energy savings benefit on a long-term basis. From the energy perspective, the prominent challenges that are to be carefully dealt with include

- The rapid depletion of primary fossil fuel–based sources
- Complexity involved in extracting fossil fuel sources
- Imminent hike in fuel prices
- Climate change and GHG emissions
- Sporadic distribution of energy being supplied to meet the actual energy demand

3.2 THERMAL ENERGY STORAGE

It has always been important among scientists, engineers, industrialists, and technologists to develop and commission energy-efficient technologies that would fulfill end-use energy requirements. In the context of confronting the aforementioned energy challenges, the search for the development of new technologies or modification of existing technologies has always been recognized as a significant measure toward accomplishing energy efficiency. A variety of energy-efficient technologies in the major energy sectors provides even more opportunity to bridge the gap between the energy supply and end-use energy demand.

Thermal energy storage (TES), or thermal storage, is the one efficient technology available that caters to end-use energy demand through energy redistribution. Energy in the form of heat or cold can be placed in a storage medium for a particular duration and can be retrieved from the same location for later usage. This is the baseline concept of TES, wherein the term *thermal* refers to either heat or cold, depending on the energy interactions between the storage medium and the energy source. The simple schematic representation as depicted in Fig. 3.1 elucidates the generic operational mode of the TES system being integrated with the available thermal energy source.

In normal operation, the critical demand heat load (cooling/heating) is met by the heat source directly, which involves the exhaustive utilization of the source energy with some useful energy being

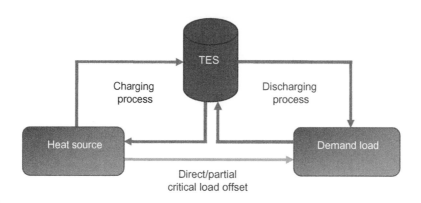

FIGURE 3.1

Simple schematic representation of TES integration and operation.

subjected to waste in the form of heat or mechanical losses. Instead, the incorporation of TES as inter-mediate between source heat energy and end-use load demand would benefit in reducing the waste of useful thermal energy and minimize GHG emissions considerably.

Typically, during the charging process (usually at part load conditions), the quantity of the heat energy that is required to offset the critical load demand (during on-peak periods) is stored in the TES. During discharging process, the stored thermal energy is retrieved from the TES and supplied to the end-use utility for the desired purpose. By integrating TES with the conventional thermal interface systems, the energy redistribution in the form of load shifting from on-peak to off-peak conditions can be effectively achieved. This in turn enables the cooling/heating plant or utility to operate at its base capacity or nominal capacity and thereby helps accomplish enhanced energy efficiency and operational performance of the thermal system.

The quantity of thermal energy that can be stored and discharged solely depends on the storage me-dium characteristics (especially on the storage material) and the associated temperature effects between the storage medium and the energy source. In other words, the amount of energy stored per unit volume (volumetric energy capacity) and the longer storage duration are considered the vital characteristics of TES. In short, the smaller the volume of the storage system and the longer the storage duration, the better would be its thermal storage performance.

3.2.1 Aspects of TES

The interesting features of adopting TES technologies for real-time applications involving the interac-tion of thermal energy, which is available in either heat or cold, are identified here:

- Increased energy efficiency of the existing cooling/heating system
- Capacity reduction of the existing cooling/heating system through load leveling or demand limiting operation
- Size reduction of the cooling/heating system achieved through volumetric energy capacity
- Operational flexibility of the plant or the system increased by storage duration
- Energy backup achieved through effective energy redistribution

- Reduction in the overall cost, including operating and initial cost
- Improved performance of the plant or system using selective heat energy storage medium, for instance sensible, latent, or thermochemical energy storage (TCES) medium

3.2.2 **Need for TES**

The concept of TES has existed for years, but is currently gaining momentum due to the increasing impact of scientific and technological advancements. The reason behind the significant attention given to TES technologies in recent years, as compared to earlier years can be due to the following needs or underlying conditions.

- A mismatch exists between the energy supply and energy demand.
- Intermittent energy sources are utilized for meeting the energy demand.
- Thermal load fluctuations occur only for a short duration.
- Cyclical thermal load conditions prevail in the system.
- An energy supply for catering the critical and part load demand is limited.
- Energy derivative equipments to meet peak load requirements are undersized or oversized.
- Cost is involved in the operation of the cooling/heating systems during day-load and off-peak conditions.
- Rebates, economic credentials, and subsidies are available for the plant or system operation through energy redistribution schemes.

For most real-world applications, the implementation of TES technologies can be considered a bright option, which would offer better energy redistribution capacity and operational flexibility to energy-consuming devices or systems. However, carrying forward such technologies often remains a question to be answered among scientific and engineering communities, despite the previously listed needs for their implementation and commissioning. The challenges and the impediments in promoting TES technologies are largely discussed in forthcoming chapters.

3.2.3 **Energy redistribution requirements**

The meritorious part of TES technologies can be experienced because of the process of energy redistribution. The cooling or heating energy requirements of a system can be shifted from peak load to off-peak load conditions by introducing the TES intermediate. For instance, in the case of a building that requires cooling/heating during its operating periods, the cooling/heating energy demand that arises in peak load conditions (daytime) can be effectively redistributed (shifted) to the part load conditions (nighttime) by TES.

The cooling/heating energy demand that is to be met during daytime can be stored in the TES medium suitably. The stored cold/heat energy can then be retrieved and supplied to the building to fulfill the requisite cooling/heating energy demand. In the case of sustainable solar-derived systems, the high-grade heat energy that is intermittent, if stored properly using the TES methodology, can benefit the end user in many ways. In fact, the heat energy obtained from the solar radiation can be redistributed (stored) during the daytime and can then be utilized during nighttime for heating. Thus, the real benefits of using TES technologies can be experienced through the process of energy redistribution.

3.3 TYPES OF TES TECHNOLOGIES

Thermal energy available in the form of heat or cold can be stored by virtue of change in internal energy of a material through sensible heat, latent heat, and thermochemical means [1–13]. Thermal energy can be either stored in the aforementioned means independently or in combination with these storage means. The three major physical principles by which the heat or cold energy can be stored is explained in forthcoming sections.

3.3.1 Sensible TES

In sensible TES (STES), the temperature of the storage material is increased by virtue of the energy being stored in that material. That is, the internal energy of the storage material is influenced by the energy being stored, which would raise the temperature of the material. In an STES, the heat capacity or the heat energy stored in the material can be directly related to the mass (m), specific heat capacity (c_p), and temperature difference (ΔT) of the material. This physical dependence can be expressed in the form of equation given by

$$Q = mc_p \Delta T \tag{3.1}$$

Equation (3.1) can be expressed in another form as:

$$Q = mc_p \left(T_h - T_l \right) \tag{3.2}$$

where T_h and T_l are the maximum and initial temperature of the material, $(T_h - T_l)$ is referred to the temperature swing.

It is interesting to note that during the TES and release processes, the relatively low thermal capacity of the STES material is usually balanced by their temperature swing. This is the reason why STES materials are largely preferred for high-temperature applications requiring large temperature swings during energy redistribution periods.

3.3.2 Latent TES

In latent TES (LTES), the heat storage material undergoes a phase transformation process for storing or discharging the heat energy. The phase change of the material either from solid to liquid or vice versa normally occurs at isothermal or near isothermal conditions. The heat or cold energy can be stored in the material by virtue of the latent heat of the material with the surrounding heat transfer medium. For instance, during the melting process, the material undergoes phase transition from solid to liquid state by absorbing the heat energy being supplied to the material.

Thus, during the phase change process, a large amount of the heat energy can be stored in the form of latent heat in the material. The case for the freezing process is similar, wherein the cold energy can be trapped from the material during its phase transformation. In general, these materials are termed phase change materials (PCMs). Like the STES materials, the heat storage capacity of the PCMs can be equal to the sum of the latent heat enthalpy at phase transition temperature and the sensible heat being stored over the temperature swing of the energy storage process. This physical process can be suitably expressed in the form as given by:

$$Q = m \left[\left(c_p \Delta T \right)_{\text{sensible}} + H + \left(c_p \Delta T \right)_{\text{latent}} \right] \tag{3.3}$$

The terminology used for referring to the heat storage and release during melting and freezing processes would differ among reference literatures. But in this book, the heat storage and release processes are referred to only with respect to the storage material, regardless of the condition of the surrounding medium (either hot or cold). For instance, freezing process refers to the storage of cold energy into the material, and melting process refers to the release of the stored energy from the material. LTES technology has become increasingly attractive in recent times, which can be attributed to the incorporation of PCMs for achieving the desired thermal storage. The thermophysical properties of the PCMs play a vital role in deciding the extent to which thermal energy can be effectively redistributed in part load conditions and reused during high demand periods. Due to the good volumetric heat storage capacity and handling of the PCMs, they are very much preferred in thermal interface systems for reducing the criticality in load capacities of the cooling/heating equipments or the plant load. LTES systems utilizing PCMs find their potential usage in a wide range of applications including industrial, residential, transport, and renewable energy integration.

3.3.3 **Thermochemical energy storage**

In the journey of TES technologies, a new and attractive energy storage concept that has recently become increasingly popular is the so-called TCES. In this concept, the chemical potential of certain materials is used as the basis for storing and releasing thermal energy with infinitesimal thermal loss. The reversible chemical interactions occurring between the reactive components of materials or chemical species is crucial for the storage and retrieval of heat energy in the TCES. Factually, the endothermic reaction of chemical constituents can be triggered through the supply of heat energy for enabling the storage and release processes to occur in chemical materials. Generically, the thermal energy that is to be stored or released during a specific time period using two chemically reactive components can be established by the following relation given by:

$$C_1C_2 + \text{Heat input} \Leftrightarrow C_1 + C_2 \tag{3.4}$$

This is an endothermic reaction, wherein the chemical species have been dissociated to individual chemical components (C_1 and C_2) by adding heat into the chemical materials. This is a typical example of how to store heat energy in TCES. Similarly, if the separated chemical components (C_1 and C_2) are combined later, the same amount of the stored heat energy can be retrieved (released) with negligible thermal losses. This is the condition of exothermic reaction occurring in the TCES, which is performed for extracting the required amount of stored thermal energy. The prototype system as proposed by the International Energy Agency (IEA) in their Task 32 SHC program works on the principle of TCES for storing thermal energy using salt hydrates as the working pair components/species. The following chemical reaction is useful to explain the working principle of the prototype system:

$$MgSO_4 \bullet 7H_2O(s) + \text{heat input} \Leftrightarrow MgSO_4(s) + 7H_2O(g) \tag{3.5}$$

Based on the aforementioned chemical process, it was estimated in the set Task 32 of IEA that the desorption-adsorption reaction of the hydrates of magnesium sulfate can store $2.8\,\text{GJ/m}^3$ of energy. Herein, the terminology of "sorption" refers to the phenomenon of trapping a gas or a vapor by a solid or a liquid substance that is available at condensed state. The gas or vapor captured is termed "sorbate," and the one that captures the sorbate is termed "sorbent." In TCES, the physics of sorption correlates to both thermophysical and thermochemical conditions, which is closely linked to the concepts of adsorption and absorption. In short, adsorption means the process of binding a gas or vapor on the surface of the porous or solid material, whereas absorption is referred to the process of causing a gas to be absorbed into the liquid medium.

The process of adsorption can be further classified into physisorption and chemisorption, wherein the former occurs mainly due to the van der Waals forces, and the latter occurs due to valence forces. Due to the involvement of surface phenomenon, the heat of sorption is mostly high for the chemisorption compared to the physisorption. In some instances, the chemisorption reaction may be irreversible. The sorption processes can also be further classified into open loop and closed loop, wherein in an open loop type, the working fluid (water vapor) is generally exhausted into the open atmosphere. In the closed loop type, the working fluid is separated from the neighborhood subsystems, thereby enabling higher heat storage capabilities.

3.4 COMPARISON OF TES TECHNOLOGIES

From the TES perspective, the storage and release of the required amount of heat energy through sensible heat and latent heat are considered direct storage techniques. Besides, the TCES, which involves the chemical interaction of chemical components by means of endothermic and exothermic reactions to store and discharge the heat energy, can be considered the indirect method of TES. The comparison between the sensible, latent, and TCES technologies is summarized in Table 3.1

TES technologies offer a spectrum of benefits for achieving a sustainable future by addressing climate change as well as the economical prospects involved in their development. The benefits of TES technologies include the following:

- The exhaustive utilization of the fossil fuels minimized
- Reduction of the GHG emissions through the energy conservative approach
- Reduction of the total operating cost of equipments/plant operation depending extensively on the fossil fuel–derived electrical energy
- Minimization of fuel pricing and costly related risks
- Extension of the existing cooling/heating systems capacity and utility service through load shifting from on-peak to part load conditions
- Reduction of the size of the cooling/heating equipments based on the thermal load profile to meet only the nominal load rather than the critical load at most of the operating periods
- Improving the energy efficiency of the cooling systems through load leveling and demand limiting schemes for acquiring energy savings potential
- Potential for integrating the renewable energy option such as solar, wind, geothermal, and underground storage for power generation and energy redistribution
- Reduction of usage of the energy consuming peaking power plants
- Reduction of redundant electrical/mechanical equipments for meeting the backup cooling capacity of essential systems, thereby reducing the installation cost
- Only energy stored and released, hence no contamination added to the environment
- Development of desalination systems to minimize the scarcity of producing fresh water

Table 3.1 Comparison of the Three Available Technologies for Seasonal Thermal Energy Storage [14]

	Sensible	**Latent**	**Chemical**
Storage medium	Water, gravel, pebble, soil...	Organics, inorganics	Metal chlorides, metal hydrides, metal, oxides...
Type	Water-based system (water tank, aquifer)	Active storage	Thermal-sorption (adsorption, absorption)
	Rock- or ground-based system	Passive storage	Chemical reaction (normally for high-temperature storage)
Advantage	Environmentally friendly cheap material	Higher energy density than sensible heat storage	Highest energy density, compact system
	Relative simple system, easy to control	Provide thermal energy at constant temperature	Negligible heat losses
	Reliable		
Disadvantage	Low energy density, huge volumes required for district heating	Lack of thermal stability	Poor heat and mass transfer property under high-density condition
	Self-discharge and heat losses problem	Crystallization	Uncertain cyclability
	High cost of site construction	Corrosion	High cost of storage material
	Geological requirements	High cost of storage material	
Present status	Large-scale demonstration plants	Material characterization, laboratory-scale prototypes	Material characterization, laboratory-scale prototypes
Future work	Optimization of control policy to advance the solar fraction and reduce the power consumption	Screening for better suited PCMs with higher heat of fusion	Optimization of the particle size and reaction bed structure to get constant heat output
	Optimization of storage temperature to reduce heat losses	Optimal study on store process and concept	Optimization of temperature level during charging/discharging process
	Simulation of ground-/soil-based system with the consideration of affecting factors (e.g., underground water flow)	Further thermodynamic and kinetic study, noble reaction cycle	Screening for more suitable and economical materials
			Further thermodynamic and kinetic study, noble reaction cycle

3.5 CONCISE REMARKS

TES can be literally delineated as the phenomenon of storing (charging) and releasing (discharging) heat energy contained by the storage material of finite mass and heat capacity while being subjected to temperature swings. The basic distinction between a heat storage and a cool storage is that in the former, the energy storage acts as a heat source, whereas in the latter, it functions as a heat sink. Thermal storage technologies can be categorized into sensible heat energy storage, latent heat energy storage, and TCES.

The three major modes of TES technologies show a huge potential in the demand-side energy management by establishing a balance between energy supply and energy demand. The success of TES technologies largely depends of several factors, including the demand load profiles, critical load to be redistributed, material properties for storing and releasing energy, mechanism/type of heat storage, short- or long-term basis, environmental conditions, cost incentives, and economic prospects. TES technologies, on the other hand, offer a wide range of opportunities and benefits to the end-use energy and demand-side management facilities primarily in terms of cost effectiveness and energy savings. It is thus expected that with the promising characteristics and energy conservative aspects, thermal storage technologies can be considered a viable measure for the development of an imminent and sustainable future.

References

[1] ASHRAE. ASHRAE handbook—HVAC systems and equipment. Atlanta, GA: ASHRAE, Inc; 2012.

[2] Silvetti B. Thermal energy storage. Encyclopedia of energy engineering and technology. Boca Raton, FL: CRC Press; 2007, pp. 1412–1421.

[3] Christenson C. Thermal energy storage. Energy management handbook. 6th ed. Lilbum, GA: The Fairmont Press, Inc.; 2007, pp. 519–537.

[4] Parameshwaran R, Kalaiselvam S, Harikrishnan S, Elayaperumal A. Sustainable thermal energy storage technologies for buildings: a review. Renew Sust Energ Rev 2012;16:2394–433.

[5] Dincer I. Thermal energy storage systems as a key technology in energy conservation. Int J Energ Res 2002;26:567–88.

[6] Hasnain SM. Review on sustainable thermal energy storage technologies, part I: heat storage materials and techniques. Energ Convers Manage 1998;39:1127–38.

[7] Hasnain SM. Review on sustainable thermal energy storage technologies. Part II: cool thermal storage. Energ Convers Manage 1998;39:1139–53.

[8] Cabeza LF, Medrano M, Castellon C, Castell A, Solé C, Roca J, et al. Thermal energy storage with phase change materials in building envelopes. Contrib Sci 2007;3(4):501–10.

[9] Baetens R, Jelle BP, Gustavsen A. Phase change materials for building applications—a state of art review. Energ Buildings 2010;42:1361–8.

[10] Sun Y, Wang S, Xiao F, Xiao F, Gao D. Peak load shifting control using different cold thermal energy storage facilities in commercial buildings: a review. Energ Convers Manage 2013;71:101–14.

[11] Seaman A, Martin A, Sands J. HVAC thermal storage: practical application and performance issues. Application Guide, BSRIA; 2000. p. 1–82.

[12] Heim D. Isothermal storage of solar energy in building construction. Renew Energy 2010;35:788–96.

[13] Han YM, Wang RZ, Dai YJ. Thermal stratification within the water tank. Renew Sustain Energy Rev 2009;13:1014–26.

[14] Xu J, Wang RZ, Li Y. A review of available technologies for seasonal thermal energy storage. Sol Energy 2013. http://dx.doi.org/10.1016/j.solener.2013.06.006.

Sensible Thermal Energy Storage

4.1 INTRODUCTION

In the classification of thermal energy storage (TES) technologies, sensible thermal energy storage (STES) is considered equally important in addressing the issues related to demand-side energy management as well as conserving primary energy consumption. STES technologies have been used in the past and have gained impetus in recent years. The archeological wonders available around the world are the real examples of STES and have their own historic significance and scientific background for their existence over the years. For instance, these structures are primarily comprised of stone or rock, which acts as the basic source for TES.

The prime intention behind the creation of huge solid structures using stone/rock as the construction material may have been to keep the indoor space at a lower temperature compared to the outside environment. This is why, if someone enters a huge and elegant structure, they would feel cooler than they felt outside the structure. The science behind this concept can be seen in the use of available sensible heat to store in or release from the stone/rock materials. The interesting fact is that the ideology of STES technologies had already been gleaned by our ancestors. Present-day STES systems being developed can be considered to be their mimetic, but with the application of advanced materials science and technology to enable them to be efficient and sustainable on a long-term basis.

4.2 SENSIBLE HEAT STORAGE MATERIALS

The core part of STES technology depends solely on the selection of suitable materials and their characteristics toward achieving the desired purpose of storing and discharging thermal energy efficiently. As the name indicates, sensible heat storage materials are those that can store or release thermal energy based on the demand requirements (long-term or short-term storage). The heat energy supply initially raises the temperature of the material considerably. Because of the change in the internal energy of the material, the supplied energy is then stored in the form of sensible heat.

The quantity of heat energy that can be stored or released by the STES material can be readily estimated using Eq. (3.1) or Eq. (3.2) (Chapter 3). STES materials can be classified into two main types, namely, (1) solid storage materials and (2) liquid storage materials. Solid storage materials include rocks, stones, brick, concrete, dry and wet earth/soil, iron, wood, plasterboard, corkboard, and so on. Likewise, water and oils, pure as well as the derivatives of alcohols, can be categorized as liquid storage materials.

4.2.1 Solid storage materials

Solid storage materials are chiefly preferred for providing thermal storage requirements in building space heating (sparingly for cooling) and high temperature (solar) heating applications. Usually, the solid storage materials utilized for these applications can be operated in temperatures ranging from 40 to 75 °C for rock beds/concrete and over 150 °C for metals. The intent behind the development of solid storage materials includes the following:

- Reduced risks related to the leakage of the heat storage material being subjected to elevated temperatures
- Viability on their usage as the storage material at very high temperatures (solar power plants)

However, these materials show certain limitations in terms of their use-related issues, including the following:

- Relatively low specific heat capacity exhibited during the heat storage (on an average $\sim 1200 \, \text{kJ/m}^3/\text{K}$)
- Reduced energy storage density compared to liquid storage materials
- Increased risks of self-discharge of thermal energy (heat losses) in long-term storage systems
- Thermophysical properties of the heat and energy transport medium
- Stratification of storage unit
- Associated cost involved in operation and maintenance of the storage unit

4.2.2 Liquid storage materials

Energy storage systems using liquid as the heat storage and transfer material have been widely preferred for applications ranging from low-temperature to medium-temperature thermal storage. In practice, water is the most common liquid material used due to its high specific heat capacity, availability, and low cost. For instance, the energy storage density of water as the heat storage material being subjected to a temperature gradient of 70 °C can be calculated to approximately $290 \, \text{MJ/m}^3$ [1]. Most of the medium-temperature applications related to solar TES make use of the essential qualities of the liquid (water) material to accomplish the desired sensible heat storage. Details pertaining to the thermal properties of a variety of solid and liquid sensible thermal storage materials are presented in Appendix II.

4.3 SELECTION OF MATERIALS AND METHODOLOGY

The selection of materials and methodology are the fundamental aspects in dealing with sensible heat storage technology for real-world applications. Numerous materials are available in literature related to materials science and engineering, and new materials are still being added every year. In this scenario, the selection of proper materials possessing the desired thermal storage characteristics is one of the most vital parts of successful development of the STES technology. On the other hand, execution of thermal storage technology using the materials that have been selected forms the other important part of successful implementation. More specifically, the essential characteristics of the selection and methodology of materials dedicated to STES applications can be identified as follows.

For the selection of high performance materials:

- The basis of design or simply the design inputs related to the thermal storage option has to be transformed into technical specifications.
- Eliminate those materials that do not fall under the requirements as detailed in the technical specifications.

- Based on thermophysical properties, segregate and grade the best and high quality materials for meeting STES requirements.
- Glean the maximum possible details about graded materials for their subsequent usage in STES systems.

For the methodology:

- Understand inherent operational features of the STES system for which high-performance materials have been selected.
- Confront the challenges related to demand-side energy management and primary energy consumption.
- Develop a framework on the best possible optimization function that would yield the optimal solution in terms of cost minimization, maximization of thermal storage performance, and so on.
- Create provision for alteration on the design parameters and variables pertaining to effective functioning of the STES system, including high-performance materials selection criterion.

The decisive step to follow in the selection of high-performance materials from the large literature database can be based on the graphs or charts framed, which would project the interdependencies between the thermophysical properties of materials. Also, one can project how such materials could perform in the long run if they are incorporated into the STES system. For instance, charts representing the values of thermal properties of various materials are shown in Fig. 4.1(a–d). Through this kind of approach, materials with suitable thermal characteristics can be sorted into groups because they would be arranged in the chart based on the property values.

For instance, in Fig. 4.1(d) polymer foam materials occupy the upper left portion in the chart, ceramics the central portion, and metals and alloys the lower right portion. Each bubble represented in the chart (Fig. 4.1(d)) pertains to a specific material. Hence, by combining the materials selection and methodology aspects, the design engineer can have a strong technical grasp to understand the behavior of high-performance materials, when preferred for their utilization into STES applications.

4.3.1 Short-term sensible thermal storage

The concept of storing heat energy to satisfy load demand peaking only for a few hours or else to make use of stored energy for meeting the load requirements based on electricity tariff ratings can be referred to as short-term TES or diurnal TES. In this method, thermal energy, which can be either renewable or nonrenewable, can be stored in the material as sensible heat and reutilized at a later stage during the day-hour operation of the STES system.

Most short-term STES systems utilize solar (renewable) energy as the heat input and store the same in solid material (usually rock bed) or use liquid material (usually water) for meeting the demand load requirements. Depending on the fluctuating load demand, design requirements, site location, and cost factors, the type of heat storage materials (i.e., either solid or liquid materials) can be selected. In this context, the variety of short-term STES technologies that are used to meet the heating demand are explained in the following sections.

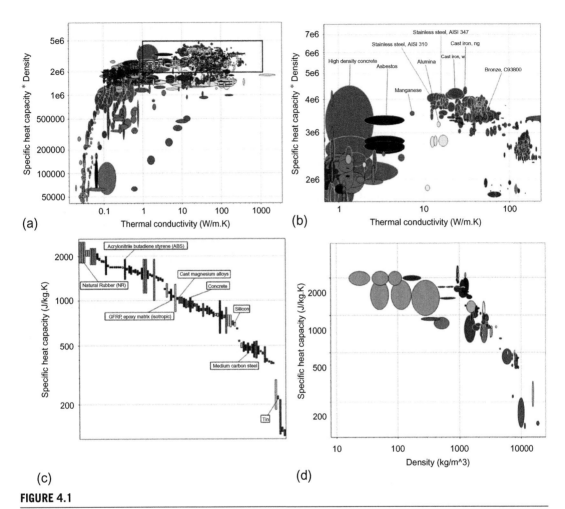

FIGURE 4.1

(a) Material property chart with combination of properties. Energy density ($c_p\rho$) vs. thermal conductivity, (b) zoom in the box area with constraints; materials that do not accomplish these requirements are not plotted, (c) bar chart of specific heat capacity for 100 of the most used materials, obtained with CES selector, and (d) specific heat capacity vs. density [2].

4.3.2 Long-term sensible thermal storage

The other form of storing thermal energy as sensible heat can be effectively achieved by using long-term STES. Heat energy can usually be stored during one time and discharged later. A long-term STES is referred to as seasonal TES because thermal energy in the form of sensible heat can be stored during one seasonal period and reutilized during other seasonal conditions.

For instance, during summer, heat energy can be trapped from the solar collectors and stored in the storage material. In winter, the same stored heat energy can be released from the storage material to meet the required heating load. Mostly, long-term STES systems utilize renewable energy as the seasonal (source) heat energy including the underground TES, earth-to-air thermal storage, energy piles, and so on. A detailed description of these TES systems appears in later chapters.

4.4 PROPERTIES OF SENSIBLE HEAT STORAGE MATERIALS

As stated earlier, a storage material should possess some essential properties to be selected for STES application. These would include [3,4] the following:

- Large thermal storage capacity per unit volume and mass of the material in terms of compactness
- High thermal conductivity within operating temperature limits
- Good packing factor (density)
- Excellent charging and discharging ability with large heat input/output without yielding large temperature differences
- Thermally reliability and stability in the long run, even after several thousand operating cycles
- Thermally as well as energy efficient
- Self-discharging very small to achieve enhanced heat storage performance
- Chemically stable for longer times without decomposition
- Nontoxic, nonexplosive, low corrosion potential or nonreactivity to heat transport medium
- Compatible with the storage construction material; mechanically stable
- Low thermal expansion coefficient
- High fracture toughness and high compressive strength
- Cost effectiveness
- Reduced carbon footprint and environmental impact

4.5 STES TECHNOLOGIES

From the perspective of energy demand and energy security worldwide, the implementation of STES technologies can pave the way to achieve energy redistribution and energy efficiency in every step of societal developments. The operational strategies of a variety of STES methods used to meet heating and cooling load requirements are described in the following sections.

4.5.1 Storage tanks using water

Using water as the storage medium in the STES system is highly regarded as an efficient way to extract and release heat energy based on demand energy requirements. The operating temperature range of water (20–80 °C) is sufficient to meet most of the heating energy redistribution needs, especially as related to residential applications [5]. Water as the liquid medium can act as both thermal storage and heat transport medium because of its high specific heat capacity and convective heat transfer characteristics.

Storage water tanks are usually fabricated of such materials as steel, aluminium, reinforced concrete, and fiberglass. Storage tanks have to be insulated with such materials as mineral wool, glass wool, or polyurethane foam to prevent self-discharge or thermal losses occurring to and from the storage tank. Water can be either statically stored in a tank or made to flow (dynamic) through the storage tank to

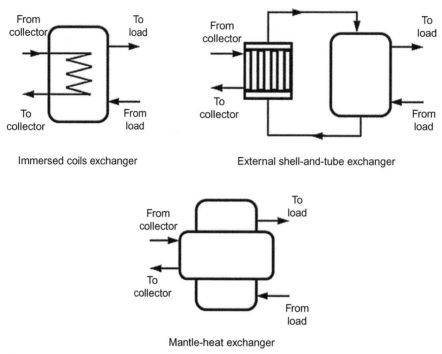

FIGURE 4.2

Different configurations of storage tanks using water as heat storage medium [5,6].

inject and extract the required quantity of heat. Various configurations of storage tanks using water as the heat storage medium are shown in Fig. 4.2.

In the immersed coils exchanger storage configuration, the heat exchanger coils are usually placed at the bottom of the storage tank. This is mainly done to allow the maximum possible heat transfer (in terms of temperature gradient) to occur between the incoming hot water from the solar collector and the heat transfer medium exiting on the load demand side. Furthermore, in this type of sensible heat storage configuration, the effects related to the temperature stratification (the difference in temperature between the upper and the lower portions of the storage tank) due to heat interactions can be greatly minimized using specialized heat exchangers. However, the option for using the specially designed heat exchangers would again be expensive and limit the adoption of such storage concepts for real-time domestic water-heating storage applications.

The external heat exchanger configuration is increasingly attractive over years due to its simple design and construction aspects as compared to the aforementioned option of heating storage. In this model, heat energy from the solar collector can be indirectly transferred to the storage tank by means of an antifreeze and water hydronic networks. The antifreeze picks up the heat from the collectors, which then transfers the same with the water entering the external shell-and-tube heat exchanger. The heated water can then be pumped into the storage tank for further usage. Thus, during nighttime the required thermal energy can be discharged from the storage tank using the heat transfer medium (air or fluid) for meeting the space-heating requirements. This type of storage tank arrangement can be made energy efficient, is cost effective, and is reliable in the long run due to the thermal stratification potential of the storage tank.

The other type of storage tank arrangement includes the mantle-heat exchanger, in which the heat transfer fluid from the solar collector can transport the heat to the water that is being stored inside the mantle or the double-walled tank structure. Due to increased heat transfer surface area, the heat exchange effectiveness and thermal storage performance of this type of arrangement can be enhanced substantially. However, due to specialized design constraints and the construction constraints involved in mantle-heat exchangers, they are generally more expensive [5].

4.5.2 Rock bed thermal storage

Storage of heat energy in rock material has been considered highly favorable for residential applications when combined with solar thermal energy as the source for heat supply. In the rock bed thermal storage method, the rock material is loosely packed in a bed-like structure as shown in Fig. 4.3. The storage system is also provided with an inlet and outlet tubing or ducting arrangement for the heat transfer medium to transfer the heat to and from the storage system.

In a typical heat charging cycle, the heat transfer medium (mostly air) carries heat from the solar collector and enters the rock bed storage through the upper airway arrangement. Due to density gradient effects in rock bed storage, hot air loses its heat, becomes warm air, and then flows back to the collector for the next charging cycle. During the discharging period (nighttime), cold air from the load space flows through the bottom air port arrangement and takes heat from the rock bed storage. Through density gradient and convection effects, the cold air is heated and reaches the required temperature, which is sufficient to meet the space heating demand. After the heat contained in the rock bed storage has been removed, the storage system is now ready for the next charging cycle to take place during daytime (sun brilliance) hours.

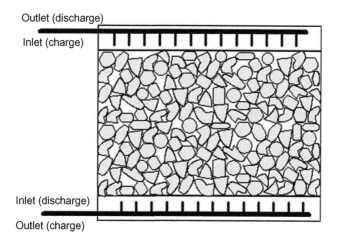

FIGURE 4.3

Schematic diagram of rock bed TES [5].

Because air is used as the heat transfer medium, the heat storage performance of the rock bed system can be expected to be lower for the same thermal load compared to a liquid (water) storage system. This is because air, as well as rock, has a lower specific heat capacity. In addition, for the same heat demand, the space occupied by the rock bed system could be three times larger (especially for large-scale seasonal thermal storage), which is not the case for a liquid heat storage system.

However, the cost involved in the construction and operation of the rock bed system is less than the water-based storage system. Also, rock bed systems can be used at elevated temperature gradients, whereas water storage systems are limited to certain temperature differentials. The rock bed system can be categorized under passive storage because the heat transfer medium (air) does not contribute to the storage process.

4.5.3 Solar pond/lake thermal storage

The storage of thermal energy using the solar pond/lake can be considered more efficient because a large quantity of the heat energy is available. The science behind this type of storage can be due to the influence of the salinity gradient of the pond/lake. The salinity gradient is the driving force for storing and extracting heat energy from a solar pond. The schematic diagram of a solar pond is shown in Fig. 4.4. As depicted in the figure, the pond water layers are divided into three major zones, namely, the upper convective zone (UCZ), middle nonconvective zone (NCZ), and lower convective zone (LCZ).

The presence of these three zones plays a significant role in heat storage and release processes. The UCZ contains little or no saline concentration, and the NCZ is usually saline in nature such that a salt concentration gradient between the UCZ and the NCZ always exists. This means that the UCZ surface water is concentrated with less salt content compared to the NCZ's higher salt concentration. However, the saline gradient between the LCZ and the NCZ is very high such that the LCZ has high salinity compared to the NCZ, and the saline concentration gradient between them is also high.

Typically, the solar radiation that enters the pond is influenced by the three zones, and the high saline concentration of the LCZ causes the heat energy to be stored in this layer. At the same time, the salinity gradient offers resistance to natural convection to take place between the water layers. Thus, the UCZ acts as an insulator for the underlying NCZ and LCZ layers [7]. Thereby, most of the heat energy can be trapped from incident solar radiation and can be stored in the bottom or LCZ of the solar pond. The stored heat energy can then be discharged (retrieved) from the LCZ with the help of a heat exchanger facility.

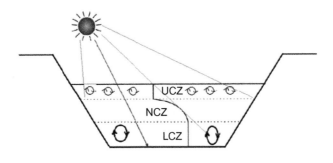

FIGURE 4.4

Schematic representation of solar pond heat energy storage [1].

The solar pond/lake thermal storage can also be artificially created by using the saline concentration gradient as the working force for storing the heat energy. The most essential factors to be considered for executing solar pond thermal storage using external salt materials can be identified as follows [8]:

- The solubility value of the salt material has to be high enough to maintain high solution densities.
- The solubility of the material has to be consistent or not vary much with respect to temperature.
- The saline solution produced must be clear and transparent to allow the incident solar radiation to transfer the heat energy to the LCZ layer.
- The used salt material has to be graded as environment friendly and be safe to handle.
- The salt material selected for the solar pond application must be readily available and affordable.

The commonly preferred salt materials for the solar pond/lake TES applications that possess good thermophysical properties would include the following [9–12]:

- Sodium chloride (NaCl)
- Natural brine of sodium, potassium, magnesium, chlorine, etc.
- Sodium carbonate (Na_2CO_3)
- Magnesium chloride (MgCl)
- Potassium nitrate (KNO_3)
- Ammonium nitrate (NH_4NO_3)
- Urea ($NH_2CO \cdot NH_2$)

Many research works have been reported pertaining to incorporation of some of the aforementioned salt materials and have established that the sodium chloride (NaCl) salt solution yielded an average temperature of 55 °C and better thermal storage performance compared to other salt materials [9,10]. Based on the experimental studies performed by various research groups, it can be inferred that the energy efficiency of solar pond thermal storage systems can be enhanced from 20 to 50% [7,13,14]. Depending on the selection of the heat exchanger module, by playing with the saline concentration gradient between the LCZ and NCZ layers, the energy storage efficiency can be improved.

4.5.4 Building structure thermal storage

Thermal energy in the form of cold or heat can be effectively stored and released with the help of certain materials that are considered integral to building structures. Building fabric materials, including concrete, brick, gypsum board, or a combination of these, can be used as an effective medium for accomplishing desired thermal storage without sacrificing the comfort cooling/heating requirements in an indoor environment. The fabric materials intended for constructing walls, floors, partitions, roofs, and ceilings are good enough to absorb and release heat or cold energy depending on the heating/cooling load fluctuations persisting in building thermal zones. The storage of heat energy in building fabrics can be classified into passive and active thermal storage depending on their functional aspects.

For instance, heat energy storage via density gradient (or natural convection) effects between the indoor environment (air medium) and the fabric material can be regarded as passive thermal storage. On the other hand, if the storage of heat energy is assisted by a mechanical component (fan, blower, or pump), it can be categorized under active thermal storage. For brevity, the concepts related to the storage of heat energy in building fabric components are discussed in the following sections.

In general, the rate of heat transfer in such storage methods depends on:

- The mass of the fabric material
- Specific heat capacity of the fabric material as well as the heat transfer medium
- Temperature gradient between the fabric material and the heat transfer medium (active) or environment (passive)
- Volume flow rate of the heat transfer fluid (in case of active system)
- Thermal conductivity of the fabric material and the heat transport medium
- Effects due to the infiltration and other thermal losses to and from the building fabric structures

Thermal storage with ceiling and floor/underfloor slab components are an attractive method that can be effectively activated through the incorporation of conventional natural ventilation and mechanical ventilation schemes. The working principle of this system is simple in the sense that during nighttime (off-peak hours), cold indoor air from building thermal zones can be made to flow over the ceiling slab component. By this, cold energy is transferred and stored in the ceiling structure. During daytime (on-peak conditions), warm indoor air can be circulated over the ceiling slab, and the required cold thermal energy can be retrieved and serve in the indoor environment to meet the cooling load demand.

The case for short-term heating storage is similar, wherein during winter daytime hours, heat energy from indoor spaces can be stored in the ceiling slab and then discharged during nighttime to offset heating demand. Generically, the temperature swing experienced between room air and the ceiling slab component can vary between 2 and 4 °C. The utilization of mechanically assisted fan or blower equipment can still improve the thermal storage performance of this system through forced convection effects and allows for better heat transfer between room air and the ceiling slab component.

The floor slab component also favors achieving heat storage by natural convection or forced convection principles. In floor slab storage, indoor air is supplied from the underfloor air diffusers, and return air is collected at the roof/ceiling level. Due to the vertical movement of air through floor level diffusers to indoor spaces, the heat energy can be effectively stored in the floor slab and the condition of temperature stratification can be maintained, which would result in comfortable conditions in the room. The concept of free cooling or economizer cycle ventilation can be coupled with the ceiling slab or floor slab components, which would still augment thermal storage performance and building energy efficiency by 5 to 10%.

TES using hollow core slab configuration has also been equally developed in recent years and is referred to as TermoDeck® [15]. This system essentially consists of a hollow core ceiling component, which functions as a heat storage facility for reducing the peaking cooling/heating load demand and improving building energy efficiency. In a typical cooling cycle operation, the cold energy contained by the indoor air or ventilation air can be trapped and stored in the hollow core ceiling component, while the air is subjected to flow through the core or voids of the ceiling component.

In practice, the pitch between the cores is designed accurately depending on the thermal load in indoor zones, thereby ensuring the maximum possible energy extraction takes place. The stored cold energy can then be released to the building spaces by circulating warm indoor air over the void cooled ceiling. The major advantages of using the TermoDeck® system include the following:

- Reduction of peak power demand up to 50%; during 4 to 5 h of operation over the midday period, this value can go up to 70 to 90%.
- Reduction in the primary energy consumption of up to 50% can be achieved.

- Mechanically assisted components including fans, chillers, ducts, water radiators, and suspended ceiling components can be eliminated.
- Savings in upper stories of up to 25% per floor is due to the absence of suspended ceilings.
- Due to design flexibility, good comfort conditions to occupants can be effectively met without sacrificing energy efficiency.
- Lower energy consumption leads to reduced carbon footprint and environmental sustainability.

TES using embedded coil elements has also been developed recently, in which the building fabric component would be embedded with coil pipe inserts during the construction stage. The heat transfer fluid (in hot or cold state) flowing through the coil element transfers the thermal energy to the building fabric through convective effects. The fabric component stores energy being transferred mainly in the form of sensible heat. The stored heat is then released to indoor spaces by making the indoor air flow over the fabric component. The exchange of heat takes place through natural convection or due to a density gradient between indoor air and the fabric component. The mass flow rate, specific heat capacity, temperature gradient, viscosity, and flow frictional pressured drop are the essential parameters to be considered during the design of embedded coil elements to ensure the system's enhanced thermal storage performance.

In a *sensible thermal storage using an electric heating* system, the hollow core bricks made up of magnetite or magnesite material suitably encapsulated in a metal container are subjected to electric resistance heating. The heated bricks store a suitable quantity of thermal energy through conduction principle. The heat energy contained by the brick component can then be transferred to the thermal zones by passing the indoor air across them. The valuable aspect of this storage system is that the temperature of the bricks can be elevated up to 760 °C by electric resistance heating during part load conditions or suitable electric tariff periods. In addition, the outer surface of the brick heater can be maintained at a temperature below 80 to 85 °C.

According to the ASHRAE Handbook 2012 [16], thermal storage capacity of brick components can be varied from 49 to 216 MJ. The charging duration of the brick component using resistance heating can range from 6 to 7 h, whereas the discharging period can be expected to be normally between 4 and 5 h. Brick heating thermal storage can enable the system to meet any intermediate heat demand due to thermal losses from indoor spaces due to change in outside environmental conditions, without compromising the level of comfort to indoor occupants. The approximate heat storage capacity being estimated for small-scale residential to large-scale industrial storage can vary from 310 to 864 MJ and 3460 MJ, respectively. On the other hand, electric power consumption of these systems varies from 14 to 46 kW and 53 to 160 kW, respectively.

4.5.5 Passive solar heating storage

The function of a passive solar thermal storage system is very similar to the one discussed earlier, but the source for heat supply can be received directly from incident solar radiation. Conceptually, the passive solar thermal storage system is driven through the thermosyphon mechanism, wherein due to the density gradient of the heat transfer medium flowing through the solar collector, the required heat energy is transferred to the storage tank. The schematic representation of the thermosyphon (passive) solar thermal storage system is shown in Fig. 4.5(a). This system is essentially comprised of a solar collector, thermal storage tank, hydronic units for heating storage side, and space load demand side.

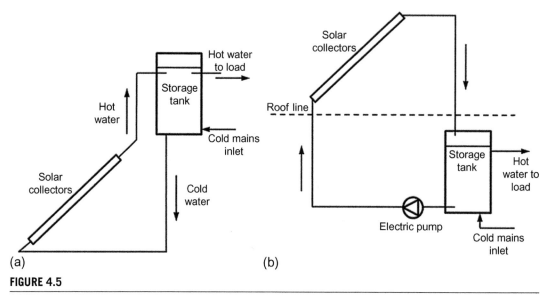

FIGURE 4.5

(a) Thermosyphon (passive) solar heating storage system, and (b) active solar heating storage system.

During the sun brilliance period, the incoming water (heat transfer medium) to the collector picks up heat from the irradiated solar energy. The hot water moves upward due to density gradient and flows through the storage tank. Due to thermal stratification effect in the storage tank, the water in the tank is heated, and the thermal energy is thus stored. The warm water then flows back to the collector inlet hydronic port by gravity, and the cycle is repeated. Depending on heating demand, the secondary heat transfer fluid (water or brine) flows through the storage tank and discharges the stored energy for meeting the heating demand.

Apart from building fabric components, sensible heat storage can also be made possible by using the glazing component. Glazed envelopes are considered to be more effective in trapping incident solar radiation and help to maintain warmth in indoor environments. By using the glazing component, dual benefits can be achieved, namely, (1) it would provide an elegant aesthetic appearance to the building, and (2) it would provide the required heat storage. Buildings with glazed envelopes arranged toward 30° of south direction are known to effectively utilize daytime solar energy, and more heat energy can be contained within building spaces through this arrangement.

Glazed structures have the ability to prevent incident solar radiation from being radiated back to the ambient as infrared radiation, thereby ensuring the highest possible means for retaining heat energy within occupied spaces. In general, the design ratio of glazing to the floor area, if followed to within 7 to 10%, would result in achieving better thermal storage in buildings without giving rise to any overheating issues in occupied spaces. Thus, passive sensible heat energy storage systems can be considered an economical and sustainable way to enhance a building's energy efficiency by 30 to 35%.

4.5.6 Active solar heating storage

By terminology, an active solar heating storage primarily consists of a mechanically assisted component for transporting and storing heat energy. As shown in Fig. 4.5(b), solar heat energy is initially trapped by the absorber plate or evacuated tube type solar collector. The flow of the heat transfer medium through the solar collector enables the transfer of heat energy from the collector to the storage tank. The warm water is then returned to the collector for the next cycle of charging process (heat storage). The heat energy that is contained in the storage tank is then extracted by secondary heat transfer fluid dedicated to meeting the space heating load demand. The circulation of the heat transfer fluid for heat transfer and energy storage is achieved through the electric pump arrangement.

To accomplish a higher rate of heat transfer and energy storage, it is much preferable to use concentrated solar collectors (parabolic or conical). This is due to the fact that the flat plate or evacuated tube collectors can be subjected to thermal stresses, convective and radiative losses, while they are subjected to elevated temperatures above 85 to 90 °C. Because of this, the solar energy that is captured during sun brilliance time may not be fully utilized and due to the inherent heat losses, only a part of which can be stored and that may not be sufficient to meet the heating demand in the buildings.

As mentioned earlier, the incorporation of concentric parabolic collector can facilitate the incident solar radiation to be focused over the receiver tubes, in order to elevate the temperature of the heat transfer medium to above 135 °C. This makes heating storage possible during the daytime, even if a switch or fluctuation takes place in the incident solar radiation between the sun brilliance periods and cloudy times. Similar to the glazing option, parabolic or conical collectors can be aligned toward 30° south corresponding to the latitude of the building and take advantage of maximum possible solar heating energy storage. Furthermore, concentrated parabolic or conical solar collectors are much efficient compared to the flat plate or evacuated tube collectors for year-round operation. The thermal efficiency of the former can be expected to be around 60%, whereas for the latter it ranges from 40 to 50%.

The *stratification of sensible TES* literally means continuous possibilities available to transfer the heat energy to the cooler regions of a fully charged thermal storage unit. In other words, it is the availability of a proper quality of energy (or exergy) from the warmer regions of a completely discharged thermal storage unit. By ensuring proper stratification in the storage unit, the thermal performance of the solar collector can be enhanced by allowing the heat transfer medium to enter the collector at a lower temperature. Numerous research studies have been performed recently on the temperature stratification of the heat storage unit [5]. On average, solar energy utilization or useful heat energy storage can be enhanced from 20 to 60% with proper stratification of heat storage tank compared to the fully mixed tank. The schematic representation of the different levels of stratification is illustrated in Fig. 4.6.

By providing suitable insulation to the storage tank, the effects related to self-discharge can be reduced significantly. Self-discharging of the storage tank refers to the thermal losses taking place from the tank to the surrounding environment due to the larger storage volume and high temperature when fully charged.

4.6 HIGH TEMPERATURE SENSIBLE THERMAL STORAGE

High temperature sensible thermal storages are largely suitable for power generation, where high temperature heat input is utilized for producing high temperature steam to drive the turbine or an engine. The heat source is basically a renewable energy and with the amalgamation of sensible heat storage materials, the overall thermal efficiency of the high temperature sensible heat storage system can be enhanced. The thermal properties of the storage materials being considered for high temperature STES are depicted graphically in Figs. 4.7–4.10.

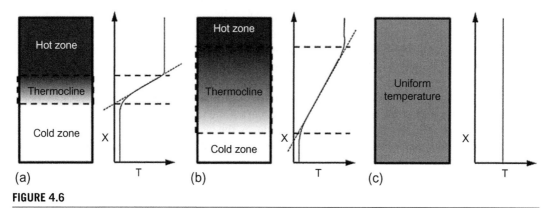

FIGURE 4.6

Differing levels of stratification within a storage tank with equivalent stored energy: (a) highly stratified, (b) moderately stratified, and (c) fully mixed (or unstratified) [5].

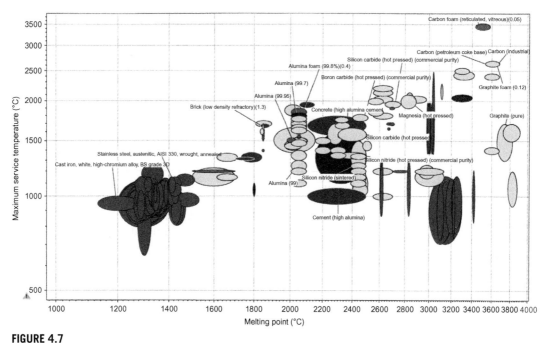

FIGURE 4.7

Maximum service temperature vs. melting point (°C) [4].

For operating temperatures ranging above 100°C, the choice of the heat storage material can be with liquids including oil or molten salts. For very high operating temperatures of above 600°C, the choice of heat storage material can be made with solid materials including concrete or ceramics. However, the technical investigation of high temperature sensible thermal storage systems still requires deep understanding of the complexity involved in their development and successful commissioning. The technical specifications of materials with sensible heat stored per kg in the temperature range of 500–750°C are summarized in Table 4.1.

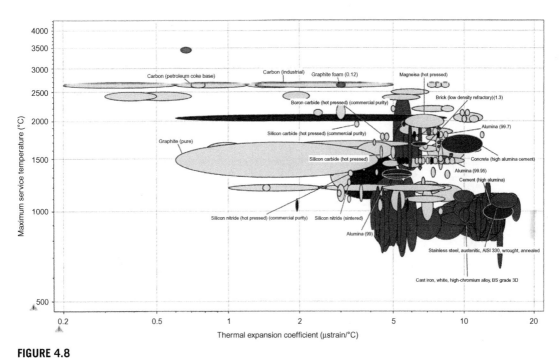

FIGURE 4.8

Maximum service temperature vs. thermal expansion coefficient (microstrain/°C) [4].

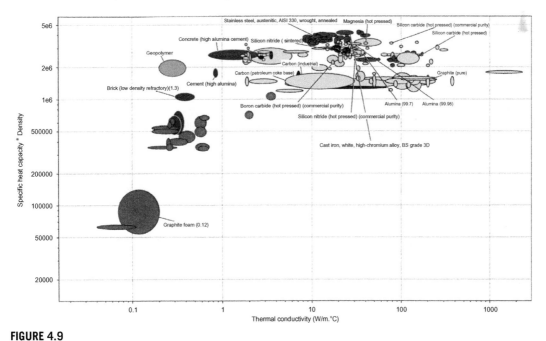

FIGURE 4.9

Energy density (J/m³ °C) vs. thermal conductivity (W/m °C) [4].

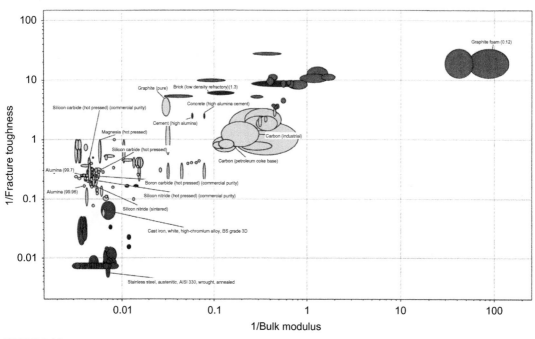

FIGURE 4.10

Fracture toughness vs. bulk modulus [4].

Table 4.1 The Summarized Technical Specifications of the Materials with Sensible Heat Stored Per Kilogram in the Temperature Range of 500–750 °C [4]

Material	Energy Stored $\Delta T = 250$ K (kJ/kg)	Energy Stored $\Delta T = 250$ K (kWh/kg)	Mass Required for 1000 kWh (kg)	Volume Required for 1000 kWh (m³)
Alumina (99.5%)	200	0.056	18,000	4.5
Cast iron	135	0.038	26,700	3.4
High alumina concrete	245	0.068	14,700	6.1
Geopolymer	298	0.083	12,100	5.1
Graphite	178	0.049	20,300	9.1
Magnesia (HP)	235	0.065	15,300	4.3
Silicon carbide (HPCP)	260	0.072	13,800	4.3

4.7 CONCISE REMARKS

TES systems using sensible heat materials (solid and water) are relatively considered to be the attractive option for reducing primary energy consumption and carbon footprint. STES systems utilize rock or water for storing and releasing heat energy, either through passive or active modes of operation. For low temperature sensible heat storage, water can be used as the selection material due to its high specific heat capacity, high density, availability, lower cost, and environmental safety. Similarly, for medium temperature sensible stores, rocks or stones may be preferred due to their high operating temperature range and compactness.

STES technologies find their potential application in residential buildings, where the implementation of passive and active thermal storages can result in the enhancement of energy efficiency and thermal efficiency by 30 to 35% and 40 to 60%, respectively. Furthermore, with a properly designed storage tank that has good stratification, the enhancement of useful heat energy utilization can be expected to achieve 20 to 60%.

References

[1] Tatsidjodoung P, Pierrès NL, Luo L. A review of potential materials for thermal energy storage in building applications. Renew Sustain Energy Rev 2013;18:327–49.
[2] Fernandez AI, Martínez M, Segarra M, Martorell I, Cabeza LF. Selection of materials with potential in sensible thermal energy storage. Sol Energy Mat Sol C 2010;94:1723–9.
[3] Abhat A. Short term thermal energy storage. Revue Phys Appl 1980;15:477–501.
[4] Khare S, Dell'Amico M, Knight C, McGarry S. Selection of materials for high temperature sensible energy storage. Sol Energy Mat Sol C 2013;115:114–22.
[5] Pinel P, Cruickshank CA, Beausoleil-Morrison I, Wills A. A review of available methods for seasonal storage of solar thermal energy in residential applications. Renew Sustain Energy Rev 2011;15:3341–59.
[6] Han YM, Wang RZ, Dai YJ. Thermal stratification within the water tank. Renew Sustain Energy Rev 2009;13:1014–26.
[7] Kurt H, Halici F, Binark AK. Solar pond conception—experimental and theoretical studies. Energy Convers Manag 2000;41:939–51.
[8] Akbarzadeh A, Andrews J, Golding P. Solar pond technologies: a review and future directions, advances in solar energy. Oxford (England): Earthscan; 2005.
[9] Hassairi M, Safi MJ, Chibani S. Natural brine solar pond: an experimental study. Sol Energy 2001;70:45–50.
[10] Kurt H, Ozkaymak M, Binark AK. Experimental and numerical analysis of sodium-carbonate salt gradient solar–pond performance under simulated solar-radiation. Appl Energy 2006;83:324–42.
[11] Ramakrishna Murthy GR, Pandey KP. Scope of fertiliser solar ponds in Indian agriculture. Energy 2002;27:117–26.
[12] Velmurugan V, Srithar K. Prospects and scopes of solar pond: a detailed review. Renew Sustain Energy Rev 2008;12:2253–63.
[13] Wang YF, Akbarzadeh A. A study on the transient behaviour of solar ponds. Energy 1982;7:1005–17.
[14] Andrews J, Akbarzadeh A. Enhancing the thermal efficiency of solar ponds by extracting heat from the gradient layer. Sol Energy 2005;78:704–16.
[15] http://www.termodeck.com/ [accessed November 2013].
[16] ASHRAE. ASHRAE handbook—HVAC systems and equipment; ASHRAE, Inc, Atlanta, GA, 2012.

Latent Thermal Energy Storage

5.1 INTRODUCTION

Energy production and consumption play a vital role in determining the energy conservation at every step of the economic development of a country. In recent years, the increased demands in the construction sector have led to the development of elegant and huge building structures worldwide. Interestingly, buildings consume one-quarter to one-third of the overall energy generated globally [1]. Incessant value-added engineering design and incorporation of the energy-efficient heating, ventilation, and air conditioning (HVAC) systems in buildings are greatly necessitated.

Although several measures are available to minimize the net energy consumption in buildings, there is still a need for an efficient system that can shift on-peak thermal load demand to off-peak conditions without sacrificing energy efficiency. From this perspective, the latent thermal energy storage (LTES) systems are primarily intended for enhancing the performance of HVAC systems in terms of storing and releasing heat energy on short-term or diurnal or seasonal basis depending on the thermal load requirements experienced in buildings.

Energy efficiency in buildings is intensely coupled with high performance materials-based energy redistribution and energy conservation, which would help in accomplishing solutions to the energy challenge and energy security. Energy redistribution requirements can be effectively met by using LTES systems integrated with the dedicated HVAC systems in buildings. In particular, the assessment of reducing energy consumption using materials that are thermally efficient and stable in the long term are vital in building cooling applications.

From this viewpoint, research interests toward developing thermal energy storage (TES) systems incorporating efficient latent heat storage materials that would offer energy redistribution requirements are increasingly popular. The nucleus of this chapter is divided into two sections: (1) to focus on apposite latent thermal storage materials having excellent thermophysical properties and (2) explore the potential opportunity available for such materials to be effectively integrated into the real-time passive and active cooling applications in buildings. The schematic representation of the heat gains in indoor environment is shown in Fig. 5.1.

5.2 PHYSICS OF LTES

The incorporation of LTES systems in buildings largely depends on the phase change characteristics of the heat storage material being considered. The inherent phenomenon that drives the operational performance of the heat storage materials is their ability to undergo phase transformation at isothermal

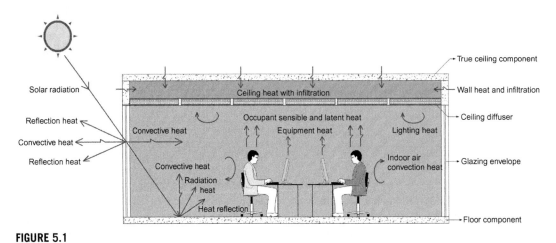

FIGURE 5.1

The schematic representation of the heat gains in indoor environment.

conditions. That is, when such materials are cooled down to below their melting temperature, they would start to freeze (or crystallize) at constant temperature.

Similar is the case for the fusion of such materials while they are subjected to temperatures higher than their melting point. The freezing or melting of the heat storage materials would characterize their phase transformation, wherein the thermal energy is stored or released in the form of differential latent heat transfer.

For a better understanding, the phase change phenomenon of a thermal storage material is shown in Fig. 5.2. It is clearly seen that the phase transition is comprised of three stages each in freezing and melting processes. In freezing, the temperature of the thermal storage material initially decreases

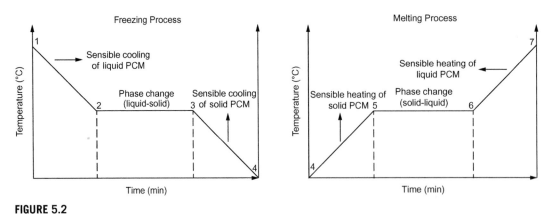

FIGURE 5.2

Heat storage and release processes of the PCM.

(process A-B), which signifies the sensible cooling of the material (in liquid state) with respect to the surrounding heat transfer (cooling) medium (HTM).

Upon continuous cooling (extraction of heat energy), the material undergoes liquid-solid phase change (process B-C), which takes place at the constant temperature. At this condition, the thermal energy (cooling energy) from the HTM is effectively transferred to the thermal storage material in the form of latent heat. The temperature again starts to decline further (process C-D), which is due to the sensible heat transfer occurring in the solid material with regard to the HTM temperature. Once the thermal equilibrium is established between the HTM and the thermal storage material, the temperature of the system remains unaltered.

Likewise, in melting, the temperature of the thermal storage material in solid state increases gradually by virtue of the sensible heat transfer (process D-E) with the surrounding HTM, which is being maintained above the fusion temperature of the thermal storage material. A similar solid-liquid phase transformation (process E-F) would take place at constant temperature leading to the release of the stored cold energy from the heat storage material to the HTM. Further increase in the temperature (process F-G) of the thermal storage material (liquid state) signifies the sensible heat transfer with the HTM, which would eventually lead to the thermal equilibrium condition.

Typically, during the process of phase transformation (B-C, E-F), the substantial quantity of the heat energy being stored and released by the TES materials would characterize their effective thermal performance while incorporated into the building structures or being integrated through a separate modular arrangement with the building cooling system. In this chapter, the convention followed for representing the freezing (charging) and the melting (discharging) processes are referenced with respect to the process of thermal energy transfer occurring between the heat storage material and the HTM.

5.3 TYPES OF LTES

The integration of TES technologies is proven to be an efficient way to confront the increasing challenges of the energy supply and energy redistribution requirements in buildings to a greater extent. In this context, the LTES methods that are primarily intended for the building cooling applications can be categorized into two types, namely, ice-thermal energy storage (ITES) and phase change material (PCM) thermal energy storage (PCM-TES).

Interestingly, ITES refers to the use of ice (solid water) in the form of crystals or slurries as the heat storage material, whereas in PCM-TES, the functional heat storage material constitutes a variety of chemical species classified under the inorganic, organic, and eutectic mixtures.

It is pertinent to note that, PCM-based LTES systems are largely preferred in practice for cooling applications in buildings, in place of ITES systems. This is due to the fact that PCMs are a class of materials with better thermophysical properties, which exhibits good phase change characteristics during repeated charging and discharging cycles.

Although, the energy storage capacity of the PCM-TES compared to the ITES is lower (because latent heat of fusion of ice: 333 kJ/kg), the former is preferred for their relatively high volumetric storage density. Furthermore, the organic PCMs reveal excellent thermophysical properties, congruent freezing and melting, thermal stability, noncorrosiveness, thermal reliability on a long-term basis at the cost of ease of operational and maintenance aspects. In subsequent sections, the incorporation of the variety of LTES methods for achieving energy efficiency and energy savings in buildings are described.

5.4 PROPERTIES OF LATENT HEAT STORAGE MATERIALS

The properties that are most essential for an LTES material (especially PCM) for them to be considered for the commercial application in buildings are summarized below [2]:

- Thermophysical properties include:
  - Fusion temperature in the range of the operating temperature
  - Latent heat of fusion higher with respect to unit volume
  - High specific heat that enables supplemental heat storage through sensible heat
  - Thermal conductivity in solid and liquid phases that must be high for enabling good heat transfer
  - Reproducibility in the congruent freezing and melting for the entire operating cycles
  - Change in the volumetric capacity and the vapor pressure of the material as low as possible with regard to the operating temperature
- Kinetic properties include:
  - Swift nucleation, growth, and dissociation of stable nucleus during freezing/melting processes at constant temperature
  - Relatively low degree of supercooling exhibited by the PCM
  - Effective heat transfer during charging and discharging cycles for storing and releasing the thermal energy at isothermal conditions
- Chemical properties include:
  - Noncorrosive with the containment or encapsulation materials
  - Chemical and thermal stability to withstand heat at elevated temperatures
  - Thermal reliability on long-term basis
  - Nontoxic, nonflammable, and nonexplosive
- Economic aspects include:
  - Ease of availability
  - Cost effectiveness
  - Ease recycling and treatment
  - Environmental friendliness

The thermal properties of some common heat storage materials are summarized in Table 5.1. The thermophysical properties of a variety of the PCMs gleaned from major research reports [3] are listed in Appendix II, which would be relevant to engineers, researchers, materials scientists, technologists, and industrial professionals.

5.5 ENCAPSULATION TECHNIQUES OF LTES (PCM) MATERIALS

The encapsulation of the PCMs paves way for the accomplishment of their maximum possible energy storage and release capabilities successfully. There are several ways by which the encapsulation of the PCMs can be performed. Some of the commonly preferred encapsulation techniques for LTES materials are

- Direct impregnation method
- Microencapsulation method
- Shape stabilization of the PCM

Table 5.1 Common Heat Storage Materials [3]

	Materials			
Property	**Rock**	**Water**	**Organic PCM**	**Inorganic PCM**
Density (kg/m^3)	2240	1000	800	1600
Specific heat (kJ/kg K)	1.0	4.2	2.0	2.0
Latent heat (kJ/kg)	–		190	230
Latent heat (MJ/m^3)	–	–	152	368
Storage mass for 10^9, avg (kg)	67,000 ($\Delta t = 15$k)	16,000 ($\Delta t = 15$k)	5300	4350
Storage volume for 10^9, avg (m^3)	30 ($\Delta t = 15$k)	16 ($\Delta t = 15$k)	6.6	2.7
Relative storage mass	15 ($\Delta t = 15$k)	4 ($\Delta t = 15$k)	1.25	1.0
Relative storage volume	11 ($\Delta t = 15$k)	6 ($\Delta t = 15$k)	2.5	1.0

5.5.1 Direct impregnation method

The simplest method of PCM encapsulation is generically referred to as the direct impregnation of heat storage material into the building construction materials. The building materials considered for direct impregnation of PCM include gypsum, concrete, plaster, vermiculite, wood, cement composite compounds, and porous materials. Each encapsulation technique has its own merits and limitations [4].

For instance, the PCM being impregnated with a porous building material would effectively occupy the pores and helps to augment heat transfer capabilities during their charging and discharging modes. However, there are more possibilities for the impregnated PCM to leak from the building material, which would generally influence the operational performance of the PCM in the long run. In some instances, the impregnated PCM would exhibit chemical interaction with the containment material, causing the deterioration of such material over periodic usage.

5.5.2 Microencapsulation method

The microencapsulation technique is rather preferred to be an effective way of encapsulating the PCM into the building materials for improving the performance of the TES system. Microencapsulation is a method of producing spherical or rod-shaped PCM particles of micron size and enclosing them into a polymeric capsule for them to be directly incorporated into the matrix materials for acquiring thermal storage in buildings.

The vital factor that has to be ensured in this technique is that the polymeric capsule containing the PCM must be stable enough to sustain the PCM inside the capsule even after many thermal cycles. In addition, the capsule coating must not exhibit chemical interaction with the matrix material in which they are incorporated.

The effectiveness or the quality of the microencapsulation process is determined based on the ratio of the mass of the hermetic capsules filled with the PCM to the total mass of the powder PCM (before processing). Interestingly, the major parameters that govern the quality of the microencapsulation process are

- Particle mean diameter (PCM)
- Thickness of the capsule shell
- Mass concentration (in percentage) of the PCM to the total mass of the capsule
- Duration of the process
- Type of reticulation agent mixed during preparation of microencapsulated PCMs.

It is noteworthy that, the microencapsulated PCM is expected to exhibit good phase transformation characteristics within the encapsulated shell when subjected to repeated cooling and heating cycles. This is due to the fact that the micron-sized PCM particles would undergo a constrained freezing and melting inside the capsule coating, thereby augmenting the possibility of lumped heat storage and discharge capabilities with the building construction material.

However, in real situations, due to the continuous phase change processes, the microencapsulated PCM is likely to exhibit hysteresis and supercooling phenomenon, which would lead to the incongruent freezing or delayed solidification of the PCM [5,6]. This in turn would directly influence the phase transition characteristics of the microencapsulated PCMs as well the desired TES aspects of the building materials.

The process of microencapsulation, scanning electron microscope (SEM) images of some microencapsulated PCMs (referred as MPCM or mPCM), the commercially available PCM microscapsules, and the factors to be considered while selecting microencapsulated PCMs for TES are presented in Figs. 5.3(a,b) and 5.4(a–d), Tables 5.2 and 5.3, respectively.

5.5.3 Shape stabilization of the PCM

As the terminology denotes, shape stabilization of the PCM is a process by which the liquid mixture of the heat storage material is combined with the supporting (matrix) material. The cooling of this prepared mixture to below the glass transition temperature of the supporting material would result in the solidification of the material. The supporting material plays a vital role in encompassing the mass proportion of the PCM into the material. Approximately, 80% of the PCM can be shape stabilized using the properly selected supporting material. The supporting material available in common includes high-density polyethylene (HDPE) and styrene-butadiene-styrene (SBS) [6].

Interestingly, the PCM being shape stabilized using HDPE and SBS shows good stabilization index, wherein the leakage of the PCM from the supporting material has not been reported [8,9]. On the other hand, the mixture of the PCM into the SBS supporting material is found better than that into the HDPE. However, the rigidity of the shape-stabilized PCM is more in the HDPE matrix, when compared to the SBS. Thus, shape-stabilized PCM is a good choice for utilizing them as potential heat storage material in the LTES applications in buildings (Fig. 5.5).

However, a limiting factor exists in using the shape-stabilized PCM extensively in the building cooling applications, which is the effective thermal conductivity of such material. Because of the relatively low thermal conductivity of shape-stabilized PCM, this would result in the insufficient heat transfer and reduced thermal storage and release capabilities during charging and discharging cycles.

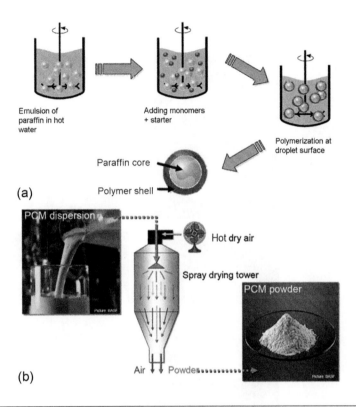

FIGURE 5.3

(a) Microencapsulation process of BASF and (b) drying process of BASF [5].

To confront this challenge, many research works have been put forth in recent years, wherein their thermal conductivity is enhanced by embedding high thermal conductivity additives during the preparation of the shape-stabilized PCM. To substantiate this, the thermal conductivity of the shape-stabilized PCM is enhanced by 53% by using the ex-foliated graphite as the thermal conductivity enhancement materials [11]. The summary of the key manufacturer of PCM worldwide in terms of product classification, range of PCM temperature, and number of PCMs in the given temperature range are presented in Tables 5.4 and 5.5.

5.6 PERFORMANCE ASSESSMENT OF LTES SYSTEM IN BUILDINGS

The successful operation of the LTES system in buildings would decisively depend on the following criterion:

- Thermal load demand persisting in the spaces of building
- Heat storage technique (passive or active)
- PCM freezing and melting characteristics (charging and discharging attributes)
- Spatial requirements for the HVAC systems
- Overall energy-cost savings potential

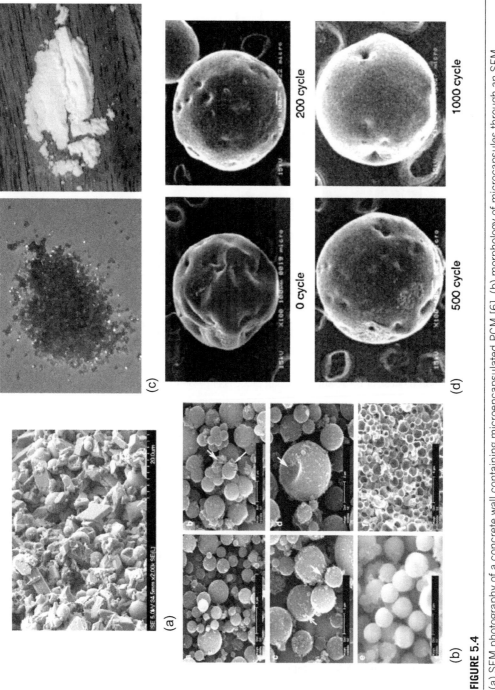

FIGURE 5.4

(a) SEM photography of a concrete wall containing microencapsulated PCM [6], (b) morphology of microcapsules through an SEM microscope [5], (c) microencapsulated PCM: left—laboratory manufactured by copolymerization; right—commercialized by BASF [7], (d) microencapsulated paraffin profile evaluated by SEM at different thermal cycles [7].

Table 5.2 Commercially Available PCM Microcapsules [5]

Manufacture	Product	Type of Product	PCM	Concentration (%)	Particle/Droplet Size (μm)	Melting Point (°C)	Latent Heat (kJ/kg)
BASF	Ds 5000	Mpcm slurry	Paraffin	42		26	45
	Ds 5007	Mpcm slurry	Paraffin	42		23	41
	Ds 5030	Mpcm slurry	Paraffin	42		21	37
	Ds 5001	Powder	Paraffin			26	110
	Ds 5008	Powder	Paraffin			23	100
	Ds 5030	Powder	Paraffin			21	90
Microtek laboratories	Mpcm-30d	Powder	N-decane		17–20	−30	140–150
	Mpcm-10	Powder	N-dodecane		17–20	−9.5	150–160
	Mpcm 6d	Powder	N-teradecane		17–20	6	157–167
	Mpcm 18d	Powder	N-hexadecane		17–20	18	163–173
	Mpcm 28d	Powder	N-octadecane		17–20	28	180–195
	Mpcm 37d	Powder	N-eicosane		17–20	37	190–200
	Mpcm 43d	Powder	Paraffin mixture		17–20	43	100–110
	Mpcm 52d	Powder	Paraffin mixture		17–20	52	120–130
Capzo	Thermusol hd35se	Powder	Salt hydrate			30–40	200
	Thermusol hd60se	Powder	Salt hydrate			50–60	160

Table 5.3 Objective Magnitudes and Influential Parameters at the Time of Selection of a PCM Emulsion or mPCM Slurry as Heat Transfer Fluid or Thermal Storage Material [5]

Influential Factors or Parameters	Objective Magnitudes	Influence When the Factor Increases	
		Positive Influence	Negative Influence
Particle diameter	Rupture of microcapsules		Rupture pressure of microcapsules decreases, higher number of ruptured capsules
	Subcooling	Greater probability of existing nucleation agents, and therefore lower subcooling	Possible nonequilibrium between PCM and water temperatures, possibility of hysteresis
	Apparent hysteresis		
	Heat transfer	Improvement in convection coefficient	
PCM concentration	Stability of emulsions Heat capacity	Increase in heat capacity, increase in transported heat	Creaming speed increases
	Pressure drop		Increase of viscosity, increase of pressure loss and pumping work. Up to PCM concentrations of 15–20% the increase is slightly superior to water
	Heat transfer	Decrease in Stefan number and therefore improvement of convection coefficient	Increase in viscosity, decrease in turbulence degree, and therefore worsening or convection coefficient
Operation temperature range	Heat transfer	The operation temperature range must fit with the phase change temperature range, and be the narrowest possible	Decrease of thermal conductivity, occasioning deterioration in heat transfer

The proper selection, experienced design, and performance assessment of the LTES system are the prime concerns for achieving energy efficiency and energy redistribution requirements in buildings. Taking the method (mode) of operation and the intended control strategies of the LTES system into account in the early stages of the design phase would result in the desired energy efficiency and energy redistribution requirements in buildings in the long run.

Moreover, the sizing of the LTES system largely depends on the thermal load profile of the building spaces, and once it has been determined, the performance assessment of the LTES system can then be performed by ascertaining load sharing capacity requirements. For example, the cooling and heating

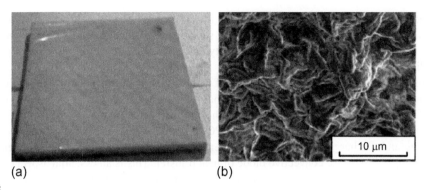

(a)   (b)

FIGURE 5.5

Pictures of the shape-stabilized PCM: (a) the PCM plate and (b) SEM picture.

Panel B: From Zhou et al. [10].

Table 5.4 PCM Manufactures Around the World [12]

Company	Country of Origin	Products	Product ID
Rubitherm GmbH	Germany	Salt hydrates/blend	SP
		Paraffins	RT
		Powder	PX
		Granules	GR
Merck KGaA	Germany	Salt hydrates	–
Climator AB	Sweden	Salt hydrates	Climsel C
Cristopia Energy Systems	France	Salt hydrates	AC
PCM Energy	India	Salt hydrates	Latest™
Mitsubishi Chemical	Japan	Salt hydrates	STL
EPS	UK	Paraffin, salt hydrates, and eutectics	PlusICE

Table 5.5 Summary of Key Manufacturer of PCM Worldwide [13]

Manufacturer and Reference	Range of PCM Temperature (°C)	Number of PCMs (Available Within the Given Temperature Range)
RUBITHERM	−3 to 100	29
Cristopia	−33 to 27	12
TEAP	−50 to 78	22
Doerken	−22 to 28	2
Mitsubishi Chemical	9.5 to 118	6
Climator	−18 to 17	9
EPS Ltd.	−114 to 164	61

load demand observed for underfloor schemes in building spaces would account for 40 and 100 W/m², respectively. Based on the thermal load flux, the selection and integration of an appropriate LTES system can help realize better load sharing capacity and energy redistribution in the occupied spaces of the building.

The case for the LTES system being integrated with the fabric component of the building interior spaces is similar, in which the net thermal load sharing capacity exclusively depends on the contact exposure of the surface area or the frontal area of the LTES module with indoor air.

For instance, the fabric wall/slab impregnated PCM module (passive system) is expected to exhibit good charging and discharging characteristics when the wallboard PCM exchanges heat energy with indoor air through the surface convective mode of heat transfer. This is an essential aspect of the LTES system, in which the frontal surface area is more likely to participate in the heat transfer process for regulating the temperature at prescribed comfort levels in indoor environments.

The cooling/heating load sharing capacity of the building can be further enhanced by using the active LTES system. The major distinction between active and the passive LTES systems is realized in terms of utilizing or minimizing mechanically assisted equipments/systems in the former and the latter, respectively.

The amalgamation of the mechanical system (fans or pumps) with the LTES module would contribute to enhanced heat transfer between the PCM and the HTM; thereby TES and release performance of the LTES system can be effectively improved. The broad classification of passive and the active LTES systems is depicted in Fig. 5.6(a,b).

On an average, about 40–50% of per day total cooling load can be effectively matched by integrating the active LTES system with the building cooling system. Hence, it is important to do a thoughtful assessment on the performance of the LTES system at every step of the design process from scheme inception to completion of an elegant building construction.

The detailed information on the assessment of various TES systems dedicated to building cooling/heating applications is presented in Chapter 12, which is beneficial to the professionals with good expertise in the design and development of systems ranging from small-scale residences to large-scale commercial applications.

5.7 PASSIVE LTES SYSTEMS

The passive LTES system, as stated earlier, utilizes natural convection as an effective mode of heat transfer for regulating temperature fluctuations in indoor spaces. PCM material impregnated within the construction material matrix is exposed to the heat source of the room or zone of a building and participates in exchanging the cold or warmth for TES, depending on seasonal variations.

5.7.1 PCM impregnated structures into building fabric components

The concept of impregnating the PCM into wallboard construction material has become increasingly attractive in recent years. Basically, the PCM (salt hydrate/inorganic/organic type) is embedded into the wallboard structure by any of the impregnation methods described in earlier sections. PCM impregnated wallboard structures are then incorporated into the building components including wall, ceiling, roof, and floor, depending on the thermal load fluctuations persisting on interior zones. The cost effectiveness of PCM impregnated wallboard structures makes them highly preferred in lightweight building construction.

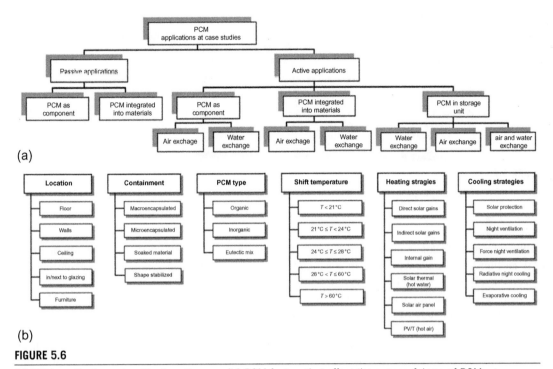

(a)

(b)

FIGURE 5.6

(a) PCM applications classification diagram. (b) PCM factors that affect the successful use of PCM applications.

The overall performance in terms of the efficiency of the wallboard structures chiefly depends on the following factors [14]:

- The method of impregnation of the PCM
- Orientation of the wall component
- Solar heat and internal heat gains of the building
- Variations in the ambient air conditions (temperature, wind speed)
- Ventilation air change rate and infiltration effects
- PCM thermophysical properties including range of the phase change (melting/freezing) temperature and the latent heat capacity per unit area of the wall component

The pictorial view of the wallboard structure comprising the microencapsulated PCM (MPCM) is shown in Fig. 5.7. The MPCM impregnated wallboard is designed for the enhancement of thermal behavior of the lightweight internal partition wall, based on the interior/exterior air temperature variations for a period of 24 h. The MPCM, with a thickness of 1 cm utilized, has a phase change temperature of 22 °C, which is capable of increasing the thermal inertia of the building component by a factor of 2.

The utilization of PCM wallboard structures in buildings would effectively reduce indoor air temperature fluctuations and the possible overheating effects. That is, the comfort conditions in an indoor environment are met without the need for supplemental energy to achieve the same. Besides, PCM

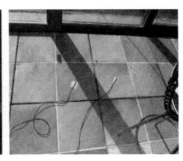

FIGURE 5.7

Dupont de Nemours PCM composite wallboard, composed of 60% of microencapsulated paraffin [15].

Photographic view of PCM tiles [16].

wallboard with a thickness of 5 mm is capable of storing the available thermal energy at a higher rate (2 times higher). Interestingly, an 8 cm concrete layer of achieves the same purpose, thus exploring the potential application of PCM wallboard in lightweight building construction.

Many research studies in terms of numerical analysis and experimentation related to the PCM wallboard application in buildings have been reported extensively in the literatures [15,17–30]. One refined conclusion that can be obtained from these studies is whether PCMs intended for TES have their phase transition temperature in the range of 20-25 °C. This is true because PCM has to exchange the heat energy with indoor air specifically through the mode of natural convection as well as to regulate the indoor air temperature to swing between the comfort temperature ranges of 23– and 26 °C.

The stabilization of indoor air temperature during winter season can be effectively achieved through the impregnation of PCM into the tile structures of the building as depicted in Fig. 5.7. The working principle of the PCM tiles is simple in the sense that during the daytime, they would absorb heat gain from irect sunlight radiation and store the required heat energy for catering the heating requirements in nighttime. The temperature of the air in the occupied space is regulated according to stored heat energy in PCM tiles. In addition, by altering PCM melting temperature, PCM tiles can be used as a heat sink during hot sunny days in summer.

The different types of PCM impregnated structures being integrated into building fabric components are shown in Fig. 5.8. These include PCM floor panel, structurally insulated panel (SIP), and vacuum insulated panel for enhancing TES capacity of lightweight building construction materials. PCM impregnated structural panels are advantageous in terms of reducing the average peak heat flux and indoor air temperature swings by 10-40% and 4-5 °C, respectively, during hot summer seasonal conditions. PCMs that are commonly preferred for wallboard applications include paraffins, eutectic mixtures of capric acid and lauric acid, polyethylene glycol (PEG 600), and so on (refer to Appendix II for PCM properties).

Double-layered PCM arrangements are considered an innovative approach for evaluating the year-round thermal performance of a building subjected to seasonal variations. The schematic representation of the LTES system is shown in Fig. 5.9. The double-layered LTES system consists of the PCM panel configured between the roof top layer and the concrete slab. As the fusion temperature of the PCM is selected to be 4-5 °C above the ambient air temperature, the PCM effectively traps the heat gain from the ambient air during early morning cold periods, thereby reducing the indoor air temperature fluctuations and helps to maintain the comfort conditions in indoor environment.

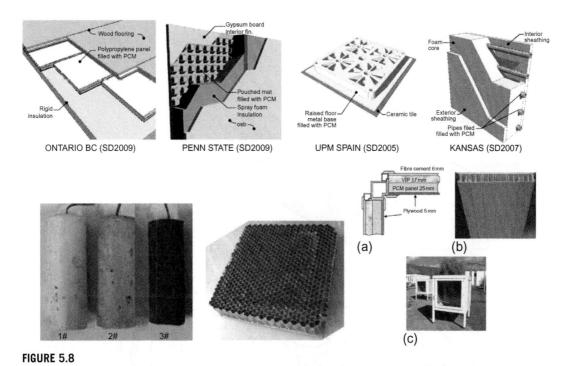

FIGURE 5.8

Passive applications: PCM as floor and wall components [31], PCM integrated in tiles and in SIP panels [31], Pictorial view of samples of: 1# – gypsum, 2# – cement, 3# – new PCM [32], Photo of honeycomb panel filled with paraffin PCM [28], (a) Sketch of test cell containing PCM, (b) photographic view of PVC panel, (c) external view of test cells [33].

In this context, the thermal gains infiltrating into the indoor zones through the ambient air can be reduced considerably by using cone frustum apertures filled with suitable LTES material as depicted in Fig. 5.9. Depending on the type of PCM being filled in the conical frustum, corresponding TES and discharge characteristics are experienced. However, whatever PCM is used for the purpose, the indoor temperature is substantially reduced compared to outdoor air temperature; thereby ensuring thermal comfort in indoor spaces.

PCMs that are usually preferred in such type of LTES system are P116, *n*-eicosane, and *n*-octadecane, whose fusion temperatures are observed to be 47, 37, and 27 °C, and their latent heat capacity values are 225, 241, and 225 kJ/kg, respectively. The conical shape of the frustum aperture contributes to augmenting the convective heat transfer between outdoor air, PCM, and indoor air, which helps to reduce the effects related to high thermal gains.

5.7.2 **PCM impregnated into building fabrics**

The spectrum of LTES materials applied to satisfy building cooling/heating requirements includes methods like those discussed in earlier sections, in which PCM are directly impregnated into building fabrics such as bricks, concrete, and trombe wall. The pictorial representation of the PCM drywall is shown in Fig. 5.10.

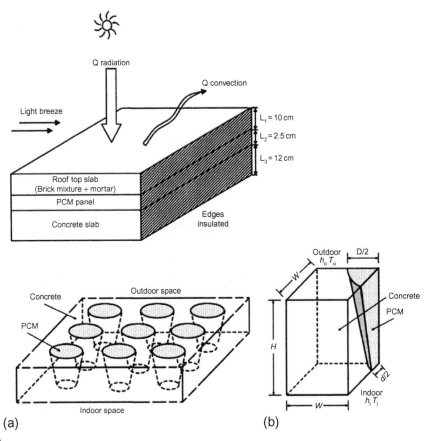

FIGURE 5.9

Sketch of building roof integrated with PCM panel [72], (a) schematic representation of the PCM-filled conical apertures on roof structure, and (b) computational domain, essential geometric parameters, and boundary conditions.

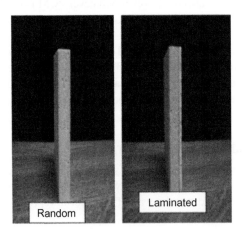

FIGURE 5.10

Samples of drywall [34].

The random drywall represents the impregnation of randomly distributed PCM into gypsum board, and the laminated drywall refers to the impregnation of the laminated profile of PCM into the same gypsum board. The phase change efficiency and the latent heat storage capacity of the gypsum board containing the laminated PCM profile are expected to achieve around 55% and 27%, respectively, when compared to randomly distributed PCM profile gypsum board.

It is pertinent to note that the thermal performance of the building construction materials is improved by embedding PCMs as an integral composite material. The photographic view depicted in Fig. 5.11 represents conventional and the alveolar brick construction containing PCM. The MPCMs utilized for this purpose are RT-27 and SP-25 A8, which are added into each of the construction cubicles and are subjected to testing [35].

The PCM-cubicle equipped with a heat pump cooling system (which helps to enhance the nighttime free cooling effect) performed better than the cubicle without the PCM. Also, energy consumption of the heat pump system during cooling mode was found to be reduced by almost 15% for the cubicle impregnated with PCM.

In a similar approach, PCM incorporated into a Portuguese clay brick masonry wall as shown in Fig. 5.11 contributes to assuage air temperature swings in indoor zones up from 5 to 10 °C to that of the outdoor conditions. In addition, the time delay of such PCM brick masonry can be increased to about 3 h.

In some cases, PCM is embedded into the hollow thermal insulation brick construction, wherein PCM bricks exhibit better thermal insulation effect during daytime direct sunlight radiation, rather than the untreated brick construction. The underside temperature of the PCM bricks is achieved at 31.7 °C, which is beneficial in terms of achieving better thermal regulation and energy redistribution requirements in buildings.

(a)   (b)

(c)   (d)

FIGURE 5.11

(a) Brick cubicle, (b) brick cubicle with polyurethane, (c) brick cubicle with RT-27 and polyurethane, and (d) alveolar brick cubicles [35]. Clay bricks with PCM macrocapsules [36].

On the other hand, PCM incorporated into concrete wall construction leads to the enhancement of thermal inertia and lower inner temperatures in buildings compared to conventional concrete [37,38]. At the same time, the concrete wall comprising PCM, if thinner, would possibly improve TES capabilities and accomplish indoor thermal comfort requirements in buildings as well [39].

In addition, the concept of providing a Trombe wall incorporated with PCM in building architectures can be considered a beneficial method of construction that would offset thermal load fluctuations in building spaces (zones) and contribute to better thermal inertia and energy savings potential.

Although Trombe walls are thought to be an ancient method of making elegant building constructions, incorporation of PCM would definitely result in the development of portable, movable, lightweight, and rotating wall-LTES systems. Through this method, the huge masonry material construction can be well reduced, and waste of available construction materials can also be minimized.

A modified design of Trombe walls has also been developed in recent times, in which rotating wall segments (pivoted around their vertical shafts) act as both absorber and radiator during the daytime and nighttime, respectively, during winter. This type of arrangement is even well suited for buildings located in cold climatic conditions, as the buildings can take advantage of such passive Trombe wall-LTES system phase change attributes and energy redistribution aspects to make them energy efficient.

5.7.3 PCM integrated into building glazing structures

In the evaluation of the total cooling/heating load demand of building envelopes, the internal heat gain due to the glazing structure is vital in determining the thermal load capacity of the building cooling/heating systems. The glazing structure possesses an inherent phenomenon by which solar radiation infiltrating through the glazing element is trapped inside the occupied space in the form of infrared radiation. Hence, the solar gain through the glazing structures must be considered an important aspect while designing the cooling/heating system for a building.

From this perspective, the integration of the PCM into glazing structures has recently gained impetus in the field of construction and building services. A typical example of such a glazing-LTES module for passive cooling/heating applications in buildings is shown in Fig. 5.12. The dividing wall

FIGURE 5.12

Dividing wall with 16 glass bricks filled with PCM [40], office with PCM sun protection system. The PCM blind was regenerated by way of tilted windows. The left opaque upper part of the facade consists of a ventilation flap for additional air flow during the night [41].

construction comprising 16 glass bricks embedded with the PCM is designed to meet both the daylight option and solar space heating simultaneously.

The operational principle of the PCM-glazing structure is quite interesting and feasible in the sense that, during daytime, visible light radiation enters the zone through the translucent PCM-glazing element and provides the daylight requirement in the building. At the same time, the infrared radiation being trapped by the PCM-glazing structure would help the PCM to undergo phase transformation, and the desired quantity of the thermal energy is stored eventually. Thus, during nighttime, PCM dissipates the stored heat energy into the indoor zone and offsets the heating load demand in buildings at the cost of reduced heat loss through the glazing.

The working principle of PCM-glazing slats is similar to that described earlier, in which PCM in the movable glazing slats absorbs (stores) the solar heat gain during daytime and releases (discharges) back into the indoor room during nighttime. Both the daylighting and heating energy requirements are equally met by this integrated system. The temperature of the occupied spaces are also maintained well within the phase change temperature of PCM (approx. 28 °C), compared to the conventional system that would have a rise in temperature of about 40 °C in summer conditions.

In addition, PCM shutters are a class of movable structure installed at the exterior of the building, which is made to function as a carrier of heat energy during daytime and radiator during nighttime. The PCM shutter is made to open during daytime such that the solar gain is trapped, which in turn would melt the PCM, enabling it to store the required heat energy through the phase transition process.

During nighttime, the PCM shutters are closed; thereby, the total heat energy that is stored by the PCM would be transferred back to the indoor zone, and the heating energy need in the building is effectively accomplished. At the same time, the heat loss occurring through the glazing wall is largely limited by the PCM shutters when they remain in the closed position.

5.7.4 PCM color coatings

The incorporation of color paint coatings containing PCM on the exterior surface of the building envelope draws special attention to maintaining the desired thermal comfort and temperature regulation in an indoor environment. The MPCM pigments with particle sizes ranging from 17 to 20 μm are mixed with the intended color paint material. The MPCM used for this passive cooling application is basically a paraffin that exhibits a latent heat of enthalpy of 170–180 kJ/kg.

Interestingly, the application of MPCM color coatings on the exterior of the building envelope has been proven to reduce the surface temperature of exterior wall structures by 7–8 °C compared to conventional color coatings. Moreover, the reduction in surface temperature is expected to be 10–12% than that of the conventional cool coatings of the same color specifications.

This will eventually help to achieve the required thermal comfort and minimize temperature swings considerably. The maximum surface temperature and temperature gradients of the MPCM-based and conventional color coatings dedicated for the passive building cooling application are summarized in Table 5.6.

These passive LTES systems integrated with building components and construction elements contribute to achieving good thermal comfort and better temperature regulating features with an improved energy savings of up to 10–15%.

Table 5.6 Max Surface Temperature and Temperature Differences of Various Color Coating Samples (°C) [42]

	Black	Blue	Green	Gray	Brown	Golden Brown
T_{max}						
Common	67.9	63.1	64.7	65.2	62.6	58.1
Cool	62.2	58.6	61.5	62.3	60.1	56.1
PCM	60.5	57.0	59.8	60.9	58.5	55.0
ΔTcommon-PCM						
Common	–	–	–	–	–	–
Cool	5.7	4.4	3.2	2.9	2.5	2.0
PCM	7.4	6.1	4.69	4.3	4.1	3.1
ΔT cool-PCM						
Common	–	–	–	–	–	–
Cool	–	–	–	–	–	–
PCM	1.8	1.7	1.6	1.4	1.6	1.1

5.8 ACTIVE LTES SYSTEMS

The energy redistribution from on-peak load conditions to part load conditions and energy efficiency of the cooling systems dedicated to provide air-conditioning requirements in building spaces can be effectively gained through incorporating active LTES systems. As pointed out earlier, an active LTES system is essentially equipped with a mechanically assisted fan/blower and pumping system as part of their design for accomplishing the aforementioned objectives.

The latent heat transformation characteristics of PCM integrated into the air distribution system (or air handling unit) can be enhanced significantly using the forced convection principle between the HTM and the PCM modules. The overall heat transfer efficiency and energy effectiveness of active systems are comparatively higher than that of passive systems, due to their dynamic performance under varying thermal load demand situations. The four major categorization of the active LTES system are as follows:

- Free cooling with the PCM-TES
- Comfort cooling with the PCM-TES
- Ice-cool thermal energy storage
- Chilled water-PCM cool thermal energy storage

5.8.1 Free cooling with the PCM TES

The concept of free cooling aided with PCM thermal storage is a preferred energy-efficient technique for buildings located in cold regions, where ambient conditions are in balance with indoor comfort criterion. The distribution of active research studies being performed on free cooling with PCM integration in buildings worldwide is depicted in Fig. 5.13. This chart shows that most of the research works in this field have been carried out in Europe (~73%), which may be due to the climatic conditions

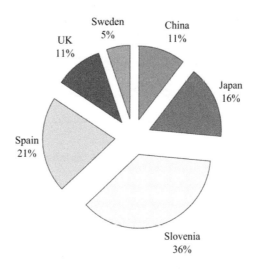

FIGURE 5.13

Research on PCM-based free cooling around the world [12].

mentioned earlier and the continuous interests shown toward climate change and developing energy-efficient and sustainable cooling systems.

The operation of free cooling-based PCM thermal storage is explained next. By forcing the low temperature ambient air over the PCM during nighttime, the charging (freezing) of takes place, and cold energy is effectively contained (stored) in the PCM. Because the PCM selected in the free cooling application has a melting temperature higher than the ambient air temperature, it would become solidified when cold air flows over the PCM.

During daytime, by passing the indoor (room) air, which is at a temperature higher than the PCM, the stored cold energy is retrieved by the warm room air, and the PCM is discharged. The return cold air circulated in the room subsequently meets the space cooling load requirements. Thus, the on-peak load (daytime) is shifted effectually to the part load conditions (nighttime) by using the PCM free cooling system. The operating strategy explained here is schematically represented in Fig. 5.14, where PCM blocks are placed on the ceiling component of the conditioned space.

It is noteworthy that thermal performance of the PCM in terms of the rate of charging and discharging depends purely on the thermophysical properties, including the thermal conductivity, phase change temperature, latent heat of fusion, degree of supercooling, ambient air temperature, and airflow rate. This type of passive cooling system exhibits good thermal storage performance and is well suited for buildings that are subjected to ambient temperature in the range of 20–23 °C during nighttime (winter season), and the same would not exceed 30 °C in daytime (summer season).

The heat exchange effectiveness of the ambient air and the indoor air during free cooling process can also be increased by using the channel (staggered) PCM configuration as shown in Fig. 5.15. Four channels of dimension 25 cm × 35 cm × 11 m are insulated with 1-cm-vacuum panels with polycarbonate profiles. The PCM utilized in these polycarbonate profiles is a salt hydrate type with a phase change temperature of 26 °C.

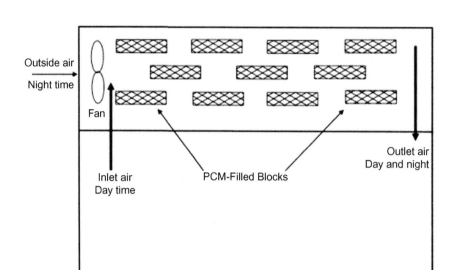

FIGURE 5.14

Night ventilation with packed-bed PCM storage system [43].

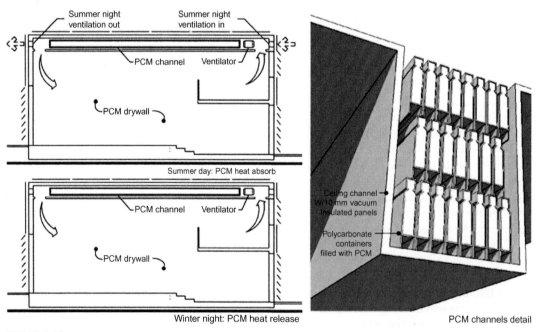

FIGURE 5.15

Active application: PCM system as component with air thermal exchange: Germany 2009 House [44,45].

In summer cooling cycles (nighttime), the ambient air transfers the cold energy effectively while flowing over the PCM channel containments and is exhausted back to the atmosphere through the variable speed ventilation fan arrangement. During on-peak hours, the indoor air that is pumped over the staggered PCM module recovers the already stored cold energy and satisfies the cooling load demand in the occupied space.

In winter heating cycles, the heat energy that is available in the indoor air acts as the source for charging the PCM (melting). In a typical operation, the room air is supplied over the PCM containment provided on the ceiling component of the building space. Through the method of forced convection using a ventilator, the warm room air transfers the heat energy to the PCM and makes it undergo a phase transition process (solid-liquid phase). The cold return air downstream of the PCM enters the indoor space and picks up the heat that is available for the next cycle of charging the PCM.

In the nighttime, the room air that is at a lower temperature flows over the PCM containment, in which the PCM undergoes discharging process (liquid-solid phase) leading to the retrieval of stored heat energy by the flowing air stream. The heated air is then allowed into the conditioned zone for meeting heating requirements. Factually, during winter night operations, ventilation air dampers are maintained in the closed position to prevent any infiltration heat loss effects due to colder outdoor air.

The concept of the PCM radiant ceiling has also been developed for cooling and heating the occupied zones as depicted in Fig. 5.16. The construction configuration shown graphically infers that cold water from the cool water tank and hot water from the hot water tank are pumped through radiant ceiling panels (radiant coils) for catering the cooling/heating load demand in indoor space. The objective of integrating PCM into the radiant cooling/heating coil elements is to ensure energy redistribution from on-peak to part load conditions of the building space.

In the cooling mode, cold water at 16 °C is pumped from the cool water tank to the PCM radiant ceiling coils, in which active heat transfer through combined radiation and convection takes place between indoor air and PCM radiant cooling coils. This in turn helps satisfy cooling load requirements in the occupied zone.

During nighttime, warm water is pumped to the solar panels being installed on the building's roof. A portion of warm water is evaporated, and the remaining cold water is then collected back into the cooling water tank. In heating mode, the hot water obtained from the solar collector, which is stored in the hot water tank, is pumped into the PCM radiant ceiling coils and ensures the heating load demand been met in the occupied zone, thereby maintaining the thermal comfort in the indoor environment.

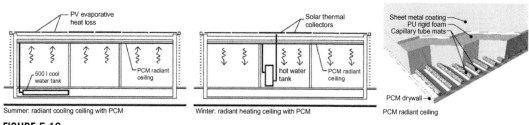

FIGURE 5.16

Active application: PCM system as component with water thermal exchange: Darmstadt 2007 House [44,45].

An interesting method of charging and discharging thermal energy using PCM combined with radiant ceiling panels shows a significant energy redistribution efficacy and energy savings potential for modern building cooling/heating applications. The schematic representation of the PCM radiant ceiling thermal storage system is illustrated in Fig. 5.17 This system essentially consists of two insulated storage boxes meant for storing cold and hot energies to meet the space cooling and the heating capacities on demand as well as to maintain a comfortable environment for indoor occupants.

Basically, the thermal storage system operates in the following manner: the daytime cooling comfort is provided by the cold water, which is available at 16 °C in the radiant ceiling panel. The excess heat in the room is transferred to the PCM system by means of the variable speed water (hydronic) system. This is referred to the discharging process (melting) of the PCM. During night hours, the ambient air at 15.6 °C is pumped into the PCM cooling box, which enables the PCM to undergo charging process (freezing).

This means that, with the help of the variable speed ventilator arrangement, latent heat transfer of the PCM from the liquid-solid phase occurs due to the thermal exchange mechanism being established between outdoor air and the PCM. The PCM considered for such application is a eutectic type, which is filled inside the HDPE blow molded containers as shown in the illustration.

Likewise, in the heating cycle, the hot water being produced by the evacuated solar thermal collector and stored in the hot water tank at the requisite temperature is pumped into radiant ceiling panels through the same hydronic circuit. Thus, space heating is achieved through radiant ceiling panels, and the desired thermal comfort is being met in the thermal zone of the building.

Meanwhile, any excess heat available in the hot water tank is routed to the PCM heating box, which facilitate utilization of PCM thermal storage for meeting the same heat duty in indoor space, in case of off-sunshine hours or partly cloudy ambient conditions. This type of PCM radiant ceiling active thermal storage system is considered energy efficient for buildings requiring both cooling and heating for a year-round operating strategy.

Other possible routes are available in the indoor space to take advantage of free cooling TES, in which the PCM module is located at the underfloor component of the building. The typical arrangements of underfloor active PCM thermal storage applications in buildings are graphically represented in Figs. 5.18 and 5.19. Conventional bricks are soaked with the PCM mixture comprising 48% butyl stearate, 50% propyl palmitate, and 2% fat acid.

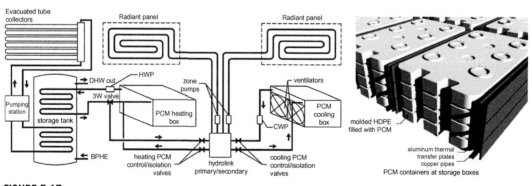

FIGURE 5.17

Active application: PCM system as component with air and water thermal exchange: Rhode Island 2005 House [46,47].

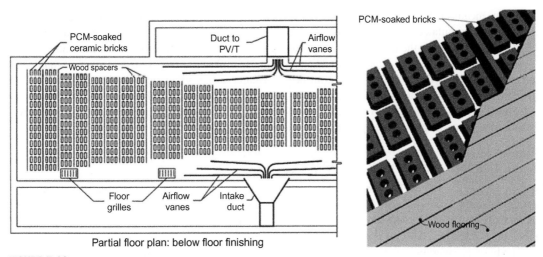

FIGURE 5.18

Active application: PCM system as component with air thermal exchange: Canada 2005 House [44].

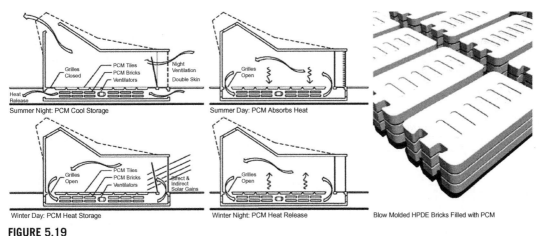

FIGURE 5.19

Active application: PCM system as component with air thermal exchange: Madrid 2005 House [48].

These soaked bricks are placed at 7 cm cavities under the floor finishing to acquire better TES during the daytime as well as night hours. The hot air for space heating is produced by means of solar PV/thermal panels, and a heat pump system is made available for meeting the space cooling demand, in case semipassive air-conditioning is insufficient to meet the same cooling requirement and indoor thermal comfort as well.

Summer day operation of the underfloor PCM thermal storage system is designed to address cooling comfort conditions in the occupied space. The warm air in the room is circulated through the underfloor component amalgamated with the PCM modules, whereby the PCM stores (charging process) the heat energy by means of forced air thermal exchange mechanism. Likewise, during night hours, the cold ambient air is circulated over PCM containments and releases (discharging process) heat energy from the PCM and facilitates the thermal storage system to be in effective operation for the next day cycle.

In addition, by using these underfloor TES strategies in buildings, energy storage capacity has been enhanced by two to three times, and overall heat exchange efficacy has been improved as well. The direction of air movement, opening and closing strategy of the grilles and dampers, and the flow path of the heat energy in indoor space represented graphically enables a better understanding of this active thermal storage concept and its application in buildings.

5.8.2 Comfort cooling with the PCM TES

In the context of active LTES systems dedicated for providing cooling comfort conditions in building spaces, the utilization of the MPCM slurry as the thermal storage material in the radiant cooling application has received increasing interest recently. The schematic diagram of the proposed cooled-ceiling amalgamated with the MPCM slurry system is represented in Fig. 5.20.

The basic refrigeration plant (chiller) is employed for producing chilled water at the designed cooling coil supply temperature, which would be between 14 and 16 °C. The MPCM used here is hexadecane (organic PCM), which has a phase transformation temperature of 18 °C and latent heat of fusion of 224 kJ/kg.

The operational performance of this system depends on the cooling load distribution between the MPCM slurry flowing through the radiant ceiling panels and the dehumidified process air being supplied into the conditioned space. Typically, during the day hours operation of this system, the MPCM slurry at the requisite temperature is pumped into the radiant ceiling panels installed over the ceiling component of the conditioned space.

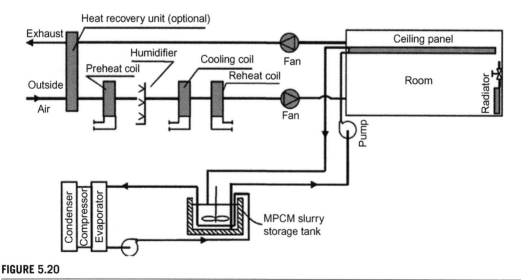

FIGURE 5.20

Schematic diagram of cooled-ceiling integrated with MPCM slurry tank [49].

The eventual phase change characteristics of MPCM helps offset the sensible load in the occupied space through the radiative heat transfer. The remaining latent heat load in the zone space is met by the cooled and dehumidified air, which is processed in the air handling unit (consisting of heat recovery elements for precooling of ventilation air) and supplied into the conditioned space.

By implementing this kind of active LTES system combined with the air-conditioning system, it is expected that about 33% of daytime electricity demand would be reduced compared to the conventional chilled ceiling system working with water as the HTM. Besides, yearly energy consumption would also be decreased with the integration of such a system for building cooling applications.

The peak load shifting of the PCM radiant ceiling integrated air-conditioning system devoted for an office building application is schematically shown in Fig. 5.21. In this system, the conventional rock wool ceiling board is replaced with the MPCM ceiling board. The MPCM considered in this system has a melting point of 25 °C, and the cooling load demand was estimated for the office space of 16 m^2.

The cold air from the air handling unit is utilized as the source for the charging (freezing) of the MPCM during nighttime operation of the cooling system. The indoor air, which is made to flow past the MPCM module, enables it to discharge (melting) the stored cold energy from the MPCM, the required cooling demand in the office space is effectively met, and indoor comfort is also equally achieved.

The incorporation of the MPCM ceiling board in place of conventional rock wool board facilitates offsetting about 85% of the total cooling load in the conditioned space. Furthermore, the peak load shaving capability of the MPCM cooling system is 25.1%, which helps the present system be cost effective in its operation by 91.6% rather than the conventional rock wool ceiling board cooling system.

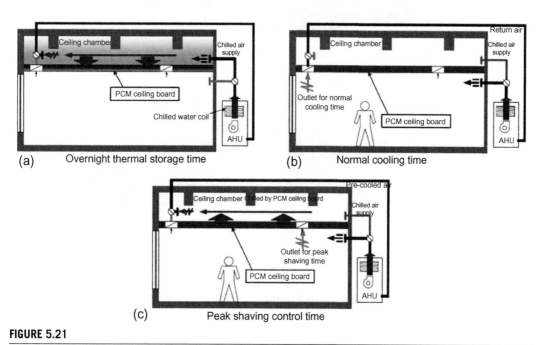

FIGURE 5.21

Outline of the proposed system (MPCM ceiling board cooling system) [50].

The development of the active MPCM assisted with the air handling cooling system has gained momentum in recent years. The configuration of organic PCM being encapsulated into the aluminum rigid slabs (weighing 135 kg approximately) that are located in parallel to the flow direction of the cold air stream in the thermal storage unit is depicted in Fig. 5.22.

The operating principle of this system is similar to that of active PCM thermal storage cooling system as described earlier and seems to be self-explanatory. It is pertinent to note that by varying the operating phase change temperature and the latent heat of fusion values of the macroencapsulated organic PCM, the rate of charging and discharging characteristics would be increased by 10–11%. This

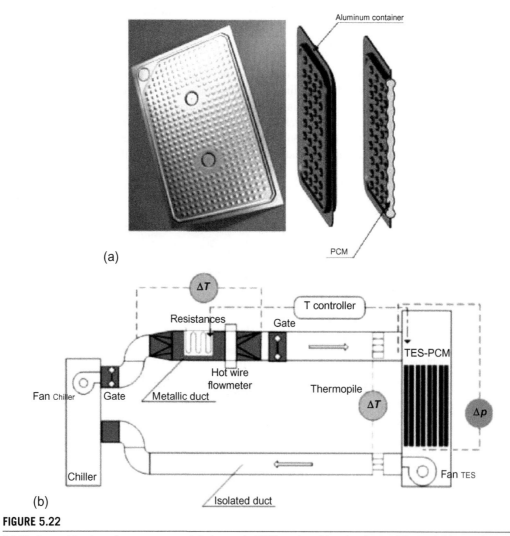

FIGURE 5.22

(a) Photographic view of macroencapsulated organic PCM enclosed by aluminum slabs and (b) experimental setup of PCM-air heat exchanger [51].

would also influence the heat transfer attributes of the PCM thermal storage system with indoor air during cooling duty cycles.

5.8.3 Ice-cool thermal energy storage

Ice-cool TES, usually referred as the ITES system, has been developed and used for many years. The ITES system, depends on the mode of operation (full or partial storage), type of storage medium, and charging and discharging characteristics to effectively match the cooling load demand and the energy redistribution requirements in buildings.

The integral part of an ITES system includes a primary chiller plant that would cool water stored in the storage tank and, at the cost of the latent heat of fusion, the cooled water changes its form into ice. The cool thermal energy is stored (charging) during the phase transition of water to ice, and the same is released (discharging) during the phase transformation of solid ice into water.

The charging and the discharging processes are activated through circulation of the HTM (brine solution or water or refrigerant) in the cooling coils embedded in the storage tank. The water present in the storage tank, which is in contact with the embedded cooling coil, absorbs cold energy from the circulating HTM and undergoes a charging process.

Likewise, during the discharging process the stored cold energy is captured from solid ice back into circulating warm HTM to satisfy the building cooling load demand. A variety of ITES systems are being developed for this purpose. Some of the major ITES systems that are considered feasible for providing cooling and energy storage in buildings are discussed in the forthcoming sections.

External melt-ice-thermal storage system usually refers to the extraction of the stored cool thermal energy from the produced solid ice by subjecting it to phase transition (melting) from the exterior surface of the primary cooling coil circuit as depicted in Fig. 5.23. It is pertinent to note that the charging process of the water (as PCM) takes place by indirect contact heat transfer, whereas during the discharging process, the direct contact heat exchange mechanism takes place between the PCM and the HTM.

The photographic illustration shown clearly explains the processing of water into ice and vice versa. As detailed earlier, during charging periods (mostly at part load conditions of the building) the HTM

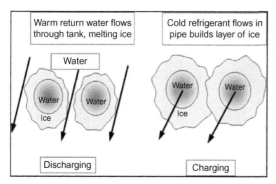

FIGURE 5.23

The charging and discharging procedure of an external melt ice storage system [52], a photograph of an external melt ice-on-coil system (subsystems of an Ice-Bear® unit) [53].

(brine or refrigerant) flows through the heat exchanger coils. Due to the heat transfer taking place between the HTM and the water that is filled in the storage tank, the phase transition of water into ice is expected to occur.

Factually, the thickness of ice that is formed purely depends on the charging temperature of the HTM. That is, for a typical application, the range of thickness of the ice that would be produced ranges from 40 to 65 mm [57]. To produce a thinner layer of ice, the charging temperature of the HTM is maintained between −7 and −3°C. Similarly, for a thicker layer of ice, the charging temperature is maintained between −12 and −9°C.

To extract the stored cold energy from the ice that is formed, the warm water from the building side is circulated inside the storage tank. Due to the direct contact heat transfer, the cool thermal energy that is stored in the ice is exchanged to the warm water for meeting the building cooling thermal load during on-peak conditions.

The design of the storage tank plays a vital role in determining the complete freezing (charging) and melting (discharging) of the water (as PCM). Generically, the ratio of water to the ice build would be kept at 70-30% for the complete freezing of water into solid ice. By maintaining this ratio, the extraction of the cool thermal energy from ice layers by the warm water from the building side during the discharging process would be facilitated further. The pictorial views of the different configurations of ITES heat exchangers produced by a variety of manufacturers are in presented in Fig. 5.24.

Internal melt-ice-thermal storage, on the other hand, is equally considered as an energy-efficient methodology for storing and retrieving cool thermal energy based on the cooling load demand in buildings. The basic demarcation of this system from the external melt system is that the HTM (brine, glycol, or refrigerant) flowing through the embedded cooling coil heat exchanger is utilized for performing the charging and the discharging processes of the water and the ice build inside the storage tank. The schematic diagram of the internal melt-ice-thermal storage system is represented in Fig. 5.25.

In a typical charging process (part load periods), the HTM (mostly glycol is preferred) at a charging temperature of −6 or −3°C (depending on the cooling load demand) is pumped through the immersed heat exchanger coils of the storage tank containing water. The relative heat exchange mechanism taking place between the HTM and the water drives the water to undergo phase change process to eventually form solid ice inside the storage tank.

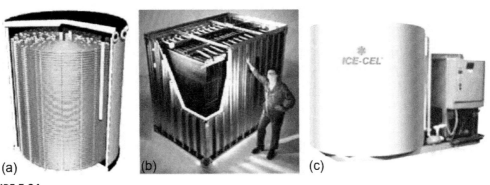

(a) (b) (c)

FIGURE 5.24

ITS heat exchangers configuration: (a) Calmac [54], (b) Fafco [55], and (c) Dunham–Bush [56].

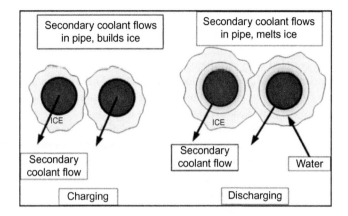

FIGURE 5.25

Charging and discharging procedure of an internal ice-on-coil storage system [53].

During the discharging mode, the solid ice build is melted using warm HTM (glycol solution) returning from the building side and that is flowing through the embedded coil elements of the storage tank. The stored cool thermal energy is thus captured by the warm HTM, and the temperature of whichin due course of time is reduced to the desired supply temperature for enabling comfort cooling in buildings.

The *ice storage using harvesting method* is a concept of producing flakes of ice combined with chilled water for meeting the fluctuating cooling load conditions in building spaces. The schematic representation of the ice storage harvesting system is shown in Fig. 5.26. The working principle of this

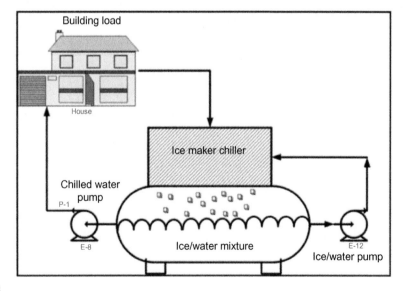

FIGURE 5.26

Schematic diagram of a typical ice harvesting ITS system [57].

cool thermal storage system is very similar to that of the external and the internal melt-ice-thermal storage systems, except for the fact that HTM (glycol) is used for producing the ice flakes during charging periods.

The refrigerant plant coupled with the storage tank containing the vertical plate surface being installed over the tank is primarily utilized for producing ice flakes. The chilled water flowing through the vertical plate (ice maker) at one point in time exhibits phase transformation, leading to the formation of ice layers on the surface of the plate heat exchanger.

The so-formed ice flakes are removed from the vertical plate by passing a hot gas stream at equal time intervals, which generally varies from 25 to 30 min. The thickness of the ice that is formed normally varies between 8 and 10 mm, which eventually depends on the cycle time for freezing process. The cool thermal energy being contained in the ice flakes together with the sensible heat of the chilled water is effectively used to satisfy the daytime cooling load demand prevailing in the conditioned space of the building envelope. Only limited installations of this type of special cool thermal storage system are identified in practice, which may be due to the complexity involved in the execution of such a system.

Ice slurry storage system stores the cool thermal energy by virtue of both sensible heat and latent heat characteristics of the HTM and water present in the storage tank. This system essentially comprises a primary cooling unit dedicated for producing ice crystals and a secondary heat exchanger coupled with the building air handling unit. The operating principle of the ice slurry storage system is depicted in Fig. 5.27.

The charging of an ice slurry storage system involves the cooling process of the HTM to a lower temperature. By mixing the low temperature HTM with the water contained in the storage tank, due to the active nucleation concept, the formation of the small ice-like structures occurs on the surface of the water present inside the storage tank. The ice slurries thus formed are then pumped to the secondary heat exchanger which is integrated with the building space conditioning unit; thereby the desired cooling effect and the energy redistribution are equally achieved.

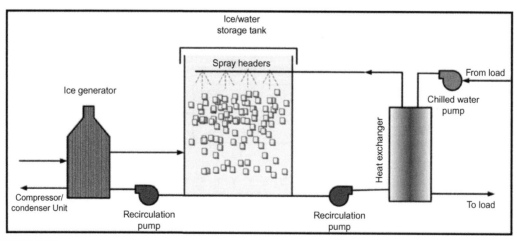

FIGURE 5.27

Schematic diagram of an ice slurry storage system [57].

It is noteworthy that for the desired cooling load, mass flow rate and the discharging temperature of the warm water returning from the building side, the supply temperature of the ice slurry storage system is mostly unaltered. In addition, the volumetric capacity ratio of the heat transfer fluid with respect to the water filled inside the storage tank would decide the effective formation of the ice slurries for them to be utilized further.

Encapsulated ice storage is a technique by which cool thermal energy is stored and released by means of the water (as PCM) being encapsulated using HDPE containments or small steel containers. The typical charging and the discharging processes of encapsulated ice storage system depicted in Fig. 5.28.

It can be seen that during the charging process, the water that is filled inside the spherical capsule is subjected to a lovw temperature circulating HTM (glycol solution between −6 and −3 °C). This in turn facilitates the phase change process of water being converted into solid ice. Similarly, during the discharging cycle, the warm HTM returning from the building side melts the ice present inside the encapsulation, and the cool thermal energy is retrieved for offsetting the cooling load demand in building spaces.

The overall heat transfer effectiveness of the encapsulated ITES purely depends on the temperature, flow rate and the thermal properties of the HTM, thermophysical properties of the water (as PCM), encapsulation material properties, storage volume, and the cooling load demand in the conditioned space.

5.8.4 Chilled water-PCM cool TES

In the quest toward developing the energy-efficient cool thermal storage systems, the concept of capturing and extracting cold energy to and from the PCM using the chilled water distribution is gaining impetus in building cooling applications. The interesting fact in developing such cool thermal storage system is to take advantage of latent heat effectiveness of the PCM to store and release thermal energy at the cost of sensible heat transfer taking place to and from the chilled water.

The basic schematic representation of the chilled water-PCM cool TES system is shown in Fig. 5.29. The operating strategy of this system during the charging and the discharging processes are obvious in the sense that the energy redistribution requirement/peak load shaving in building can be effectively addressed. The fundamental distinction between this system and the ITES system is realized by the following factors:

- The charging temperature of the chilled water (HTM) is kept between 4 and 5 °C.
- The discharging temperature of the chilled water (warm HTM) is maintained between 12 and 13 °C.

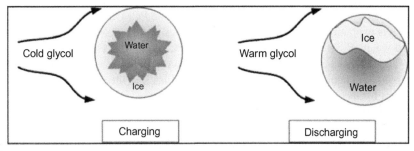

FIGURE 5.28

Charging and discharging procedure of an encapsulate ice storage [57].

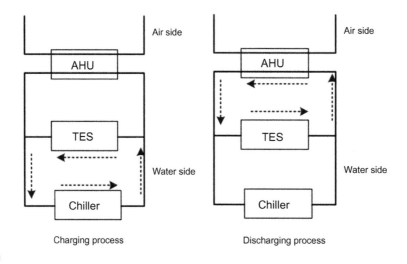

FIGURE 5.29

Schematics of charging and discharging processes using thermal energy storage system [58].

- The phase transition temperature of the PCM is selected in the range from 7 to 10 °C.
- The PCM is usually encapsulated within HDPE capsules.
- Only the chilled water circulating in the secondary loop of the chiller is utilized in the heat transfer process during the charging and the discharging periods of the PCM.
- The volumetric storage capacity of the chilled water-PCM system is comparatively less than the ITES system.

The chiller plant is coupled with the building side air handling unit through the TES interface as shown graphically. The thermal storage interface plays a significant role in enhancing the cooling load shaving from on-peak to part load conditions. Basically, the hydronic circuit of chilled water-PCM thermal storage system is divided in two parts during the charging and discharging processes: (1) a portion of the chilled water is routed to the building cooling loop and (2) a portion is diverted to the TES interface.

During the charging process, the chilled water produced from the chiller plant at the requisite temperature (as mentioned earlier) is pumped to the building side air handling unit for catering the cooling load demand and comfort conditions in the occupied space. Meanwhile, a portion of the chilled water enters the packed-bed thermal storage tank, which contains the PCM encapsulation.

For the reasons explained earlier, the charging (freezing) of the PCM takes place due to low temperature chilled water (HTM) being circulated over the PCM capsules (shown by the direction of the arrow heads). The warm water is then collected and mixed with the warm water returning from the building side in the secondary loop, which is then pumped back to the primary side (processing side) of the chiller plant for the next cycle of cooling process.

In the discharging process, a portion of the warm water returning from the building air handling unit (cooling unit) is pumped into the storage tank, wherein the stored cold energy is transferred from the PCM to the water (shown by the direction of the arrow heads). This in turn enables it to acquire the requisite temperature to satisfy the cooling demand in the conditioned space. The chilled water flowing out of the storage tank is mixed with the incoming chilled water from the secondary loop (production side) of the chiller plant. The combined flow of the chilled water is allowed through the building side cooling unit, and the process is repeated again.

The modes of operation of the cool TES (including ITES as well as chilled water-PCM) systems are summarized in Table 5.7 and Fig. 5.30. In case of full storage, the system would meet the on-peak load requirement entirely from the storage unit itself. But a partial storage system takes advantage of meeting the cooling load demand partly from storage and the remaining from the chiller plant as explained earlier.

In particular, the chilled water-PCM-TES system is usually applied for meeting the cooling demand and energy redistribution needs under the partial storage strategy. This is because this system operates on the sensible heat transfer of chilled water with the PCM, which mainly depends on the supply and the return temperature differential of the chilled water preferred for cooling the occupied spaces in buildings. However, the ITES storage system can be preferred to be operated under both full storage and partial storage strategies, depending on the type of building cooling application.

The advanced version of the chilled water-PCM-TES system as applied for commercial building cooling and air-conditioning needs is shown in Fig. 5.31 [60]. The cool TES system being proposed works on the partial storage strategy as explained in the earlier sections and helps to achieve 28–47%

Table 5.7 Modes of Operation of Active LHES System

Modes of Operation	Cooling Cycle	Heating Cycle	Presence of Heat Transfer Fluid (HTF)	Type of Storage
Charging process	Cooling of storage system by means of separate cooling unit to remove heat from the storage	Supplying heat to the storage system using separate heating unit	Yes	Full storage
Simultaneous charging process and thermal load balancing	Cooling of storage system by means of separate cooling unit to remove heat from the storage and offsetting cooling load from building directly	Providing necessary heat to the storage system using separate heating unit as well as catering heating load in buildings directly	Yes	Partial storage (either for load leveling or demand limiting operation)
Discharging process	Meeting cooling load demand entirely using stored cold energy from storage system only	Operating only the storage system to completely retrieve heating load demand	Yes	Full storage
Instantaneous discharging process and thermal load balancing	Sharing of cooling load demand by retrieving stored cold energy from storage system and operating cooling unit in parallel cycles	Retrieving heat load demand by activating storage system and heating unit in combined manner	Yes	Partial storage (either for load leveling or demand limiting operation)

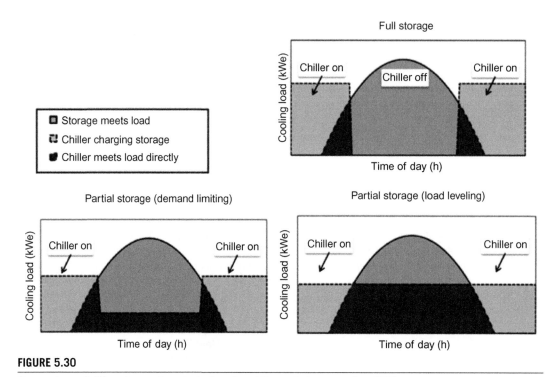

FIGURE 5.30

Comparison of different operating strategies of cool thermal storage system [59].

of energy savings potential when combined with the variable air volume (VAV) system and the energy conservative ventilation techniques.

Many other techniques are available for enhancing thermal storage performance of the LTES systems dedicated for building cooling applications. One such method by which the heat transfer mechanism could be augmented in the PCM is the inclusion of thermally conductive materials directly into the PCM or through the insertion of such materials in the form of extended heat transfer surfaces as shown in Fig. 5.32. The detailed discussion on how these thermally conductive materials would help improve the thermal properties of the PCM is provided in Chapter 12. The essential aspects of major types of cool TES systems are summarized in Table 5.8.

5.9 MERITS AND LIMITATIONS
5.9.1 Merits of LTES materials

Organic (paraffins and nonparaffins)

- Available in a large range of temperature
- High latent heat of enthalpy values (fatty acids have much higher latent heat of fusion than the paraffins)

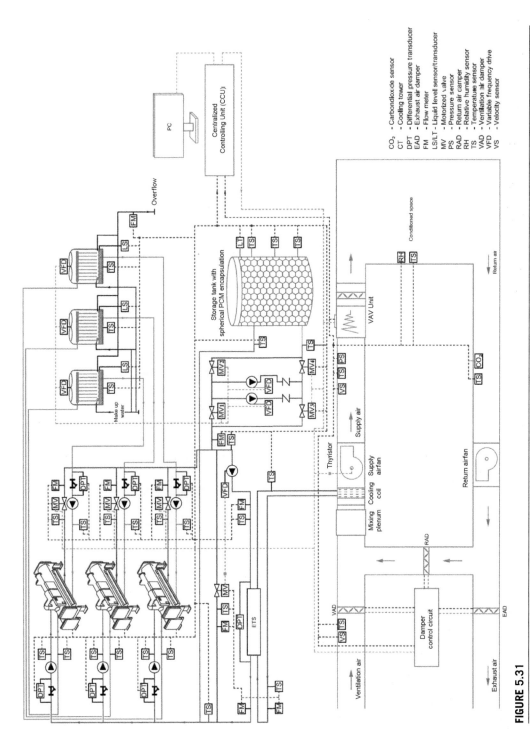

FIGURE 5.31

Schematic representation of the VAV–TES air conditioning system [60].

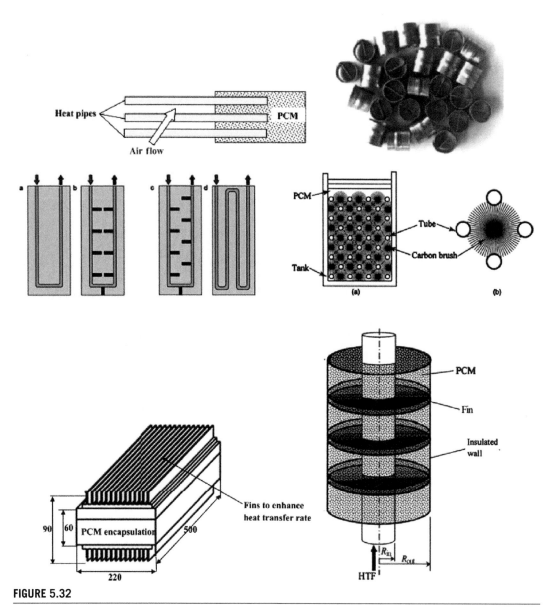

FIGURE 5.32

Heat transfer enhancement techniques for PCMs [61–69].

- Freezing with low supercooling degree or without surfusion effect
- Congruent phase transition process
- Self-nucleation and growth rate properties
- Relatively low segregation even after several thermal cycles (thermal reliability)
- High thermal stability

Table 5.8 Essential Aspects of Active CTES and LTES Systems [70,71,73]

	Chilled Water Storage	Ice Storage	Eutectic Salt Storage	PCM Storage
Specific heat (kJ/kg K)	4.19	2.04	–	2-4.2
Latent heat of fusion (kJ/kg)	–	333	80–250	130–386
Heating capacity	Low	High	Medium	Medium
Type of chiller	Standard water	Low temperature secondary coolant	Standard water	Standard water
Volume of storage tank (m³/kWh)	0.089–0.169	0.019–0.023	0.048	–
Storage charging temperature (°C)	4–6	–6 to –3	4–6	–10 to 6
Storage discharging temperature (°C) (higher than charging temperature)	1–4	1–3	9–10	5–8
Ratio of cooling capacity	20–30	More than 50	15–40	20–50
Performance coefficient of chiller	5.9–5	4.1–2.9	5.9–5	5.9–5
Fluid for discharging storage	Standard water	Secondary coolant/brine solution	Standard water	Standard water
Tank interface	Open system	Closed system	Open system	Closed system
Space requirements	More	Less	Less	Less
Flexibility	Existing chiller usage, fire protection duty	Modular tanks suitable for small/large installations	Existing chiller usage	Existing chiller usage
Maintenance	High	Medium	Medium	Medium

Inorganic salt hydrates

- High volumetric latent heat storage capacity
- Possess low vapor pressure in the melt state
- Noncorrosive, nonreactive, nonflammable, and not dangerous (fatty acids may be corrosive)
- Have good compatibility with the conventional construction materials
- Recyclable, cost effective, and readily availability
- Exhibits high latent heat of enthalpy and sharper phase transformation
- High thermal conductivity with lower volumetric changes during phase change
- Safe to the environment in terms of handling and disposing compared to the paraffins

Eutectics

- Exhibits application specific sharp phase change temperature (melting temperature)
- Slightly higher volumetric thermal storage density than the organic compounds
- Congruent phase transition with good thermal reliability (low or no segregation)

5.9.2 Limitations of LTES materials

Organic (paraffins andnonparaffins)

- Density, thermal conductivity, and latent heat of fusion inherently lower
- Inflammable, less compatible with plastic containments
- Expensive by nature
- Larger volumetric changes possible during charging and discharging (applicable to some grades of organic compounds)

Inorganic salt hydrates

- Relatively high supercooling properties
- Low degree of nucleation (requires nucleating additives and thickening constituent materials)
- Incongruent phase change and dehydration occurs during freezing and melting cycles
- Decomposition associated with phase separation
- Compatibility with some building materials is limited
- Exhibits corrosion properties when subjected to most metals
- Slightly toxic in nature

Eutectics

- Analysis of eutectics for TES applications limited due to insufficient and nonavailability of thermophysical property data
- In some cases, fatty acid eutectics evolve pungent odor making them less suitable for PCM wallboard TES applications in indoor environments.

5.9.3 Merits of LTES systems

External melt-ITES system

- The mechanism of heat transfer between the water/ice and the HTM/warm water can be expected to be appreciable due to the uniqueness of the medium used during the discharging process.
- By estimating the cooling load demand or fluctuations persisting in the building spaces, the rate of discharging of the ice build can be adjusted in one stretch or on timely basis.
- The energy consumption at the primary chiller can be minimized considerably during the on-peak load conditions; thereby the overall cooling system efficacy can be enhanced.

Internal melt-ITES system

- The closed loop configuration helps to achieve effective control on the HTM and the built-in-ice charging temperature and discharging temperature, respectively.

- The coefficient of performance of this system is better compared to the external melt system.
- The energy redistribution requirements from on-peak to the part load conditions in buildings is effectively matched with this system, due to the time-dependent charging and discharging operating attributes.
- The water present in the storage tank only would undergo phase transition process, which facilitates for accomplishing reduced pumping power consumption.

Chilled water-PCM TES system

- The cooling load demand in the building spaces can be effectively distributed between the cooling/air conditioning facility and the thermal storage system.
- The chiller plant can be operated at its nominal capacity at most of the operating periods on a cooling design day cycle.
- Peak load shaving from on-peak to part load can be accomplished through the charging process of the PCM.
- Energy efficiency and energy savings potential on the chiller side is achieved by making use of the PCM discharging cycle during on-peak load conditions.

5.9.4 Limitations of LTES systems

- The thermal performance of the cooling system is influenced by the lower charging temperature and the thermal losses occurring during the charging and the discharging processes.
- The pressure drop and frictional losses would limit the overall heat transfer effectiveness of the HTM with the water/ice during the freezing and melting cycles.
- The inherent operational and maintenance cost of such thermal storage system is generally affordable only beyond several hundred tons of cooling capacity requirements in buildings.

5.10 SUMMARY

Success in achieving the energy redistribution, energy savings potential, and energy management strongly depends on the ways by which the total energy demand has been matched with the energy supply in buildings. The focus on value-added technology implementation has a significant role to play at every step of the design and development of building envelopes.

From this perspective, TES systems offer a wide range of opportunities to bridge the gap between energy supply and energy demand in several ways. The TES systems integrated with HVAC system are capable of efficiently shifting and leveling as well as limiting the cooling/heating load demand in buildings.

The acceptable heat storage and release characteristics of the PCMs allow them to be pronounced as potential candidates for TES systems meant to satisfy the cooling/heating requirements in buildings.

In total, the energy performance of the existing and the new building envelopes can be enhanced with the application of LTES technologies, without sacrificing energy efficiency and environmental sustainability.

References

[1] Parameshwaran R, Kalaiselvam S, Harikrishnan S, Elayaperumal A. Sustainable thermal energy storage technologies for buildings: a review. Renew Sustain Energy Rev 2012;16:2394–433.

[2] Oro E, Gracia AD, Castell A, Farid MM, Cabeza LF. Review on phase change materials (PCMs) for cold thermal energy storage applications. Appl Energy 2012;99:513–33.

[3] Dutil Y, Rousse DR, Salah NB, Lassue S, Zalewski L. A review on phase-change materials: mathematical modeling and simulations. Renew Sustain Energy Rev 2011;15:112–30.

[4] Regin AF, Solanki SC, Saini JS. Heat transfer characteristics of thermal energy storage system using PCM capsules: a review. Renew Sustain Energy Rev 2008;12:2438–58.

[5] Delgado M, Lazaro A, Mazo J, Zalba B. Review on phase change material emulsions and microencapsulated phase change material slurries: materials, heat transfer studies and applications. Renew Sust Energ Rev 2012;16:253–73.

[6] Kuznik FDR, David D, Johannes K, Roux J-J. A review on phase change materials integrated in building walls. Renew Sustain Energy Rev 2011;15:379–91.

[7] Cabeza LF, Castell A, Barreneche C, de Gracia A, Fernández AI. Materials used as PCM in thermal energy storage in buildings: a review. Renew Sustain Energy Rev 2011;15:1675–95.

[8] Xiao M, Feng B, Gong K. Preparation and performance of shape stabilizes phase change thermal storage materials with high thermal conductivity. Energy Convers Manage 2002;43:103–8.

[9] Sari A. Form-stable paraffin/high density polyethylene composites as a solid–liquid phase change material for thermal energy storage: preparation and thermal properties. Energy Convers Manage 2004;45:2033–42.

[10] Zhou G, Zhang Y, Wang X, Lin K, Xiao W. An assessment of mixed type PCM–gypsum and shape-stabilized PCM plates in a building for passive solar heating. Sol Energy 2007;81:1351–60.

[11] Zhang Y, Ding J, Wang X, Yang R, Lin K. Influence of additives on thermal conductivity of shape-stabilized phase change material. Sol Energy Mater Sol Cells 2006;90:1692–702.

[12] Waqas A, Din ZD. Phase change material (PCM) storage for free cooling of buildings—a review. Renew Sustain Energy Rev 2013;18:607–25.

[13] Tyagi VV, Kaushik SC, Tyagi SK, Akiyama T. Development of phase change materials based microencapsulated technology for buildings: a review. Renew Sustain Energy Rev 2011;15:1373–91.

[14] Soares N, Costa JJ, Gaspar AR, Santos P. Review of passive PCM latent heat thermal energy storage systems towards buildings' energy efficiency. Energy Build 2013;59:82–103.

[15] Kuznik F, Virgone J, Noel J. Optimization of a phase change material wallboard for building use. Appl Therm Eng 2008;28:1291–8.

[16] Ceron I, Neila J, Khayet M. Experimental tile with phase change materials (PCM) for building use. Energy Build 2011;43:1869–74.

[17] Heim D, Clarke JA. Numerical modelling and thermal simulation of PCMgypsum composites with ESP-r. Energy Build 2004;36(8):95–805.

[18] Heim D. Isothermal storage of solar energy in building construction. Renew Energy 2012;35(4):788–96.

[19] Zhang Y, Lin K, Jiang Y, Zhou G. Thermal storage and nonlinear heat transfer characteristics of PCM wallboard. Energy Build 2008;40(9):1771–9.

[20] Stovall TK, Tomlinson JJ. What are the potential benefits of including latent storage in common wallboard. J Sol Energy Eng Trans ASME 1995;117(4):318–25.

[21] Diaconu BM, Cruceru M. Novel concept of composite phase change material wall system for year-round thermal energy savings. Energy Build 2010;42(10):1759–72.

[22] Neeper DA. Thermal dynamics of wallboard with latent heat storage. Sol Energy 2000;68(5):393–403.

[23] Kuznik F, Virgone J, Roux JJ. Energetic efficiency of room wall containing PCM wallboard: a full-scale experimental investigation. Energy Build 2008;40(2):148–56.

[24] Ahmad M, Bontemps A, Sallee H, Quenard D. Experimental investigation and computer simulation of thermal behaviour of wallboards containing a phase change material. Energy Build 2006;38(4):357–66.

[25] Kuznik F, Virgone J. Experimental investigation of wallboard containing phase change material: data for validation of numerical modeling. Energy Build 2009;41(5):561–70.

[26] Schossig P, Henning HM, Gschwander S, Haussmann T. Microencapsulated phase-change materials integrated into construction materials. Sol Energy Mater Sol Cells 2005;89(2–3):297–306.

[27] Koo J, So H, Hong SW, Hong H. Effects of wallboard design parameters on the thermal storage in buildings. Energy Build 2011;43(8):1947–51.

[28] Chen C, Guo H, Liu Y, Yue H, Wang C. A new kind of phase change material (PCM) for energy-storing wallboard. Energy Build 2008;40(5):882–90.

[29] Ahmad M, Bontemps A, Sallee H, Quenard D. Thermal testing and numerical simulation of a prototype cell using light wallboards coupling vacuum isolation panels and phase change material. Energy Build 2006;38(6):673–81.

[30] Athienitis AK, Liu C, Hawes D, Banu D, Feldman D. Investigation of the thermal performance of a passive solar test-room with wall latent heat storage. Build Environ 1997;5:405–10.

[31] Rodriguez-Ubinas E, Ruiz-Valero L, Vega S, Neila J. Applications of phase change material in highly energy-efficient houses. Energy Build 2012;50:49–62.

[32] Borreguero AM, Sanchez ML, Valverde JL, Carmona M, Rodríguez JF. Thermal testing and numerical simulation of gypsum wallboards incorporated with different PCMs content. Appl Energy 2011;88(3):930–7.

[33] Hasse C, Grenet M, Bontemps A, Dendievel R, Sallée H. Test and modelling of honeycomb wallboards containing a phase change material. Energy Build 2011;43:232–8.

[34] Alawadhi EM, Alqallaf HJ. Building roof with conical holes containing PCM to reduce the cooling load: numerical study. Energy Convers Manage 2011;52:2958–64.

[35] Darkwa K, Kim JS. Dynamics of energy storage in phase change drywall systems. Int J Energy Res 2005;29:335–43.

[36] Castell A, Martorell I, Medrano M, Pérez G, Cabeza LF. Experimental study of using PCM in brick constructive solutions for passive cooling. Energy Build 2010;42(4):534–40.

[37] Silva T, Vicente R, Soares N, Ferreira V. Experimental testing and numerical modelling of masonry wall solution with PCM incorporation: a passive construction solution. Energy Build 2012;49:235–45.

[38] Cabeza LF, Castellon C, Nogues M, Medrano M, Leppers R, Zubillaga O. Use of microencapsulated PCM in concrete walls for energy savings. Energy Build 2007;39(2):113–9.

[39] Cabeza LF, Medrano M, Castellon C, Castell A, Solé C, Roca J, Nogués M. Thermal energy storage with phase change materials in building envelopes. Contrib Sci 2007;3(4):501–10.

[40] Chandra S, Kumar R, Kaushik S, Kaul S. Thermal performance of a non-airconditioned building with PCCM thermal storage wall. Energy Convers Manage 1985;25(1):15–20.

[41] Bontemps A, Ahmad M, Johannes K, Sallée H. Experimental and modelling study of twin cells with latent heat storage walls. Energy Build 2011;43(9):2456–61.

[42] Weinlaeder H, Koerner W, Heidenfelder M. Monitoring results of an interior sun protection system with integrated latent heat storage. Energy Build 2011;43(9):2468–75.

[43] Santamouris M, Synnefa A, Karlessi T. Using advanced cool materials in the urban built environment to mitigate heat islands and improve thermal comfort conditions. Sol Energy 2011;85:3085–102.

[44] Yanbing K, Yi J, Yinping Z. Modeling and experimental study on an innovative passive cooling system—NVP system. Energy Build 2003;35:417–25.

[45] http://www.solardecathlon.gov/history.html (last accessed 08.04.11).

[46] Hegger M. Sunny Times: The surPLUShome of Team Germany for Solar Decathlon 2009. Wuppertal: Verlag Muller+Busmann KG; 2010.

[47] Lipinski D. Winning Teams and Innovative Technologies from the 2005 Solar Decathlon. U.S. House of Representatives, Committee on Science; 2005 [November 2, 2005].

[48] PCM Products, Solar House Thermal Energy Storage Systems, Phase Change Material Products Limited, http://www.pcmproducts.net/files/solar_house1_1_.pdf [last accessed 08.04.11].

[49] Neila González FJ, Acha Román C, Higueras García E, Bedoya Frutos C. Materiales de Cambio de Fase (MCF) empleados para la acumulación de energía en la arquitectura. Su aplicación en el prototipo Magic Box. Mater Constr 2008;58(251):119–26.

[50] Wang X, Niu J. Performance of cooled-ceiling operating with MPCM slurry. Energy Convers Manage 2009;50(3):583–91.

[51] Kondo T, Ibamoto T. Research on thermal storage using rock wool phase change material ceiling board. ASHRAE Trans 2006;112:526–31.

[52] Dolado P, Lazaro A, Marin JM, Zalba B. Characterization of melting and solidification in a real scale PCM-air heat exchanger: numerical model and experimental validation. Energy Convers Manage 2011;52:1890–907.

[53] www.ice-energy.com.

[54] ASHRAE. Thermal storage. ASHRAE Handbook. HVAC Applications. Atlanta: American Society of Heating, Refrigerating and Air Conditioning Engineers. Inc.; 2007 [chapter 34].

[55] www.calmac.com.

[56] www.Fafco.com.

[57] www.dunham-bush.com.

[58] Yau YH, Rismanchi B. A review on cool thermal storage technologies and operating strategies. Renew Sust Energy Rev 2012;16:787–97.

[59] Sun Y, Wang S, Xiao F, Gao D. Peak load shifting control using different cold thermal energy storage facilities in commercial buildings: a review. Energy Convers Manage 2013;71:101–14.

[60] Dorgan CE, Elleson JS. Design guide for cool thermal storage. Atlanta, GA: ASHRAE, Inc.; 1994.

[61] Parameshwaran R, Harikrishnan S, Kalaiselvam S. Energy efficient PCM-based variable air volume air conditioning system for modern buildings. Energy Build 2010;42:1353–60.

[62] Turnpenny J, Etheridge D, Reay D. Novel ventilation cooling system for reducing air conditioning in buildings. Part I: testing and theoretical modeling. Appl Therm Eng 2000;20:1019–37.

[63] Turnpenny J, Etheridge D, Reay D. Novel ventilation system for reducing air conditioning in buildings. Part II: testing of prototype. Appl Therm Eng 2001;21:1203–17.

[64] Kurnia JC, Sasmito AP, Jangam SV, Mujumdar AS. Improved design for heat transfer performance of a novel phase change material (PCM) thermal energy storage (TES). Appl Therm Eng 2013;50:896–907.

[65] Velraj R, Seeniraj RV, Hafner B, Faber C, Schwarzer K. Heat transfer enhancement in a latent heat storage system. Sol Energy 1999;65:171–80.

[66] Fukai J, Hamada Y, Morozumi Y, Miyatake O. Effect of carbon-fiber brushes on conductive heat transfer in phase change materials. Int J Heat Mass Transfer 2002;45:4781–92.

[67] Fukai J, Kanou M, Kodama Y, Miyatake O. Thermal conductivity enhancement of energy storage media using carbon fibers. Energy Convers Manage 2000;41:1543–56.

[68] Mosaffa AH, Talati F, Basirat Tabrizi H, Rosen MA. Analytical modeling of PCM solidification in a shell and tube finned thermal storage for air conditioning systems. Energy Build 2012;49:356–61.

[69] Stritih U. Heat transfer enhancement in latent heat thermal storage system for buildings. Energy Build 2003;35:1097–104.

[70] Butala V, Stritih U. Experimental investigation of PCM cold storage. Energy Build 2009;41:354–9.

[71] Wang J-J, Zhang C-F, Jing Y-Y, Zheng G-Z. Using the fuzzy multi-criteria model to select the optimal cool storage system for air conditioning. Energy Build 2008;40:2059–66.

[72] Pasupathy A, Velraj R. Effect of double layer phase change material in building roof for year round thermal management. Energy Build 2008;40:193–203.

[73] Hasnain SM. Review on sustainable thermal energy storage technologies. Part II: cool thermal storage. Energy Convers Manage 1998;39:1139–53.

Thermochemical Energy Storage

6.1 INTRODUCTION

Thermal energy can be effectively stored and retrieved by means of sensible heat and latent heat principles. The other way of storing and releasing thermal energy can be performed through chemical reaction principles. The reversible chemical reactions occurring between working reactants or reactive components help to store and release the required heat energy. By supplying heat energy to definitive chemical material pairs, the intermolecular bonding between them can be broken, and they can be separated into individual reactive components. This would eventually allow the material to store heat energy.

On the other hand, by recombining the same individual reactive components, the stored heat energy can be effectively recovered and used to meet the heating/cooling load demand. Most thermochemical energy storage systems are developed for space heating applications in buildings rather than cooling applications. This could be because, for heating applications, high grade heat energy is available from solar radiation, which is a renewable source of energy and can be easily trapped through solar collectors for further usage. Likewise, the combination of a thermochemical energy storage system with a long-term seasonal TES system can also be an advantageous approach to reduce the carbon footprint and greenhouse gas emissions and contribute to maintaining environmental sustainability. In this context, the concepts and inherent operational characteristics of various thermochemical energy storage systems are discussed in the following sections.

6.2 PHENOMENA OF THERMOCHEMICAL ENERGY STORAGE

Sorption storage and chemical storage are the two broad areas necessary for performing active and potential research work for effectively charging and discharging available energy. The phenomena in the background for sorption storage can be categorized into adsorption and absorption storage. Likewise, chemical storage can be made effective through the electrochemical, electromagnetic (photochemical or photosynthesis), thermochemical without sorption, chemical adsorption (chemisorption), and chemical absorption phenomena. The structures of sorption storage and chemical storage are presented in Fig. 6.1.

Basically, the phenomenon of capturing a gas or a vapor (sorbate) by a substance (sorbent) existing in the condensed state is known as sorption. The sorbent substance can be in either solid or liquid form. The sorption phenomena includes both thermophysical and thermochemical features, and in general it is referred to as adsorption and absorption. From the perspective of energy storage, absorption refers to the process of capturing a gas or a vapor by a liquid (absorbent). Adsorption, on the other hand, refers to the phenomenon of binding a gas or vapor to the surface of a solid or a porous media. As the surface

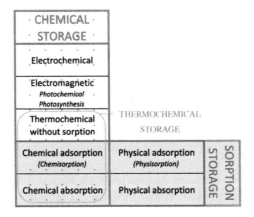

FIGURE 6.1

Classification of chemical storage and sorption storage [1].

phenomena drives the adsorption, if the binding takes place due to van der Waals forces, then it is termed physical adsorption (or physisorption), and if the binding is governed by valency forces, then it is called chemical adsorption (or chemisorptions). Generically, the chemisorption process is considered to yield larger heat of sorption than physisorption. As pointed out in Chapter 3, the reversible chemical reactions occurring between reactive components or reactant pairs are largely responsible for charging and discharging the heat energy. The reactants or reactive components being considered in veracity related to sorption and chemical processes for the last two decades are shown in Fig. 6.2. The comparison between sensible heat, latent heat, and thermochemical energy storage phenomena is shown in Fig. 6.3.

It can be seen from Fig. 6.3 that to meet the annual storage requirement for an energy-efficient passive building, the volume needed to fulfill the requirement is much less for thermochemical storage compared to the other two phenomena of energy storage. Likewise, promising features of thermochemical energy storage would facilitate its effective utilization in building applications. The chemically reactive materials used in thermochemical storage are capable of yielding high heat storage capacity, significant increase of temperature, and storage of reactant materials at ambient conditions without undergoing self-discharge (heat losses during charging and discharging in the long term).

6.3 THERMOCHEMICAL ENERGY STORAGE PRINCIPLES AND MATERIALS

In principle, thermochemical energy storage utilizing sorption material would release water vapor by virtue of supplied heat energy and would release heat energy while the water vapor is being adsorbed or absorbed. Specifically, the purpose of the sorption material is that it facilitates a shift from the vapor-liquid phase equilibrium of pure working fluid to the other vapor-condensed phase equilibrium of the same working fluid, but in the presence of the sorption material.

In a typical charging phase, the heat energy supplied tends to overcome the reversible bonding energy between the molecules of the working fluid and the sorption material as well. By this, the gaseous working fluid is released to the environment (in the case of an open system) or else is condensed (in the case of a closed system). This clearly indicates that the energy and the entropy flux are separated

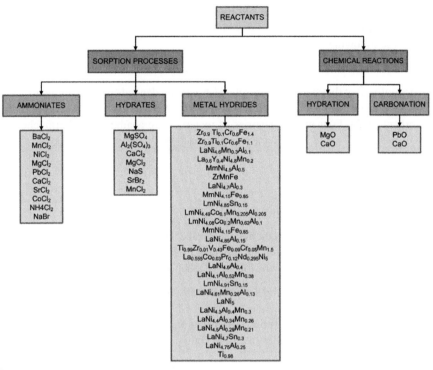

FIGURE 6.2

Overview working pairs tested in prototypes under practical conditions [2].

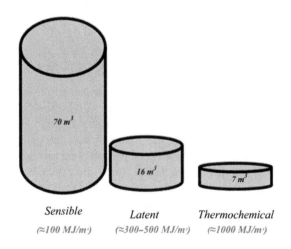

FIGURE 6.3

Comparison of heat storage volume requirement for an energy-efficient passive house (6480 MJ) [3].

through the thermochemical storage principle. By recombining the separated reactive components/working fluid pairs, discharge of the entropy occurs. The energy of adsorption/absorption can be expressed through the following relation [4],

$$\Delta H_a = \Delta H_v + \Delta H_b \tag{6.1}$$

The equation infers that the energy of adsorption/absorption depends on the summation of the energy required for phase change evaporation of the working fluid (ΔH_v) and the bonding energy of the working fluid to the sorbent (ΔH_b). Four main categories of materials are available for thermochemical energy storage, each of which possess its own characteristics based on thermophysical properties, and they are

- Adsorption materials
- Absorption materials
- Pure thermochemical materials
- Composite thermochemical materials

The thermal properties of a variety of thermochemical energy storage materials have been reviewed, and their properties are gleaned and presented in Appendix II. The vital parameters that have to be considered while selecting the proper materials for thermochemical energy storage are listed here [5–9]:

- High affinity of the sorbent toward the sorbate, thereby influencing reaction kinetics (rate of the reaction)
- Volatility of the sorbate during absorption better than that of the sorbent
- Thermal conductivity and heat transfer rate high enough with the heat transfer medium during adsorption process
- The temperature of desorption ensured as low as possible
- Eco-friendliness, less toxicity, low carbon footprint, low global warming potential, and ozone depletion potential
- Noncorrosive to the storage containers or heat exchange materials
- Possess good thermal and molecular stability under the stated operating temperature and pressure
- Excessive pressure and high vacuum conditions avoided

6.4 THERMOCHEMICAL ENERGY STORAGE SYSTEMS

The functional aspects of the thermochemical energy storage and sorption systems purely depend on the chemical reaction rates, which are quite reversible. The energy absorption and release occurring during the reversible chemical reactions are characteristic to the reaction kinetics and coordination of the chemical components (constituents) involved. Storage of heat energy is said to take place during the dissociation (desorption) of reactive components into individual components, and by reversing the reaction by combining the individual components, the stored energy can be released. The former reaction is said to be endothermic, and the latter is designated as an exothermic reaction.

The classification of chemical and thermochemical energy storage systems is shown in Fig. 6.4. Through performing the desorption reaction, the resulting individual components can be stored separately, and during the peak load demand conditions, the separated components can be combined for the reason stated earlier to form the parent components again. The thermal properties and reaction potential of some important materials utilized for thermochemical energy storage is summarized in Table 6.1.

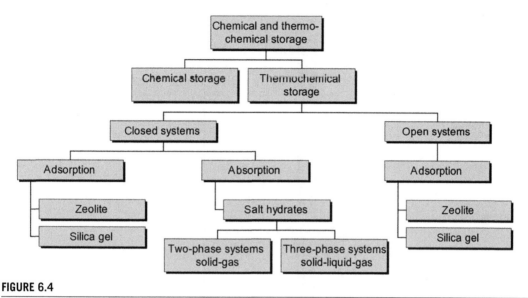

FIGURE 6.4

Classification of chemical and thermochemical processes for heat storage applications [4].

Reproduced with permission of PERGAMON in the format reuse in a book/textbook via Copyright Clearance Center.

As shown in Fig. 6.4, thermochemical energy storage systems can be classified into two major types, namely, open system and closed system. In an open system, the working fluid in the gaseous state is released to the environment (or space) directly (releasing the entropy). In the case of a closed system, the working fluid is not released directly, but the entropy is released to the environment through a heat exchanger interface. The inherent operational characteristics of some major thermochemical energy storage systems are described in the following sections.

6.4.1 Open adsorption energy storage system

Open adsorption energy storage is an attractive scheme in the sense that it can be used as a heat store for meeting peak load demand and energy redistribution requirements. This system has a vast opportunity to be incorporated as thermal storage in buildings. The main source for carrying out desorption reaction can be obtained from the largely available solar (renewable) energy. The Institute of Thermodynamics and Thermal Engineering (ITW) in the University of Stuttgart (Germany) has been involved in the development of an adsorption energy storage system, and they have also recently proposed an open adsorption energy storage system [1]. The operating principle of this type of system is schematically represented in Fig. 6.5, and the schematic representation of how the heating demand can be met using the open adsorption thermochemical energy storage system in building applications is shown in Fig. 6.6.

Typically, during the regeneration cycle, the hot air leaving the solar collector at a temperature ranging from 180 to 190 °C is allowed to flow across the energy storage unit. The energy storage unit is filled with zeolite 4A material. Due to the desorption reaction taking place in the energy store, the water content present in the zeolite 4A material is trapped by the hot air flow. At the same time, the hot air releases its heat to the material, thereby enabling the heat energy to be stored in the material. The warm air flowing out of the thermal store can then be used for primary heating purposes.

Table 6.1 Potential Materials for Chemical Reaction Storage Identified During IEA SHC Task 32 [1,10]

| Material Name | Dissociation Reaction | | | Energy Storage Density of AB GJ/m³ | Turnover Temperature °C | Realization Potential % |
	AB ⇔	B +	A			
Magnesium sulfate	$MgSO_4 \cdot 7H_2O$	$MgSO_4$	H_2O	2.8	122	9.5
Silicon oxide	SiO_2	Si	O_2	37.9	4065+HF:150	9.0
Iron carbonate	$FeCO_3$	FeO	CO_2	2.6	180	6.3
Iron hydroxide	$Fe(OH)_2$	FeO	H_2O	2.2	150	4.8
Calcium sulfate	$CaSO_4 \cdot 2H_2O$	$CaSO_4$	H_2O	1.4	89	4.3

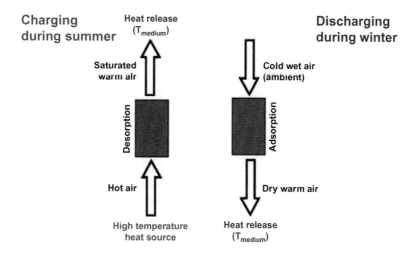

FIGURE 6.5

Operating principle of an open adsorption energy storage system [1,11].

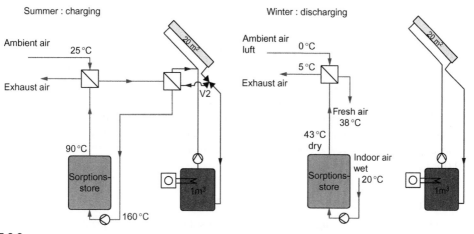

FIGURE 6.6

Open adsorption thermal storage system for application in buildings [1,11].

During the discharging cycle, by passing the wet/moist air over the zeolite 4A material, the stored heat energy is transferred to the flowing air stream, where the moisture from the air is removed by the heat storage material at the cost of heat energy release. The heated air can then be utilized for meeting the demand load requirements. The honeycomb-like design structure of the heat storage system allows more heat energy to be stored and released by virtue of enhanced adsorption and reaction kinetics with reduced pressure drop occurring across the process length. The major restraint in coupling the solar energy as the direct heat source is that desorption temperature is elevated with the utilization of high grade solar heat energy.

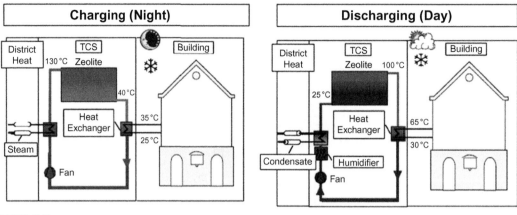

FIGURE 6.7

Open adsorption energy storage system on district heating net in Munich [1,11].

The other type of open adsorption system (Fig. 6.7), which has been developed and studied by the ZAE Bayern (Center for Applied Energy Research) in Germany, intends to provide the buffer or cushioning effect for the district heating facility. The district heating network is dedicated to meeting the heating demand in buildings, and this adsorption store is not specifically meant as long-term sorption storage for the heating application. When using this system for a school building, the energy that can be stored would account for 1300 kWh for 14 h of operating period and 130 kW of maximum power consumption.

This system makes use of heat energy from the district heating network as the source for performing desorption reaction during off-peak load conditions and is almost independent throughout its discharging operating periods. The heat storage material utilized in this system is zeolite 13X, and the operating principle is almost identical to that of the ITW heat storage system explained earlier. Furthermore, the storage density of this system can be accomplished up to 124 kWh/m³ and 100 kWh/m³ for the heating and cooling applications, where the coefficient of performance (COP) of this system can be estimated to be 0.9 (for heating) and 0.8 (for cooling).

6.4.2 Closed adsorption energy storage system

The closed adsorption energy storage system also utilizes solar energy as the prime source for completing the dissociation reaction and to enable heat storage to the highest possible extent for building applications. The Modestore (modular high energy density heat store) prototype system was first developed by the AEE INTEC in Austria. The Modestore system has been coupled with solar collectors of 20.4 m² in area, which is capable of receiving maximum heat energy from incident solar radiation [1].

This system is intended to meet space heating and domestic hot water production applications. The working pair or reaction constituent materials used in this system are silica gel/water pairs. In this system, silica acts as the adsorbent, and water is the sorbate. The choice of silica as the adsorbent means that it can adsorb the moisture from the vapor. The operating cycle of this storage system is depicted in Fig. 6.8.

During the charging cycle, the heated air from the solar collector is made to flow over the silica gel by means of the dedicated heat exchange arrangement. The temperature of the heated air can be increased to about 90°C. Due to the heat interaction between the air and the silica gel, desorption

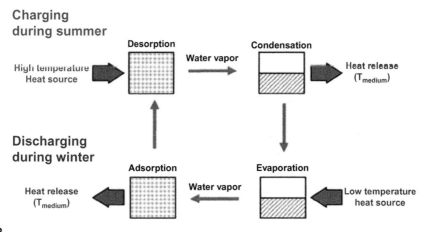

FIGURE 6.8

Operation principle of closed adsorption system [1].

reaction takes place, resulting in the release of water vapor from the silica gel. The released water vapor is then cooled in the condenser, and the dry silica gel and the water vapor are stored separately for use during the discharging cycle. In the discharging cycle, depending on the energy demand, the low temperature heat source evaporates the collected water, and the vapor is subjected to flow over the silica gel present in the adsorber unit. Due to the adsorption process, the water vapor is adsorbed into the silica gel, and the heat energy can then be released from the adsorbent to satisfy the required load demand.

However, the storage density of this system is estimated to drop to only 50 kWh/m³, which is almost 30% less efficient compared to the water storage system [1,12,13]. This may be due to the fact that the temperature lift is not sufficient to complete the process in which the silica gel water content is about 13%. Likewise, the desorption could not be possible using the flat plate solar collectors due to the temperature levels being achieved and the water content of silica gel at below 3%. The optimal water content of the absorbent material must be in the range of 3 to 13%.

6.4.3 Closed absorption energy storage system

The closed absorption energy storage combined with the solar seasonal heat storage system has been investigated in recent years by the EMPA (Swiss Federal Laboratories for 793 Materials Testing and Research in Switzerland) [1,14]. This system was basically developed for long-term seasonal thermal storage application, where the commonly used inexpensive sodium hydroxide (NaOH) or caustic soda [15] is preferred as the absorbent, and water acts as the sorbate.

The operating principle of this system is very similar to the closed adsorption system. The solar heat is supplied directly to the regenerative heat exchanger containing the low concentration solution (NaOH solution). The heat energy supplied helps to separate the water content from the NaOH solution, and thus the desorption process is effectively accomplished. The highly concentrated caustic soda (without water content) is stored separately for further use. The water vapor formed in the desorption process is then cooled, condensed, and stored suitably in the sorbate storage tank. Any

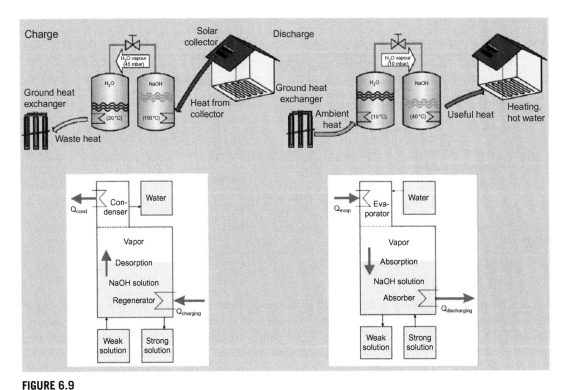

FIGURE 6.9

Operation principle of closed absorption storage [14].

excess heat that would have been produced is supplied back to the ground heat exchanger for storage and reutilization during the winter season. In the discharging cycle, the low temperature heat energy obtained from the ground heat exchanger is used to evaporate the water that is stored in the sorbate tank. The resulting vapor is then fed to the absorbed tank, where the concentrated caustic soda lye releases the stored heat.

By integrating this system for a single-family home in accordance with the passive house standard of 120 m², demand heating requirement of 15 kWh/m² at 35 °C, domestic hot water production of 50 l/day (approx.) at 60 °C, and evaporator temperature of 5 °C would yield a total storage volume of 7 m³ including the tanks and heat exchangers. Also, this system is enabled for the maximum absorber temperature and the lowest condenser temperature of 95 and 13 °C, respectively, for the concentration of lye of 62 wt.% (that is 7% higher than the expected value). The schematic representation of the closed absorption energy storage system is shown in Fig. 6.9.

6.4.4 Solid/gas thermochemical energy storage system

The solid/gas thermochemical energy storage system can be categorized under the short-term or diurnal storage scheme dedicated to air conditioning applications in buildings. This system also makes use of high grade solar energy as the heat input for its effective functioning. The schematic representation and pictorial view of this system are shown in Fig. 6.10.

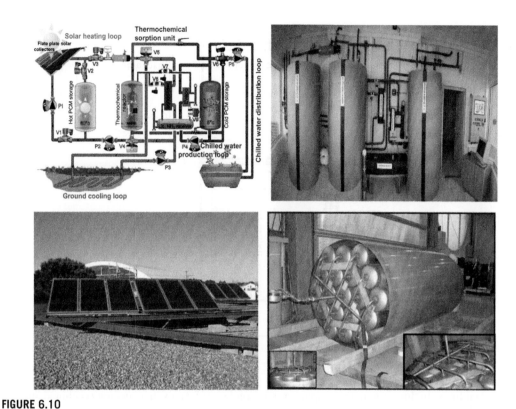

FIGURE 6.10

Schematic description and pictorial view of the solar air-conditioning thermochemical pilot plant [16].

During the daytime cycle, the reactor is heated by energy received from the solar collectors. The supplied heat energy helps desorb the reactive gas (usually ammonia) from the reactor to the condenser unit. In the condenser, the evolved ammonia gas is cooled, condensed, and then stored in the condensate tank. The gas undergoes condensation in the condenser because the outdoor temperature (diurnal temperature) is higher than the nocturnal (nighttime) temperature. The condensation temperature is attained by the reactive gas, which decides the maximum possible operating pressure of the system.

During nocturnal periods, as the reactor is in the cooled state, it starts absorbing the reactive gas from the condensation unit. Because of the pressure gradient being established between the reactor and the evaporator (both are integrated), the ammonia gas starts to boil and evaporate in the evaporator unit. The cooling effect realized in the evaporator can be suitably removed and stored in a separate storage facility integrated to this system. The evaporated ammonia gas is then absorbed by the reactor and produces heat of absorption at the nocturnal ambient temperature.

6.4.5 Thermochemical accumulator energy storage system

The significance of combining the absorption storage and the solar thermal technology with minimum system inconsistency has been demonstrated by the Solar Energy Research Centre, Sweden, and their industrial partner, ClimateWell AB. This system is basically intended for cooling applications in

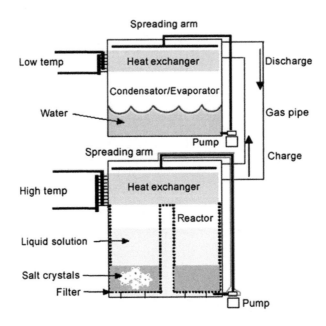

FIGURE 6.11

Schematic representation of the single unit TCA [1,17].

buildings, but sometimes it can also be extended to provide heating and cooling energy redistribution needs in buildings. In this system, the evaporator and condenser heat exchangers are combined similar to how the absorber and desorber reactors are combined in the TCA system. The schematic representation of the single unit TCA is illustrated in Fig. 6.11.

In the charging phase, the solution with a weak concentration (poor solution) is initially pumped to the heat exchanger, where the solution is evaporated. Once the saturation point is achieved in the heat exchanger, the vapor produced by the desorption process is then transferred to the condensator/evaporator unit, where the solid crystal remains are dropped to the bottom of the tank by gravity. This enables storage of heat energy due to the heat of condensation and binding energy release. The heat thus produced can be supplied to the indoor thermal zones or can be transferred to the ground-coupled heat exchanger for later usage.

During the discharging phase, the low temperature heat source either from the building spaces or ground-coupled heat exchanger supplied to the condensator/evaporator unit produces water vapor. The condenser acts as the evaporator, and the water vapor thus produced is transferred back to the reactor heat exchanger for the next cycle of operation. The heat storage density of this system with LiCl salt is estimated to be 253 kWh/m^3 [17,18]. The economics involved in the system configuration limits its market development for long-term thermal energy storage (TES) applications. The performance index of the TCA system is presented in Table 6.2.

6.4.6 Floor heating system using thermochemical energy storage

Thermochemical energy storage can also be applied to floor heating in buildings. The system developed by CWS-NT (Chemische Wärmespeicherung—Niedertemperatur) utilizes a combination of

Table 6.2 Claimed Performances of TCA ClimateWell 10 Machine [1,18]

Mode	Storage Capacity[a] (kWh)	Maximum Output Capacity[b] (kW)	Electrical COP[c]	Thermal Efficiency
Cooling	60	10/20	77	68%
Heating	76	25	96	160%

[a]Total storage capacity (i.e., including two barrels).
[b]Cooling capacity per barrel: 10 kW cooling is the maximum capacity. If both barrels are used in parallel (double mode) the maximum cooling output is 20 kW, and the heating output is 25 kW.
[c]Coefficient of performance (COP) = cooling or heating output divided by electrical input.

zeolite and salt materials for energy storage and release. The solar thermal combisystem coupled with this system is considered to be the vital inclusion for enhancing overall thermal efficiency and performance of the system. The collector loop heat exchanger connects the solar thermal combisystem with the thermochemical energy storage.

The thermochemical energy store is equipped with a reactor and a material store, in which the heat and mass transfer of the reacting components takes place in the reactor during the charging and discharging processes. The material store and the reactor are the two major provisions provided for efficient operation of the system, and it has to be carefully designed to withstand high temperature that may arise during the regeneration process. The schematic representation of the CWS-NT concept is depicted in Fig. 6.12. The schematic sketch of the reactor design and the laboratory prototype of

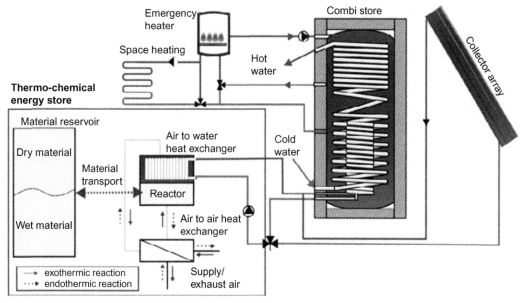

FIGURE 6.12

Schematic representation of the CWS-NT concept [19].

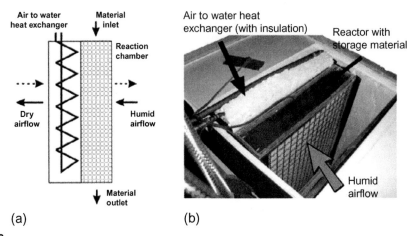

(a)   (b)

FIGURE 6.13

(a) Sketch of the cross-flow reactor (dotted line: material regeneration, solid line: heating mode), and (b) photo of the laboratory cross-flow reactor (view from top) [19].

the cross-flow reactor are shown in Fig. 6.13(a) and (b). An external reactor provision can help to separate the material store and the reactor. By doing this, only the required quantity of the material is used per unit time of the total storage volume of the material. Also, as the required quantity of the storage material is transported between the material store and the reactor through a vacuum conveyer, the process efficiency can be improved with reduced energy consumption. In a typical heating cycle, the required quantity of the material from storage is fed to the cross-flow reactor from the top, and by means of the gravitational force, the material flows into the reactor.

The ambient (outside) air that enters the reactor in the lateral direction transfers the heat energy and moisture (mass transfer) to the reactor. The heat that is adsorbed by the reactor is then transported to the water loop of the air-to-water heat exchanger. On the other hand, the regeneration of the reactor is said to take place where the heat storage material (combination of zeolite and salt) is heated by the flowing hot air being supplied from the solar collector heat exchanger.

For the effective operation of the derived reactor, the crucial factors to be considered during the design stage include the following:

- For attaining reduced pressure drop/losses and fan power consumption, the cross-flow sectional area for the flow or air and the material width has to be kept large and minimum, respectively.
- Transport of the reactive material has to be optimized in terms of reliability and economics, which ensures that low stress develops in the material during transportation.
- The distance between the heat source and heat storage has to be kept as minimum as possible to accomplish good heat storage with reduced heat losses to the ambient.

6.4.7 Thermochemical energy storage for building heating applications

The combination of thermochemical energy storage and seasonal TES has always been an attractive scheme because the heat energy required to drive storage is renewable and sustainable. This system takes advantage of both sensible heat and thermochemical reactions for charging and discharging thermal energy based on the load demand in buildings.

During the charging process, the heat energy trapped from the solar collector is transferred to the thermochemical reactor, where the reaction of desorption takes place on the reactive salt component, resulting in the release of water vapor and storage of heat energy. The vapor is then condensed, evaporated, and recombined with the concentrated salt, making it unsaturated. This facilitates the discharge of heat energy from the salt component.

The water flowing through the floor heating loop (System A-dotted line) takes advantage of the heat energy being dissipated from the thermochemical storage for meeting the space heating demand of the building. The recently developed thermochemical energy storage combined with the aquifer sensible heat storage facility is shown in Fig. 6.14 [20].

The reversible chemical reactions involved in the proposed storage system are given as:

$$SrBr_2 \; 6H_2O + Heat \rightarrow SrBr_2 \; H_2O + 5H_2O \, (\text{charging process}) \qquad (6.2)$$

$$SrBr_2 \; H_2O + 5H_2O \rightarrow SrBr_2 \; 6H_2O + Heat \, (\text{discharging process}) \qquad (6.3)$$

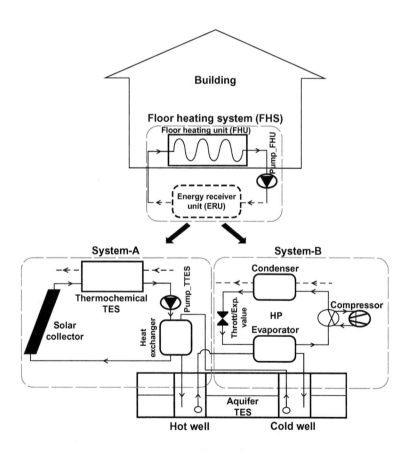

FIGURE 6.14

Schematic layout of the proposed thermochemical energy storage combined with the aquifer sensible heat storage system [20].

From the reversible reactions presented, it can be seen that during the charging and discharging processes, the reactive component H_2O remains in vapor phase and $SrBr_2 \cdot 6H_2O$ and $SrBr_2 \cdot H_2O$ stay in the solid phase. It is noteworthy that the aquifer TES system can be simultaneously charged through this combined system, where any excess heat being produced during the chemical reactions can be transferred to the aquifer by means of the heat exchanger interface. In case the heating demand in the building could not be entirely met by the thermochemical storage, the aquifer TES can help supplement the required heat energy. That is, the water from the hot well can be transferred to the cold well through the evaporator (in System B) of the heat pump.

By operating the heat pump, the required heat energy is exchanged between the heat transfer medium and the cold water flowing through the floor heating loop (System B-dashed line) in the condenser section. This in turn effectively supplements the desired heating demand in buildings. The major accomplishments of Task 32 of the IEA program related to advanced storage concepts for solar and low energy building applications are presented in Table 6.3.

As is expected, the sorption and thermochemical storage methods or systems do have their inherent limitations with respect to successful implementation in reality, which includes the following:

- Availability of the low temperature heat source at all times erects a barrier for the heat storage option because the heat source has to be maintained between 5 and 10 °C for evaporation to take place, which is an energy consuming task.
- Specialized designs are required for storage reactor beds, particularly for the solid sorbents, where an adequate rate of sorption is needed to produce usable delivered power.
- Risks are involved in the solidification/freezing of the working pair materials containing water as sorbate while flowing through the hydronic piping network.
- A possible chance exists for thermal losses during the starting and completing stages of operation of the storage due to the high discharge temperature requirement from storage.
- In closed loop systems, maintaining the vacuum for longer times carries a huge risk in the operating efficiency of heat storage.
- The dense materials used in the thermochemical or sorption storage are most suitable for heat pump or cooling applications due to their storage density and power density.
- It is very expensive in terms of both storage system and reactive materials.

6.5 CONCISE REMARKS

Thermochemical energy storage can be considered an energy-efficient approach that offers a wide opportunity for conserving primary energy sources as well as reducing greenhouse gas emissions. When compared to sensible heat and latent heat storage, thermochemical energy storage can yield the highest heat storage capacity without producing any thermal losses during the storage period. The working pairs of materials incorporated in thermochemical energy storage system including silica gel/water, magnesium sulfate/water, lithium bromide/water, lithium chloride/water, and NaOH/water have been considered the most prominent materials for achieving increased heat storage capacity.

By using the high porous structured carrier materials dispersed with the reactive material inside, the heat and mass transfer processes can be improved. Besides, the incorporation of chemical heat pumps working on the principles of chemical sorption processes can help acquire enhanced heat storage capacity even at very high temperatures, where the conventional heat pump does not need

Table 6.3 Main Achievements in "Task 32: Advanced Storage Concepts for Solar and Low Energy Buildings" [1]

Parameter	TCA, 80–100°C	NaOH, 95°C Test, 150°C Calculated	Modestore, 88°C	SPF, 180°C	Monosorp, 130°C	ECN, 150°C
Type of technology	Closed three-phase absorption	Closed two-phase absorption	Closed adsorption	Closed adsorption	Open adsorption	Closed thermo-chemical
Cost of material	3600 €/m³	250 €/m³	4300 €/m³	2–3000 €/m³	2500–3500 €/m³ᵃ	4870 €/m³
Storage materials weight	LiCl salt 54kg Water 117kg Steel 47kg	NaOH 160kg Water 150kg	Silica gel 200kg Water 30kg Steel 100kg Copper 50kg	Zeolite 13X 7kg	Zeolite 4A 70kg Steel 10kg	MgSO$_4$·7H$_2$O
Storage capacity for heat	35kWh	8.9kWh	13kWh	1kWh	12kWh	–
Floor space required for prototype	0.46m²	2m²	0.4m²	0.09m²	0.4m²	–
Energy density of material (NRJ4.1) (ratio to water 25/85°C)	253kWh/m³ (3.6)	250kWh/m³ (3.6)	50kWh/m³ (0.71)	180kWh/m³ (~3)	160kWh/m³ (2.3)	420kWh/m³ (6.1)
Energy density of prototype (NRJ4.1) (ratio to water 25/85°C)	85kWh/m³ (1.2)	5kWh/m³ (0.07)	33.3kWh/m³ (0.48)	57.8kWh/m³ (~1)	120kWh/m³ (1.7)	–
Energy density of prototype—cold (ratio to water 7/17°C)	54kWh/m³ (4.7)	–	–	–	–	–
Charge rate	15kW	1kW	1.0–1.5kW	–	2.0–2.5kW	–
Discharge rate	8kW	1kW	0.5–1.0kW	0.8/1.8kW	1.0–1.5kW	–
Estimated size for 70kWh (ratio to water 25/85°C)	0.64m³ (1.6)	1.3m³ (0.75)	1.7m³ (0.59)	1.2m³ (0.59)	0.54m³ (1.9)	0.4m³ (2.5)ᵇ
Estimated size for 1000kWh (ratio to water 25/85°C)	5.3m³ (2.7)	5m³ (2.9)	23m³ (0.62)	17m³ (~1)	7.7m³ (1.9)	5.6m³ (2.5)ᵇ

ᵃCost for large quantity of extruded material is unknown and is estimated for zeolite 4A.
ᵇEstimations are based on experimental storage density of ~420kW/h/m³ for reaction MgSO$_4$·6H$_2$O+heat ⇔MgSO$_4$·0.2H$_2$O+5.8H$_2$O.

to be suited for the desired purpose. However, the thermochemical energy storage if blended with long-term seasonal TES techniques can still result in enhanced thermal performance of the storage system without sacrificing energy efficiency and environmental sustainability.

References

[1] N'Tsoukpoe KE, Liu H, Pierrès NL, Luo L. A review on long-term sorption solar energy storage. Renew Sustain Energy Rev 2009;13:2385–96.

[2] Cot-Gores J, Castell A, Cabeza LF. Thermochemical energy storage and conversion: a-state-of-the-art review of the experimental research under practical conditions. Renew Sustain Energy Rev 2012;16:5207–24.

[3] Tatsidjodoung P, Pierrès NL, Luo L. A review of potential materials for thermal energy storage in building applications. Renew Sustain Energy Rev 2013;18:327–49.

[4] Thermal properties of materials for thermo-chemical storage of solar heat. A Report of IEA Solar Heating and Cooling programme—Task 32 "Advanced storage concepts for solar and low energy buildings". Report B2 of Subtask B: 2005, p. 1–20.

[5] American Society of Heating Refrigerating and Air-Conditioning Engineers (ASHRAE). Handbook: fundamentals. I–P edition, 2005.

[6] Bales C, Gantenbein P, Hauer A, Henning H-M, Jaenig D, Kerskes H. Thermal properties of materials for thermo-chemical storage of solar heat: Report no. B2-Task 32. 2005. 20p; Available from: http://www.iea-shc.org/publications/downloads/task32-Thermal_Properties_of_Materials.pdfS.

[7] Gantenbein P, Brunold S, Flückiger F, Frei U. Sorption materials for application in solar heat energy storage: Report no.SPF-01. 2001. 10 p; Available from: http://www.solarenergy.ch/fileadmin/daten/publ/sorption01.pdfS.

[8] Visscher K. Simulation of thermo chemical seasonal storage of solar heat-material selection and optimum performance simulation. In: Proceedings of 2nd workshop Matlab/Simulink for building simulation, CSTB; 2004.

[9] Wongsuwan W, Kumar S, Neveu P, Meunier F. A review of chemical heat pump technology and applications. Appl Therm Eng 2001;21:1489–519.

[10] Visscher K. Energy research Center of the Netherlands (ECN). Simulation of thermo chemical seasonal storage of solar heat—material selection and optimum performance simulation. In: Proceedings of workshop 2004 2nd workshop Matlab/Simulink for building simulation, CSTB. Paris, France; 2004.

[11] Kerskes H, Institut für Thermodynamik und Wärmetechnik (ITW). Seasonal sorption heat storage. DANVAK seminar (solar heating systems—Combisystems—heat storage). Lyngby, Denmark: DTU; 2006.

[12] Bales C. Laboratory tests of chemical reactions and prototype sorption storage units: 2008, www.iea-shc.org.

[13] Gartler G, Jähnig D, Purkarthofer G, Wagner W. Development of a high energy density sorption storage system. In: Proceedings of the Eurosun. Freiburg, Germany; 2004 aee-intec.at.

[14] Weber R, Dorer V. Long-term heat storage with NaOH. Vacuum 2008;82:708–16.

[15] Bales C. Final report of Subtask B "Chemical and Sorption Storage" the overview: 2008. p. 23, www.iea-shc.org.

[16] Stitou D, Mazet N, Mauran S. Experimental investigation of a solid/gas thermochemical storage process for solar air-conditioning. Energy 2012;41:261–70.

[17] Bales C, Nordlander S. TCA evaluation—lab measurements, modelling and system simulations. Borlänge, Sweden; 2005 www.serc.se.

[18] ClimateWell AB. Product description ClimateWell™ 10: 2008, http://www.climatewell.com/.

[19] Mette B, Kerskes H, Drück H. Concepts of long-term thermochemical energy storage for solar thermal applications—selected examples. Energy Procedia 2012;30:321–30.

[20] Caliskan H, Dincer I, Hepbasli A. Energy and exergy analyses of combined thermochemical and sensible thermal energy storage systems for building heating applications. Energy Buildings 2012;48:103–11.

Seasonal Thermal Energy Storage

7.1 INTRODUCTION

In the rapidly expanding demand for end-use energy, the penetration of thermal energy storage (TES) technologies has proven to be a viable option for confronting the challenges related to climate change, greenhouse gas (GHG) emissions, and energy security. In this context, the search for an energy storage technology that can use renewable sources of energy (as the primary source) for charging and discharging thermal energy has become increasingly attractive in recent years. Plentiful thermal energy (cold or heat) available in natural reservoirs including underground, lakes or ponds, and sea water can be extracted and stored for further usage depending on seasonal conditions (summer or winter).

For instance, in summer, cold energy from underground can be trapped and transferred to dwelling spaces to meet the prevailing cooling load requirements. Similarly, heat energy contained in the earth's subsurface can be retrieved and transmitted to buildings to offset the space heating demand. Seasonal TES (SeTES) technologies fundamentally differ from other TES methods in that thermal energy in the form of heat or cold is largely available as the *source* naturally. Other TES options including sensible, latent, or thermochemical energy stores using separate materials require thermal energy (not as a source) derived from primary energy sources to enable energy transfer and storage.

However, from the perspectives of energy front, SeTES as well as the other TES technologies are equally advantageous because they contribute to conserving energy on seasonal occasions or year-round. A variety of SeTES technologies are available in practice, including the aquifer TES (ATES), borehole TES (BTES), cavern thermal storage, earth-to-air thermal storage, earth piles heat storage, sea water TES, rock thermal storage, and roof pond energy storage. The description and operational characteristics of these SeTES methods are discussed in the following sections.

7.2 SEASONAL (SOURCE) TES TECHNOLOGIES

Storing heat or cold energy as the source can be considered an effective and efficient way to harness the useful energy that is available in the form of aquifer, geothermal, earth, rocks, sea water, and solar pond. It is appropriate to mention that the development of SeTES technologies is a relatively more efficient way to tackle climate change and reduce GHG emissions, as well as to satisfy the demand-side management of useful energy.

The SeTES technologies currently available and functioning in the energy market are mostly based on sensible thermal storage concepts. This is because source energy is typically renewable energy, which is intermittent and requires large installations for capturing and storing the desired heat energy to meet the demand load for district cooling/heating systems. Some of the most prominent sensible thermal storage technologies are explained in the following sections:

7.2.1 **Aquifer thermal storage**

The principle of storing the heat energy during summertime and extracting it during wintertime is usually referred to as underground TES (UTES), and ATES can be categorized under the UTES. The storage of heat from and to the ground/earth using a UTES largely depends on the local geological environment, type of storage technology adopted (open or closed loop), cooling/heating applications, and operating temperature range.

The ATES can be classified under the open loop system, wherein groundwater from the earth's subsurface is generically used for storing and extracting the thermal energy. The enthalpy of thermal energy contained in low temperature groundwater is utilized to meet the cooling/heating load demand. The schematic representation of the ATES system is illustrated in Fig. 7.1. The ATES system is essentially comprised of aquifer well, heat exchangers, extraction pumps, and space cooling/heating system (represented as systems).

In a typical operation of the ATES during summer season, the cold groundwater that is available in temperatures ranging from 5 to 10 °C is pumped from the extraction well and supplied to the heat exchanger component. The warm air or fluid (brine) coming from the space (or building side) transfer the heat energy to the cold groundwater in the heat exchanger and gains the cold energy in turn. The cooled air or brine solution is then supplied to the building for meeting the requisite cooling load demand. The warm groundwater is then pumped back to the injection well through the earth layers to maintain the groundwater level consistently. However, in some site locations, the groundwater

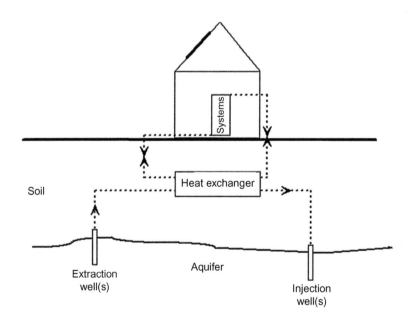

FIGURE 7.1

Aquifer thermal energy storage system [1].

temperature may also vary from 5 to 30 °C, depending on how deep the extraction and injection wells are established.

During winter conditions, the warm groundwater from the earth extraction well is pumped to the heat exchanger and returned to the earth through the injection well. Similar to that of the cooling cycle, the entering cold air or brine from the building side picks up heat energy from the warm groundwater flowing through the heat exchanger element. Heat energy is then transferred to the building, and thus the space heating load demand can be effectually met. The major factors that decide the efficient operation of the ATES system are listed here:

- The site or location preferred for the establishment of an ATES system should be checked for low natural groundwater flow conditions.
- The installation of the ATES system should be made between the impervious layers of the earth or stratigraphy (sequence of layers).
- The earth layers containing sandstone, limestone, gravel, or water-filled permeable sand with high hydraulic conductivity are also equally preferable for executing the ATES for district cooling/heating applications.
- Distribution of grain size is primarily applied to the porosity aquifers.
- Geometry and construction are related to the hydraulic boundaries.
- Storage coefficient is seen in terms of hydraulic storage capacity.
- Leakage factor is due to vertical hydraulic effects.
- Consider hardness or degree of consolidation.
- Look at temperature difference with respect to depth of extraction and injection wells.
- Check groundwater flow direction, chemical concentration, and static head.

The ATES systems are further classified into three major types depending on the thermal load demand persisting in the building side, namely single source (mono), double source (doublet), and recirculation (year-round). On the basis of the system configurations and components, including the extraction well, injection well, heat exchangers (heat pumps), and directional flow between groundwater and heat transfer fluid (using circulation pumps), the ATES systems are recognized as feasible for medium to large-scale TES installations. The TES capacity of ATES systems can go up to 15 kWh/m³ [3].

The ATES system established by Schmidt and Müller-Steinhagen [4] and Schmidt et al. [5] in Rostock, Germany, is capable of handling the heating load demand for 108 apartments of heating area accounting to 7000 m². The ATES system is combined with a 1000 m² solar collector installed on the roof of the dwelling. The ATES system is primarily meant for meeting the domestic hot water and space heating needs of the apartment building with the help of the solar collector (as source). This ATES system can be considered the first large-scale SeTES project in Germany.

The depth of the shallow ATES system is about 30 m, and the operating temperature range varies between 10 and 50 °C. To counteract a high return temperature (not advisable for the ATES system), a low temperature heating system coupled with radiator and a supply temperature that does not exceed 45 °C is equipped. By storing solar thermal energy during the summer season, the heating energy demand arising during winter can be effectively matched. Furthermore, this long-term thermal storage system offers a maximum solar fraction value of 62%. The details of the critical hydrogeological parameters of the ATES systems application most relevant to the site locations are presented in Table 7.1 (Fig. 7.2).

Table 7.1 Hydrogeological Parameters of the Aquifer Thermal Energy Stores in Germany [6]

Site	Dresden (Field Test)	Rostock-Brinckmanshohe	Buildings of the German Parliament in Berlin		Neubrandenburg
Geological formation	Quaternary	Quaternary	Hettangian	Quaternary	Upper postera
Depth (m)	7-10	13-27	285-315	30-60	1234-1274
Porosity (%)	~25	~20	30	~30	25
Permeability (μm^{2})	>2	8	2.8-4.2	>1	>1
Mineralization	Freshwater	Freshwater	29g/L	Freshwater	133g/L
Store temperature (initial; °C)	8	10	19	10	54

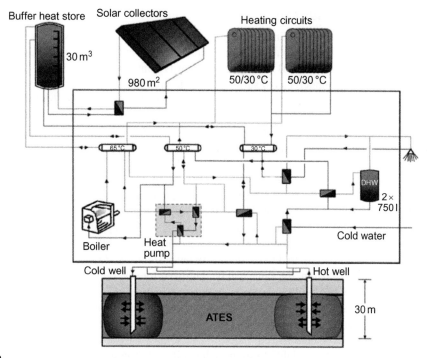

FIGURE 7.2

Scheme of a heat supply system (DHW, domestic hot water) [4,5].

It is interesting to note that the ATES system coupled with a conventional chiller plant operating between the temperature limits of 6 and 7 °C (supply and return water temperature differential) can achieve the coefficient of performance (COP) up to 4.0. For a natural cooling (free cooling) process enabled ATES, the COP can be expected to be accomplished between 15 and 35. The average COP of the heat pumps integrated ATES systems normally vary from 4.5 to 6.5 during the heating process. Furthermore, the ATES system can be energy efficient on a long-term basis with low operational and maintenance costs involved. The life expectancy of the ATES system can be expected to be around 20 to 25 years. The implementation of the ATES system can help to achieve a 50 to 60% reduction in the primary and natural gas consumption compared to conventional systems.

In practice, the ATES systems offers certain limitations related to their functional aspects as described here:

- Destabilization of the quality of groundwater in the underground levels of the earth
- Formation of deep scaling in the extraction and injection wells
- Susceptible conditions for the growth of algae, parasitic, fungal, and bacterial agents in the heat exchange component surfaces
- Penalty in the pumping capacity and pumping power due to the frictional pressure losses created in the heat pump or heat exchanger
- Possible creation of fluctuations/disturbances in the water table

7.2.2 Borehole thermal storage

BTES or duct heat storage can be put in the category of a closed loop UTES system, where the operational strategy of the BTES system is very similar to the ATES system except for the design configuration and installation procedures. The BTES system primarily contains high-density polyethylene pipe structures to be embedded in underground surfaces through predrilled boreholes. The presence of heat exchanger (heat pump), heat transfer fluid (brine solution), and other accessories also forms an integral part of the BTES system. The schematic diagram of the Drake Landing Solar Community SeTES [7,8] bore field configuration and boreholes details are shown in Fig. 7.3.

In a typical operation, the heat transfer fluid (glycol or antifreeze brine solution), which is made to flow through the embedded pipe structures, is utilized for extracting thermal energy from underground. With this configuration, overall thermal performance and heat transfer effectiveness can be improved. During summer, cold energy from underground is effectively extracted by the heat transfer fluid and transfered to the building cooling element. The warm brine solution transfers the heat energy captured from the building space to underground, and the cooling cycle is then repeated. Because this is a closed loop configuration, the more the borehole structures, the higher the BTES performance in terms of better heat transfer output as compared to the conventional system. In addition, as the brine solution is contained within the BTES system, the same can be reutilized for several hundred to thousand thermal cycles with minimum flow and heat losses in the hydronic circuit.

The closed loop BTES system can be further categorized into horizontal, vertical, and slinky loop systems. These configurations have their own TES characteristics, but they take advantage of ATES systems due to design aspects and efficient mode of operation year-round. For instance, during summer the high grade heat energy available from the solar collectors can be stored underground using the

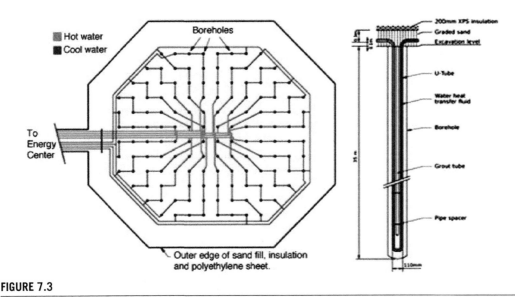

FIGURE 7.3

Drake Landing Solar Community SeTES [7] bore field configuration (left) and boreholes details (right).

closed loop BTES system. The stored heat energy can then be retrieved and supplied to the building through the closed loop hydronic network to satisfy the space heating load demand.

The schematic of the horizontal BTES system is depicted in Fig. 7.4. The characteristic values of a BTES system to be considered for seasonal thermal storage application are summarized in Table 7.2.

Similar to the ATES systems, these systems also exhibit some limitations for their successful implementation on medium to large-scale expansion projects, which include

- Huge capital investment (roughly 20 to 25% of total storage systems cost) required for drilling deep boreholes through the hard earth subsurface
- Heat imbalances developing through the underground thermal masses
- Creation of thermal fluctuations/disturbances in the underground hydrogeological structures, etc.

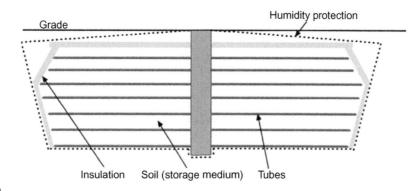

FIGURE 7.4

Schematic representation of the Vaulruz system (horizontal tubes) [8].

Table 7.2 Characteristic Values of a BTES System for Seasonal Thermal Storage Application [2]

Parameter	Values
Borehole diameter (mm)	100-150
Borehole depth (m)	30-100
Distance between boreholes (m)	2-4
Thermal ground conductivity (W/m K)	2-4
Flow rate in U-pipes (m/s)	0.5-1.0
Average capacity per meter of borehole length (W/m)	20-30
Minimum/maximum inlet temperature (°C)	−5/>+90
Typical cost of BTES storage per meter borehole length (€/m)	50-80

7.2.3 Cavern thermal storage

The cavern TES method as applied to meeting seasonal cooling/heating requirements in dwellings has gained momentum in recent years. This is also a type of UTES system, where heat or cold energy can be stored and distributed to the end-use load demand through the water reservoirs available on a large landscape. Two types of cavern TES systems are known commercially—hot water storage and gravel/water storage. The schematic layout of the hot water cavern TES system is presented in Fig. 7.5. The configuration of a hot water cavern TES system is comprised of a huge insulated underground built-in cavity or pit-like structure (storage tank) that is filled with the designed capacity of water. The seasonal storage tank is integrated with the building cooling/heating equipment for making use of the underground heat or cold depending on seasonal conditions.

During summer, the cold energy present underground is transferred into the storage tank structure, where the energy is captured in the water in the form of sensible heat. Through the hydronic network established between the underground storage tank and the building cooling heat exchanger, the cold water transfers cooling energy to the building cooling equipment, and thereby the space cooling requirements are effectively met. Similarly, during winter, the heat energy present in the earth's subsurface is captured and stored in the water tank and then transferred to the building to satisfy the space heating load demand.

The hot water cavern TES shown in Fig. 7.5 was built in Friedrichshafen, Germany [9], with a storage volume of $12,000 \, m^3$ and an additional stainless steel inner liner provided to prevent heat loss effects during operation. Heating energy for the underground built storage tank is provided through $3515 \, m^2$ integrated solar roof collectors, which were constructed over the top of a multifamily dwelling. This hot water storage was connected to a district heating facility and was commissioned for full operation in 1996. Due to the reduced solar fraction (20–30%) and the higher heat losses (about 40%) of this seasonal storage system, the net output TES has been affected due to the high return temperature. However, it is suggested that by admixing the return flow to the supply flow, heating losses can be minimized to the possible extent.

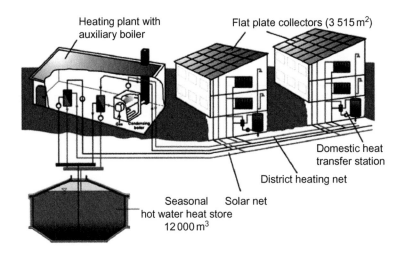

FIGURE 7.5

Schematic layout of hot water cavern TES system [9].

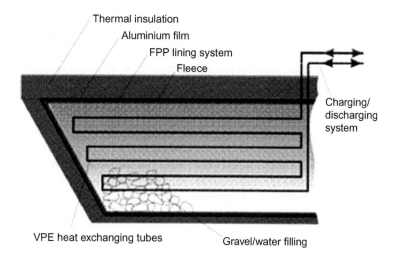

Thermal insulation
Aluminium film
FPP lining system
Fleece

Charging/
discharging
system

VPE heat exchanging tubes   Gravel/water filling

FIGURE 7.6

Schematic representation of a gravel/water cavern TES system [10].

On the other hand, the gravel/water cavern TES system shown in Fig. 7.6 offers great potential for storing and retrieving heat or cold energy. The 1500 m³ capacity gravel/water cavern TES system being constructed in Steinfurt [10] is comprised of a sealed and modified double polypropylene liner and is well insulated with granulated recycling glass. The operating temperature of this system can be extended up to 90 °C. This system on operation in line with design conditions yields 34% of the annual heating demand by using solar energy, which is quite appreciable in terms of seasonal storage of thermal energy.

The influence of the capillary and thermal conductivity aspects of the gravel, rocks, or pebbles used can play a significant role in acquiring the desired cooling/heating thermal storage characteristics in dwellings. From an economic point of view, gravel/water cavern TES systems are considered highly expensive compared to other SeTES systems. This could be due to constructional features, operational strategy, site location, and selection being graded on a long run.

From this perspective, the additional limitations encountered in the successful installation of UTES systems are summarized here:

- Capital investment for excavating the ground soil or earth is higher.
- The effects related to the soil pressure are considerable, which leads to the maintenance of static in the UTES that is more complex.
- For a moist soil condition, the construction of the storage tank or pit structure must be highly resistive toward the wetting nature of the insulation provided.
- Thermal losses are quite natural and cannot be deprived in UTES systems and those for which the surface to volume ratio of the built-in structures must be at a minimum.
- For the constructional geometry of storage tanks, utilization of concrete, or prefabricated materials can be the viable option compared to pit-like structures.
- Pit-like construction is highly prone to geometric constraints, including the slope angles of the pit by virtue of the coefficient of friction of the earth/soil, leading to the restriction on its depth.

Thermal insulation plays a crucial role in increasing the net TES output of the UTES systems. For instance, utilizing porous materials for insulation, increasing the temperature, and increasing the moisture content would in turn increase their thermal conductivity. The list of pilot and research UTES projects incorporated with different types of the thermal insulation materials are given in Table 7.3.

Similar to the thermal insulation materials, the liners or the lining material used in the UTES system must also possess certain qualities to achieve better thermal storage capabilities, and they are listed in Table 7.4.

Several types of lining materials are available for the UTES systems, including stainless steel, polyolefins (polypropylene, high-density polyethylene, low-density polyethylene), elastomers, asphalt, bitumen, clay, resin, and high-performance concrete. The most important characteristics that a lining material should possess are resistance to temperature variations, permeation (water vapor or moisture) proof, durability, and reliability on a long-term basis.

Table 7.3 List of Pilot and Research UTES Projects Incorporated with Different Types of Thermal Insulation Materials [11]

Project Location	Type of Thermal Insulation
Vaulruz, Lyngby, Chemnitz, Augsburg, Attenkirchen	XPS
Marstal (HW), Rottweil, Egenhausen	EPS
Ottrupgaard, Sjökulla, Herlev Växjö, Lombohov, Studsvik, Illmenau	PUR
Särö, Rottweil Friedrichshafen, Sjökulla, Lyngby, Marstal (HW)	Mineral/rock wool
Berlin, Lisse	Foam glass
Steinfurt, Hannover, Crailsheim, Munich	Expanded glass granules
Eggenstein, Munich	Foam glass gravel

Table 7.4 Essential Requirements and Qualities of Lining Materials [11]

Essentials	Remarks	Qualities	Remarks
Waterproof	—	Proof to water vapor	<0.001 g/m² d at 95 °C
Temperature resistant	Up to 95 °C	Ultraviolet resistance	During the construction stage
Hydrolysis resistant	Contact to hot water	Robustness	Tensile strength, endurance limit, tear, and wear resistance
Durability on a long-term basis	20–30 Years (minimum)	Economics	Installation and leakage proof checks
Availability	Relatively in small quantities	Processable	Flexible at all ambient conditions
Weldability	Hot air or wedge and extrusion	Service/maintenance	Weldable even after many years/time

The most common reasons for experiencing high thermal losses of UTES systems, which are in some cases incomparable to the design values, may be due to the following:

- Frequent fluctuations in the thermal load demand in buildings requiring high mean storage temperature
- High return temperature of the hydronic network resulting in higher thermal losses to the earth/ground, particularly occurring at the bottommost section of the storage tank (usually not insulated at the bottom section)
- Underestimation of the thermal conductivity of the lining and insulation materials during the design as well as construction stages, thereby leading to erratic thermal losses (typical example can be for the porous material as explained earlier)
- Destitute construction and workmanship adopted for the establishment of the thermal storage facility or network.

The comparison of hot water storage, gravel/water thermal storage, BTES, or duct heat storage ATES concepts are summarized in Table 7.5.

7.2.4 Earth-to-air thermal storage

This type of energy storage applies to small residential and commercial buildings, where the earth's temperature is stable at specific depth from the ground surface. This storage comprises of array of plastic pipes or a single lengthy pipe, room air handling unit, or heat exchanger with necessary acces-

Table 7.5 Comparison of Storage Concepts [12]

Storage Concept	Hot Water	Gravel-Water	Duct (or BTES)	Aquifer
Storage medium	Water	Gravel-water	Ground material (soil/rock)	Ground material (sand/water-gravel)
Heat capacity (kWh/m3)	60-80	30-50	15-30	30-40
Storage volume for 1 m^3 water equivalent (m^3)	1	1.3-2	3-5	2-3
Geological requirements	– Stable ground conditions – Preferably no groundwater – 5-15m deep	– Stable ground conditions – Preferably no groundwater – 5-15m	– Drillable ground – Groundwater favorable – High heat capacity – High thermal conductivity – Low hydraulic conductivity (k f < 1 × 10^{-10} m/s) – Natural groundwater flow < 1 m/a – 30-100m deep	– Natural aquifer layer with high hydraulic conductivity (k f > 1 × 10^{-5} m/s) – Confining layers on top and below – No or low natural ground water flow – Suitable water chemistry at high temperatures – Aquifer thickness 20-50m deep

sories. It operates by charging the earth soil material by tubes buried 3 to 4 m beneath the ground level. Early morning or nighttime air is fed into the room during summer season for cooling the room, and during daytime, if the outdoor air temperature is sensed to be greater than the room air temperature, the indoor air is routed through the pipe structures that are embedded in the earth. The air that circulates through the pipe traps the cold energy possessed by the earth and thus serves the cooling load demand effectively. The outdoor air temperature variations during seasonal conditions govern the options of diverting indoor air through the earth pipes.

7.2.5 Energy piles thermal storage

In this type of seasonal thermal storage system, the heat or cold energy from the ground is trapped by coils or pipes embedded in building pile structures. During the construction stage of the dwelling, the pile structures are formed as major part of the foundation work. At this stage, the thermally conductive coil or pipe elements are amalgamated within the pile structures. The working principle of this storage system is very similar to other UTES systems described earlier.

In summer, cold energy from underground is captured by the coil/pipe element through building pile structures. The sensible heat that is stored in the coil/pipe material is then transferred to the heat pump arrangement (heat exchanger) equipped in the building side. The stored cold energy is thus exchanged to the dwelling for meeting the space cooling requirements. During winter, the required stored heat energy from underground is transferred to building spaces for meeting the heating demand.

The energy piles thermal storage system can be considered a closed loop system, which is very similar to the BTES system, wherein a brine solution or heat transfer fluid is used for effective thermal energy transfer process to take place. This storage system can be installed in geological locations that have a scarce underground water supply because the energy efficient heat pumps loaded with the heat transfer fluid or brine solution can help extract more thermal energy than the one produced by groundwater.

7.2.6 Sea water thermal storage

The other possible way to extract sensible heat (mainly cold energy) can be achieved by establishing the sea water thermal storage for offsetting the desired cooling energy demand, especially in buildings. Logically, the temperature of sea water at a depth of few meters beneath its surface would be lower compared to the temperature measured nearer its surface. By using this temperature gradient, a heat pump system can be blended with sea water, and the required thermal energy can be stored and retrieved for future usage.

In normal operation of the sea water thermal storage system, the low temperature sea water from the required depth of the sea is pumped into the heat pump (or chiller plant) installed in the district cooling facility. The low temperature sea water (around 25 °C) is then cooled to the requisite temperature (around 5–6 °C), which is then supplied to the respective air handling units for satisfying the cooling load demand in the buildings. The return sea water coming out of the air handling units (around 12–15 °C) is then pumped back to the chiller for the next cycle of cooling process to be performed. The sea water quantity in terms of discharge or flow rate can be maintained by supplying only the required quantity of the sea water, and the remainder is routed back to the seabed. Using the sea water thermal storage system, the cooling capacity of the district cooling facility can be fetched from 30 to 50 MW. However, as mentioned for the other SeTES systems, the sea water thermal storage system also posses certain limitations related to its successful implementation, which include:

- Corrosion/erosion of the piping network and storage facility (including heat pumps/chiller)
- Salt and scale deposition on the hydronic circuit delivering the sea water for cooling storage
- Clogging of filter elements due to aquatic sediments and other ecological factors
- Increase of frictional pressure drop and pumping power due to the scale formation
- Total cost, repair, and maintenance factors involved leads to economic issues for implementation of the storage system

7.2.7 Rock thermal storage

The interesting part of storing sensible heat on a large scale can be accomplished using the rock TES method. Herein, the thermal energy (cold or heat) can be effectively stored or released from and to the heavy rock structures with the help of a heat transfer medium. Typically, for storing the heat energy being captured from the solar collector, the heat transfer medium (air or water or brine solution) is made to flow through the drilled pipes being inserted into the rock structures.

Due to temperature swing, heat capacity, and mass of the rock, heat energy is transferred to rock structures sensibly. Similarly, to extract heat energy (cold or heat) during peak seasonal conditions (summer or winter), the same heat transfer fluid is pumped through the rock storage, and the desired thermal energy is then utilized for offsetting the cooling/heating demand in dwellings through the operation of a room heat exchanger or heat pump system. The essential criteria for enabling a high rate of thermal energy from the rock storage system are that the rocks must be impermeable, buckle-free, robust, and durable on long-term basis.

The schematic layout of a solar air heating system dedicated to space heating and hot water supply for a two-story building located in Qinhuang Island, China [13], is shown in Fig. 7.7. This rock thermal storage facility has been recognized as a large-scale solar air heating project in China and was put into full operation in December 2010.

The heating spaces are comprised of two different thermal zones requiring $717\,m^2$ for dormitory area and $2602\,m^2$ for cafeteria. The heating demand for these areas is different during the heating season, as the former requires 24 h and the latter one 5 h, respectively. To meet the heating loads of the two different thermal zones, solar collectors of $473.2\,m^2$ were coupled with the thermal storage facility. The thermal storage facility was constructed with a $300\,m^3$ pebble bed configuration, which is capable of storing collected heat energy during the daytime and releasing it during night hours.

The heat transfer medium used for the energy transfer is air, which picks up heat from the rock store and releases it to indoor spaces. Based on experimental measurements, about 19.1% of the mean solar fraction value has been achieved, with a highest value of 33% obtained during the second half of December 2010. This system was optimized with the help of TRNSYS simulation and exhibited 53.03% of average annual solar fraction value. As pointed out earlier, the implementation of a rock thermal storage system shows certain limitations in terms of the cost factors and thermal storage properties including low energy density. For the same capacity requirement, the rock thermal storage typically calls for larger storage volumes compared to a water thermal storage facility.

7.2.8 Roof pond thermal storage

Roof pond TES takes advantage of the natural evaporation of water molecules to create a cooling effect in dwelling spaces. The roof structure of the dwelling spaces filled with the water is evaporated, by which the indoor space located beneath the roof structure can be effectively cooled. The rate of

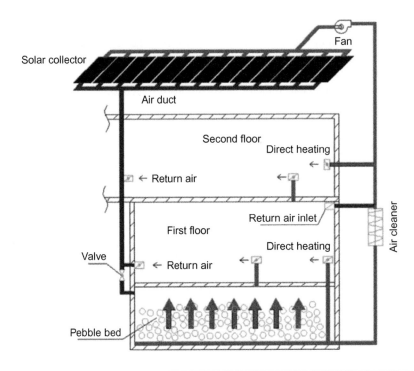

FIGURE 7.7

Schematic diagram of the solar air heating system on Qinhuang Island, China [13].

evaporation of water from the roof surface can be enhanced by forced convection principle or else by filling the water pond using pebbles or rock structures.

Because of the capillary forces exerted between the water molecules and the pebbles, the water can be raised above the pond surface leading to an effective evaporation process. The energy storage potential of the roof pond storage can be expected to achieve 1 kW per square meter of the pond area, wherein the cooling temperature ranges from 12 to 15 °C. The issues related to water penetration, leakage, and growth of microbes, fungal, and bacterial agents can be normally expected in a roof pond thermal storage, which if properly maintained gives promising cooling TES in the long run. The review of major large-scale seasonal sensible TES projects is presented in Table 7.6.

7.3 CONCISE REMARKS

From the standpoint of energy front, seasonal, or source TES technologies present a wide scope for conserving the primary energy and the possible reduction of GHG emissions. This is due to the fact that the backbone of the SeTES technologies is basically the natural resource (chiefly underground or underwater energy), which enables them to be operated longer without sacrificing energy efficiency. The SeTES enables the end-use energy demand to be equally offset through the extraction of heat or cold energy on a large scale.

Table 7.6 Review of Major Large-Scale Seasonal Sensible Heat Storage Projects [14]

Project	Heated Living Area (m²)	Demand by District Heating (GJ/a)	Solar Collector Area (m²)	Storage Volume (m³)	Solar fraction (%)	References
Hot water						
Hamburg, DE	14,800	5796	3000	4500	49[a]	[5]
Friedrichshafen, DE	39,500	14,782	5600	12,000	47[a]	[5]
Hannover, DE	7365	2498	1350	2750	39[a]	[5]
Munich, DE	24,800	8280	2900	5700	47[a]	[15]
Eggenstein, DE	12,000		1600	4500	37	[17]
Pit						
Rise, DK	115 Buildings		3575	5000		[15,16]
Marstal, DK	1300 Houses	104,400	26,000	70,000	29	[17]
Herlev, DK		4520	1025	3000	35	[13]
Ottrupgard, DK		1630	560	1500	16	[18]
Hoerby, DK				500		[15]
Ingelstad, SE	50 Houses		1320	5000		[15]
Lambohov, SE	50 Houses		2700	10,000		[15]
Lyckebo, SE			4320	100,000		[15]
Neuchatel, CH	Office		1120	1000		[15]
Calabria, IT	1750	111	91.2	500	28.2	[19]
Lisse, NL	Agriculture		1200	1000		[20]
Charlestown, US	Historic park		5700			[21]
Aquifer						
Rostock, DE	7000	1789	1000[b]	20,000	62[a]	[5]
Berlin, DE		57,600 (heat) 18,000 (cold and heat)			77 (heat) 93 (cold and heat)	[6,22]
Rastatt, DE		18,345	6780	23,000	41	[17]
Neubrandenburg, DE					46	[23]
2 MW, NL			2900			[15]

(Continued)

Table 7.6 Review of Major Large-Scale Seasonal Sensible Heat Storage Projects—cont'd

Project	Heated Living Area (m²)	Demand by District Heating (GJ/a)	Solar Collector Area (m²)	Storage Volume (m³)	Solar fraction (%)	References
Westway Beacons, UK	130 Apartments					[15]
Richard Stockton, US	College with 7000 (cold) students					[15]
Balcali, TR	Hospital	50,400				[24]
Çukurova, TR	360 m² Greenhouse		Use greenhouse as solar collector			[25]
Antwep, BE	Hospital				81	[26]
Gravel						
Chemnitz, DE	4680	4450	2000	8000	42[a]	[5]
Steinfurt, DE	3800	1170	510	1500	34[a]	[5]
Stuttgart, DE		360	211	1050	62	[27]
Augsburg, DE				6500		[22]
Duct						
Neckarsulm, DE	20,000	5987	2700	20,000	50[a]	[5]
Attenkirchen, DE	30 Homes	1386	836	500+10,500 (hot water+duct)		[28]
Crailsheim, DE	School and gymnasium	14,760	7300	37,500	50[a]	[15]
Anneberg, SE	50 Residential units	1980	2400	60,000	70[a]	[29]
Lidköping, SE		3528	2500	15,000	70	[17]
Groningen, NL			2400			[15]
DLCS, CA	52 Homes		2313	33,657	80	[7]
Kerava, FI			1100			[15]

DE, Germany; DK, Denmark; SE, Sweden; CH, Switzerland; IT, Italy; NL, Netherlands; US, United States of America; UK, United Kingdom; TR, Turkey; BE, Belgium; CA, Canada; FI, Finland.

[a]Calculated values for long-time operation.
[b]Combined heat and power plant (waste heat and ambient cold).

In the spectrum of a variety of SeTES options available, the major kinds of SeTES systems generally preferred for charging and discharging thermal energy are the hot water pit (tank) cavern thermal storage, gravel/water thermal storage, ATES, and BTES or duct heat storage. Although the constructional features of these systems differ by geological locations and design, the basic concept of storing and discharging thermal energy remains almost consistent among these storage technologies. Based on storage concepts and large-scale implementation comparisons, the SeTES technologies posses certain limitations, but can be considered potential candidates contributing to the development of the sustainable energy future.

References

[1] Pinel P, Cruickshank CA, Beausoleil-Morrison I, Wills A. A review of available methods for seasonal storage of solar thermal energy in residential applications. Renew Sustain Energy Rev 2011;15:3341–59.

[2] Schmidt T, Miedaner O. Solar district heating guidelines. 2012. Fact Sheet 7.2, www.solar-district-heating.eu, p. 1-13.

[3] Lemmens B, Desmedt J, Hoes H, Patyn J. Haalbaarheidsstudie naar de toepassing van koude-opslag met recirculatie bij Kaneka te Westerlo. Mol: Vlaamse Instelling voor Technologisch Onderzoek; 2007, p. 1-48.

[4] Schmidt T, Müller-Steinhagen H. The central solar heating plant with aquifer thermal energy store in Rostock—results after four years of operation. In: The 5th ISES Europe solar conference. Germany; 2004.

[5] Schmidt T, Mangold D, Müller-Steinhagen H. Central solar heating plants with seasonal storage in Germany. Sol Energy 2004;76:165–74.

[6] Seibt P, Kabus F. Aquifer thermal energy storage—projects implemented in Germany. In: ECOSTOCK '2006. 10th International conference on thermal energy storage. USA; 2006.

[7] Drake Landing Solar Community. Borehole thermal energy storage (BTES). Available from: http://www. dlsc.ca/borehole.htm [accessed January 2011].

[8] Chuard P, Chuard D, Van Gilst J, Hadorn JC, Mercier C. IEA task VII Swiss project in Vaulruz—design and first experiences. In: International conference on subsurface heat storage in theory and practice; 1983.

[9] Raab S, Mangold D, Heidemann W, Müller-Steinhagen H. Solar assisted district heating system with seasonal hot water heat store in Friedrichshafen (Germany). In: The 5th ISES Europe solar conference. Germany; 2004.

[10] Pfeil M, Koch H. High performance—low cost seasonal gravel/water storage pit. Sol Energy 2000;69:461–7.

[11] Ochs F, Heidemann W, Müller-Steinhagen H. Seasonal thermal energy storage: a challenging application for geosynthethics, Eurogeo4. In: 4th European geosynthetics Conference. Edinburgh; 2008. p. 1–8.

[12] Novo AV, Bayon JR, Fresno DC, Hernandez JR. Review of seasonal heat storage in large basins: water tanks and gravel–water pits. Appl Energ 2010;87:390–7.

[13] Zhao DL, Li Y, Dai YJ, Wang RZ. Optimal study of a solar air heating system with pebble bed energy storage. Energ Convers Manage 2011;52:2392–400.

[14] Xu J, Wang RZ, Li Y. A review of available technologies for seasonal thermal energy storage. Sol Energy 2013;. http://dx.doi.org/10.1016/j.solener.2013.06.006.

[15] Dalenbäck J-O. Available from: http://www.solar-district-heating.eu/SDH/LargeScaleSolarHeatingPlants. aspx; 2012.

[16] SOLARGE, Available from: http://www.solarge.org/index.php?id=1631

[17] Fisch MN, Guigas M, Dalenbäck JO. A review of large-scale solar heating systems in Europe. Sol Energy 1998;63:355–66.

[18] Heller A. 15 Years of R&D in central solar heating in Denmark. Sol Energy 2000;69:437–47.

[19] Oliveti G, Arcuri N, Ruffolo S. First experimental results from a prototype plant for the interseasonal storage of solar energy for the winter heating of buildings. Sol Energy 1998;62:281–90.

[20] Bokhoven TP, Kratz J, Van Dam P. Recent experience with large solar thermal systems in The Netherlands. Sol Energy 2001;71:347–52.

[21] Breger DS, Michaels AI. A seasonal storage solar heating system for the Charlestown, Boston Navy Yard National Historic Park. In: First E.C. conference on solar heating. Amsterdam; 1984. p. 858–63.

[22] Schmidt T, Mangold D, Müller-Steinhagen H. Seasonal thermal energy storage in Germany. In: ISES solar world congress. Schweden; 2003.

[23] Kabus F, Wolfgramm M. Aquifer thermal energy storage in Neubrandenburg—monitoring throughout three years of regular operation. In: EFFSTOCK'2009. 11th International conference on thermal energy storage. Sweden; 2009.

[24] Paksoy HO, Andersson O, Abaci S, Evliya H, Turgut B. Heating and cooling of a hospital using solar energy coupled with seasonal thermal energy storage in an aquifer. Renew Energy 2000;19:117–22.

[25] Turgut B, Paksoy H, Bozdağ Ş, Evliya H, Abak K, Dasgan HY. Aquifer thermal energy storage application in greenhouse climatization. In: ECOSTOCK '2006. 10th International conference on thermal energy storage. USA; 2006.

[26] Vanhoudt D, Desmedt J, Van Bael J, Robeyn N, Hoes H. An aquifer thermal storage system in a Belgian hospital: long-term experimental evaluation of energy and cost savings. Energ Buildings 2011;43:3657–65.

[27] Hahne E. The ITW, solar heating system: an oldtimer fully in action. Sol Energy 2000;69:469–93.

[28] Reuss M, Beuth W, Schmidt M, Schoelkopf W. Solar district heating with seasonal storage in Attenkirchen. In: ECOSTOCK'2006. 10th International conference on thermal energy storage. USA; 2006.

[29] Lundh M, Dalenbäck JO. Swedish solar heated residential area with seasonal storage in rock: initial evaluation. Renew Energy 2008;33:703–11.

Nanotechnology in Thermal Energy Storage

8.1 INTRODUCTION

In recent years, the importance of nanoscience and nanotechnology has attracted the interest of a variety of research communities involved in scientific and engineering fields, where the scientific principles meet engineering endeavors. The underlying science explored at the nanoscale level can create a significant change in the final outcome of an engineered product. Basically, nanoscience is the study of fundamental properties of matter or materials produced in the range of 1 to 100 nm.

Nanotechnology, on the other hand, deals with the technological aspects of transforming the science explored on matter or materials at the nanoscale level into real-time applications [1]. The summary of nonexhaustive applications of nanoscience and nanotechnology in the spectrum of energy and the environment has been gleaned and is presented in Table 8.1.

The properties of the matter or material found at the nanoscale level can be significantly different from the same material in its bulk state. By exploring changes in the properties of materials at the nanoscale level, their potential to be embedded with other materials available either at the nano or micro or macro scale level can be made possible. This can improve the performance of a bulk material to the desired extent in real-time applications. The applications of nanomaterials are numerous; in particular, their penetration into the field of thermal energy storage is inexorable. The thermophysical property changes of heat storage materials by virtue of embedding nanostructured materials are functional and energy efficient.

8.2 NANOSTRUCTURED MATERIALS

The term *nanostructured materials* or *nanomaterials* refers to a class of materials in which the formation of the particles or structures occurs at the nanoscale level with sizes ranging from 1 to 100 nm. The morphology in terms of the shape and size of the nanostructured materials is relatively vital, which determines the ultimate performance of bulk materials that contain them.

The high surface-to-volume ratio of nanomaterials allows them to establish excellent thermophysical property changes in bulk materials and thereby enhances their energy performance. The broad categorization and the global production scenario of nanomaterials are shown in Fig. 8.1 and Table 8.2, respectively.

8.2.1 Preparation and characterization of nanomaterials

The fundamental approach in nanomaterials starts from their preparation (or synthesis) stages to the characterization needed for them to be incorporated into bulk materials for real-time energy applications. The commonly preferred approaches for manufacturing nanoparticles or nanomaterials in various shapes and

Table 8.1 Various areas where nanoscience and nanotechnology (N&N) are expected to have a significant impact on energy and environmental systems [2]

Area				Nanoscience and nanotechnology (N&N) aspects
Energy sources	Renewable and/or unlimited	Solar	Solar thermal	Materials
			PV	Light harvesting, materials
			PEC, solar hydrogen	Light harvesting, materials, catalysis
			Solar power satellites	Materials
			Wind	Materials
			Wave	Materials
			Hydro	Materials
			Biomass	Catalysis
		Nuclear fusion		Materials, phase (fuel) separation
		Geothermal		Sensors, materials
		Gravitational		Materials
		Extraterrestrial feedstock		Materials
	Non-renewable	Nuclear fission		Materials, phase (fuel) separation, nanofluids
		Oil		Catalysis, materials, sensors
		Gas		Catalysis
		Coal		Catalysis
		Unconventional fossil fuels		Catalysis
Energy storage	Water dams			Materials
	Super giro, pneumatic			Materials
	H_2 storage			Materials, catalysis, kinetics
	Batteries			Materials
	Supercapacitors			Materials, nanolayered structures, electrolytes
Energy conversion	Combustion engines			Materials, diagnostics
	Turbines			Materials, diagnostics
	Fuel cells			Catalysis, diagnostics, electrolytes
	Magneto-hydrodynamic generators			Materials
	Thermo/piezoelectric materials			Nanolayered structures
Energy use	Industrial/mining			Sensors, phase separation
	Residential			Nonthermal light sources, materials (smart windows), nanofoams
	Transportation			Materials
Emission cleaning	Gas/air			Catalysis
	Water			Catalysis

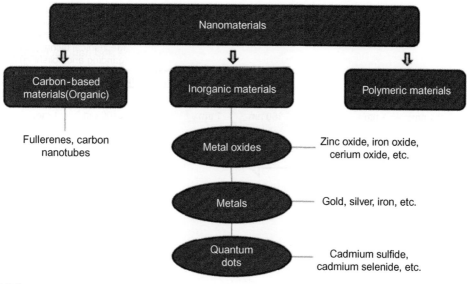

FIGURE 8.1

Classification of nanomaterials [3].

Table 8.2 Global production of nanostructure materials [4,5]		Estimated Global Production (Tons/Year)		
Application	**Nanomaterial Device**	**2003/04**	**2010**	**2020**
Structural application	Ceramics, catalysts, film & coating, composites, metal	10	10^3	10^4-10^5
Sink care products	Metal oxides (e.g., TiO_2 and ZnO)	10^3	10^3	10^3
Information and communication technologies	SWCNT, nanoelectronics and optoelectronics materials (excluding CMP slurries), organic light emitters	10	10^2	>10^3
Biotechnology	Nanocomposite, encapsulates, target drug delivery, diagnostic marker, biosensors	<1	1	10
Environmental	Nanofiltration membranes	10	10^2	10^3-10^4

sizes are depicted in Fig. 8.2. As shown in this figure, two major routes are used for the preparation of nanomaterials, which includes the top-down strategy (approach or method) and the bottom-up strategy (approach or method).

In the top-down strategy, nanomaterials can be formed by breaking up or fragmenting bulk materials into much smaller sizes ranging from micro to nanometer scale in steps of manufacturing processes as depicted in Fig. 8.2. This is referred to as the top-down strategy because the nanomaterials

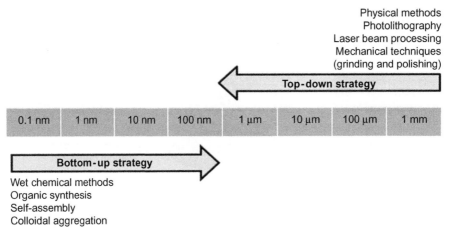

FIGURE 8.2

Strategies for manufacturing nanomaterials [3,6].

fabrication takes place from the macro scale or bulk state (top) to the nanoscale (down to nanometer) level. Likewise, in the bottom-up strategy, the nanomaterials of required shapes and sizes can be prepared atom by atom (or self-assembly of atoms) in the range of nanometer scale.

The shape and the size of the nanomaterial can be effectively controlled using the bottom-up strategy, which is not so in the case of top-down strategy. Self-assembly of the atoms or molecules occurs within the nanometer scale from the bottom range (0.1 nm) up to 100 nm as illustrated in Fig. 8.2. The comparison between nanostructures on the basis of chemical composition and dimensionality is shown in Figs. 8.3 and 8.4, respectively. It is pertinent to note that a basic difference exists between the terminologies *nanostructures* and *nanostructure materials*.

Nanostructures are characterized either by form or dimensionality, whereas nanostructure materials exemplify the composition of addition. More clearly, the inner-crystallite grain boundaries and free surfaces play a major role in demonstrating the essential behaviors of the nanostructures and nanostructure materials. That is, the grain boundaries can give rise to the inner classical size effects including enhancement of diffusion, melting point reduction, or changes in the lattice parameters.

On the other hand, the surface aspects take part in determining the shape, form, dimensionality and class of nanostructures as represented in Fig. 8.4. To give a better understanding of the preparation of different types of nanomaterials and nanoalloys, a detailed comparison is presented in Table 8.3, which highlights the merits and demerits of a variety of synthesis procedures. This comparative analysis (nonexhaustive) can be expected to be beneficial to the scientific community for selecting the appropriate method for the preparation of nanomaterials.

Based on the comparison provided in Table 8.3, it can be observed that the chemical reduction method offers good potential for the preparation of nanomaterials of desired physicochemical properties, easier handling, equipment configuration, and cost factors associated with the chemical processes. Using strong reducing agents, nanomaterials of relatively smaller size and desired shape can be produced.

At the same time, by utilizing stabilizing agents (or surfactants), the dispersion stability of the nanomaterials being embedded into the parent material (in liquid or colloidal form) can be established

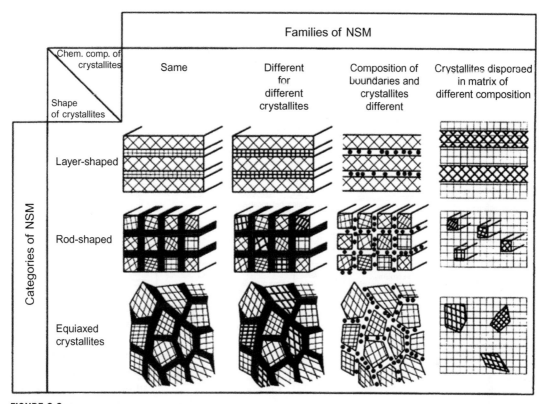

FIGURE 8.3

Gleiter's classification schema for nanostructured materials according to chemical composition and dimensionality of the crystallites forming the nanostructures [3,7].

without forming any sort of aggregation between the nanoparticles. Moreover, depending on the shape of the nanoparticles, they can be used for different applications.

After preparation of the nanomaterials, extensive analyses have to be performed to characterize whether their physicochemical and thermophysical properties are suitable for real-time applications. A variety of characterization methods are available to explore the properties of nanomaterials. Some of the major characterization techniques along with their strengths and limitations are summarized in Tables 8.4 and 8.5 for ready reference, which gives better insight about the characterization of nanomaterials.

8.2.2 Hybrid nanomaterials

The interesting aspects of the nanomaterials of different morphologies and functional properties have paved for the development of hybrid nanomaterials. In particular, the interest shown toward the fabrication of hybrid nanomaterials in addition to the preparation of individual nanomaterials have been gaining impetus in recent times.

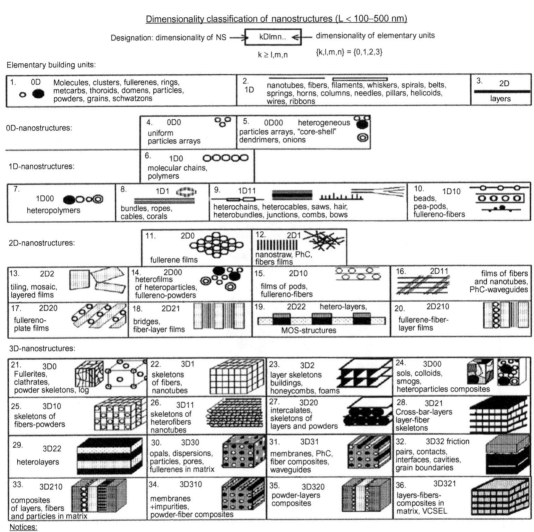

FIGURE 8.4

Dimensionality classification of nanostructures [3,8,9].

Hybrid nanomaterials are referred to as the combination of one or more dissimilar materials into a singular and multifunctional material that can exhibit synergetic properties not found in constituent materials. The formation of the hybrid nanomaterials depends on several factors, including elemental composition, morphology, surface effects, density of materials, and so on.

Factually, the chemical reactivity of nanoparticles is governed by internal and external interfacing abilities as well as the size distribution of the nanoparticles on the supporting matrix of the nanostructures.

Table 8.3 Mainstream methods for synthesizing nanoparticles and nanoalloys [3]

Method	Description of Method	Advantages	Disadvantages	References
Molecular beams	Beams are directed toward specific metal targets using a variety of methods; laser vaporization, pulsed arc, ion and magnetron sputtering. This creates clusters of metallic nanoparticles including nanoalloys	(1) Any type of nanoparticle or nanoalloy can be created from metallic/alloy targets. (2) Nanoparticle synthesis is quick and not lengthy	(1) Process is expensive and requires equipment setup in most cases	[10,11]
Chemical reduction	Use of precursor salts, reducing agents, and stabilizer to synthesize nanoparticles. In most cases a catalyst and some heating is used.	(1) Can readily produce bulk quantities of nanoparticles and nanoalloys. (2) Process can be easily scaled up to meet mass manufacturing needs. (3) Process enables synthesis of particles close to 1 nm, and this can easily be controlled. (4) Process is relatively cheaper compared to other synthesis methods because the technology is quite standard.	(1) Mass use of chemicals, and some may be harmful to the environment. (2) Processing is time consuming and depends on many parameters.	[11–21]
Thermal decomposition of metals	Thermal decomposition of metal or metal complexes (for nanoalloys) is produced using high temperature mediums or solvents.	(1) Nanoparticles can be created at relatively low temperatures. (2) Process can create nanoparticles in a wide range of sizes.	(1) Requires use of chemicals and solvents, which may be harmful to the environment.	[11]
Ion implantation	This method is used to create nanoalloys by implanting two or more metal ions into a specific matrix. This generates metallic/bimetallic clusters.	(1) Metallic ions can be implanted into exact positions in a matrix. (2) Various combinations of ions can be used to yield nanoalloy clusters.	(1) Requires equipment setup, which is relatively expensive compared to chemical reduction.	[11]
Electro-chemical synthesis	Using an electrolysis cell and two electrodes of metallic elements, bimetallic alloys/nanoalloys can be created in solution. Core–shell structures have also been created via this method.	(1) Various nanoalloy combinations can be synthesized. (2) Cell setup is rather easy and does not need extensive equipments.	(1) Use of chemicals as electrolytes, which may yield harmful/toxic gases as by products from the process.	[11–15]

Continued

Table 8.3 Mainstream methods for synthesizing nanoparticles and nanoalloys—cont'd

Method	Description of Method	Advantages	Disadvantages	References
Radiolysis	Radiolysis of an aqueous solution of metal ions to produce nanoparticles. This method has also been used to create nanoalloys/bimetallic particles.	(1) Irradiation of molecules is able to create nanoparticles with a wide range of sizes, as well as very narrow sizes.	(1) Requires expensive equipment setup. (2) Radiation is harmful to the health of living organisms, including humans. (3) The use of this method requires extensive clearance from concerned authorities.	[11]
Sonochemical synthesis	Irradiation of metal salt solutions using ultrasound to create nanoparticles and nanoalloys.	(1) Sonic wave irradiation is able to create narrow particle sizes.		[11,16–18]
Biosynthesis	Biological means are used to synthesize nanoparticles and nanoalloys using microorganisms or plants. Synthesis of nanoparticles using these mediums can make the nanoparticles more bio-compatible.	(1) This method is cheap and uses sources from nature. (2) The method does not produce waste detrimental to human beings. (3) Does not require extensive equipment setup.	(1) This method is slow and takes time.	[11,22,23]

Through manipulation of the precursor to the seed or the reducing agent ratio (either in the presence of surfactant or not), inducing heterogeneous nucleation and growth kinetics, and coordinated organic linkages one can achieve the synergetic properties of hybrid nanomaterials.

The two major factors influencing the formation of hybrid nanomaterials are the controlled synthesis procedure and knowledge of particle-to-particle interfacial interactions. Generalized growth models for the preparation of hybrid nanostructures of different configurations are illustrated in Fig. 8.5.

In a typical synthesis of colloidal hybrid nanostructures, the molecular precursors in solution phase react with reducing agents in the presence of organic surfactants. The role of surfactants is highly regarded to influence growth control of the nanoparticles species generated in the colloidal solution. On activation of the controlled synthesis procedure, monomers are generated, which in turn facilitate achieving the nucleation and the growth kinetics of the nanoparticles in the colloidal solution.

The synthetic routes through which the formation of the hybrid nanomaterials can be accomplished can be categorized into:

• Direct growth of secondary precursors over the surface or tip of the preformed seed support structure by heterogeneous means

Table 8.4 Analytical modalities for evaluation of the physicochemical characteristics of nanomaterials [24–107]

Techniques	Physicochemical Characteristics Analyzed	Strengths	Limitations
Dynamic light scattering (DLS)	Hydrodynamic size distribution	Nondestructive/invasive manner Rapid and more reproducible measurement Measures in any liquid media, solvent of interest Hydrodynamic sizes accurately determined for monodisperse samples Modest cost of apparatus	Insensitive correlation of size fractions with a specific composition. Influence of small numbers of large particles Limit in polydisperse sample measures Limited size resolution. Assumption of spherical shape samples
Fluorescence correlation spectroscopy (FCS)	Hydrodynamic dimension Binding kinetics	High spatial and temporal resolution Low sample consumption Specificity for fluorescent probes Method for studying chemical kinetics, molecular diffusion, concentration effect, and conformation dynamics	Limit in fluorophore species Limited applications and inaccuracy due to lack of appropriate models
Zeta potential	Stability Referring to surface charge	Simultaneous measurement of many particles using electrophoretic light scattering (ELS)	Electro-osmotic effect Lack of precise and repeatable measurement
Raman scattering (RS) Surface enhanced Raman spectroscopy (SERS) Tip-enhanced Raman spectroscopy (TERS)	Hydrodynamic size and size distribution (indirect analysis) Conformation change of protein–metallic NP conjugate Structural, chemical, and electronic properties	Complementary data to IR No requirement of sample preparation Potential of detecting tissue abnormality Enhanced RS signal (SERS) Increased spatial resolution (SERS) Topological information of nanomaterials (SERS, TERS)	Relatively weak single compared to Rayleigh scattering Limited spatial resolution (only to micrometers) Extremely small cross section Interference of fluorescence Irreproducible measurement (SERS)

Continued

Table 8.4 Analytical modalities for evaluation of the physicochemical characteristics of nanomaterials—cont'd

Techniques	Physicochemical Characteristics Analyzed	Strengths	Limitations
Near-field scanning optical microscopy (NSOM)	Size and shape of nanomaterials	Simultaneous fluorescence and spectroscopy measurement Nano-scaled surface analysis at ambient conditions Assessment of chemical information and interactions at nano-scaled resolution	Long scanning time Small specimen area analyzed Incident light intensity insufficient to excite weak fluorescent molecules Difficulty in imaging soft materials Analysis limited to the nanomaterial surface
Circular dichroism (CD)	Structure and conformational change of biomolecules (e.g., protein and DNA) Thermal stability	Nondestructive and prompt technique	Nonspecificity of residues involved in conformational change Less sensitive than absorption methods Weak CD signal for nonchiral chromophores Challenging for analysis of molecules containing multiple chiral chromophores
Mass spectroscopy (MS)	Molecular weight Composition Structure Surface properties (secondary ion MS)	High accuracy and precision in measurement High sensitivity to detection (a very small amount of sample required)	Expensive equipment Lack of complete databases for identification of molecular species Limited application to date in studying nanomaterial-bioconjugates
Infrared spectroscopy (IR) Attenuated total reflection (ATR) Fourier transform infrared (ATR–FTIR)	Structure and conformation of bioconjugate Surface properties (ATR–FTIR)	Fast and inexpensive measurement Minimal or no sample preparation requirement (ATR–FTIR) Improving reproducibility (ATR–FTIR) Independence of sample thickness (ATR–FTIR)	Complicated sample preparation (IR) Interference and strong absorbance of H_2O (IR) Relatively low sensitivity in nanoscale analysis

Table 8.4 Analytical modalities for evaluation of the physicochemical characteristics of nanomaterials—cont'd

Techniques	Physicochemical Characteristics Analyzed	Strengths	Limitations
Scanning electron microscopy (SEM) Environmental SEM (ESEM)	Size and size distribution Shape Aggregation Dispersion	Direct measurement of the size/size Distribution and shape of nanomaterials High resolution (down to sub-nanometer) Images of biomolecules in natural state provided using ESEM	Conducting sample or coating conductive materials required Dry samples required Sample analysis in non-physiological conditions (except ESEM) Biased statistics of size distribution in heterogeneous samples Expensive equipment Cryogenic method required for most NP-bioconjugates Reduced resolution in ESEM
Transmission electron microscopy (TEM)	Size and size distribution Shape heterogeneity Aggregation Dispersion	Direct measurement of the size/size distribution and shape of nanomaterials with higher spatial resolution than SEM Several analytical methods coupled with TEM for investigation of electronic structure and chemical composition of nanomaterials	Ultrathin samples in required Samples in nonphysiological condition Sample damage/alternation Poor sampling Expensive equipment
Scanning tunneling microscopy (STM)	Size and size distribution Shape Structure Dispersion Aggregation	Direct measurement High spatial resolution at atomic scale	Conductive surface required Surface electronic structure and surface topography unnecessarily having a simple connection
Atomic force microscopy (AFM)	Size and size distribution Shape Structure Sorption Dispersion Aggregation Surface properties (modified AFM)	3D sample surface mapping Sub-nanoscaled topographic resolution Direct measurement of samples in dry, aqueous, or ambient environment	Overestimation of lateral dimensions Poor sampling and time consuming Analysis in general limited to the exterior of nanomaterials

Continued

Table 8.4 Analytical modalities for evaluation of the physicochemical characteristics of nanomaterials—cont'd

Techniques	Physicochemical Characteristics Analyzed	Strengths	Limitations
Nuclear magnetic resonance (NMR)	Size (indirect analysis) Structure Composition Purity Conformational change	Nondestructive/ noninvasive method Little sample preparation	Low sensitivity Time consuming Relatively large amount of sample required Only certain nuclei NMR active
X-ray diffraction (XRD)	Size, shape, and structure for crystalline materials	Well-established technique High spatial resolution at atomic scale	Limited applications in crystalline materials Only single conformation/binding state of sample accessible Low intensity compared to electron diffraction
Small-angle X-ray scattering (SAXS)	Size/size distribution Shape Structure	Nondestructive method Simplification of sample preparation Amorphous materials and sample in solution accessible	Relatively low resolution

- Adsorption of the metal clusters on oppositely charged support surface through photo-irradiation
- Adsorption of reduced precipitation of the secondary precursors on hydrophilic-surfaced nanohybrids by means of the ion-exchange deposition
- Growth of secondary precursors on the support surface after being activated through chemical means
- Self-assembled controlled growth of the secondary precursors on the support surface through one-pot synthesis procedure

By conceptualizing and following the aforementioned synthetic routes, promising opportunities are available for the successful preparation of hybrid nanomaterials with excellent physicochemical and thermophysical properties for energy and environmental applications.

8.3 NANOMATERIALS EMBEDDED LATENT HEAT STORAGE MATERIALS

The incorporation of nanomaterials into latent heat storage materials (i.e., phase change materials, PCMs) has been well acquainted in modern times. The potential opportunities available for enhancing the thermal storage properties of the PCM can be effectually accomplished by embedding nanostructured materials into the pure (or base) PCM. By having other kinds of heat storage materials

Table 8.5 Physicochemical characteristics of nanomaterials and suitable evaluation modalities [108–164]

Nanomaterial Characteristics	Techniques
Size/size distribution	DLS, FCS, RS, NSOM, SEM, TEM, STM, AFM, NMR, TOF-MS, XRD, SAXS, FS, UV–visible, AUC, GE, CE, FFF
Surface charge	Zeta potential (ELS), ATR–FTIR, GE, CE
Shape	NSOM, SEM, TEM, STM, AFM, XRD, SAXS, AUC
Structure	TERS, CD, MS, IR, STM, AFM, RS, NMR, XRD, SAXS, FS, DSC, AUC
Composition	MS, NMR
Purity	MS, NMR, HPLC, HDC
Stability	Zeta potential measurement, CD, TGA, DSC, ITC, thermophoresis, HPLC, HDC
Dispersion	ESEM, TEM, STM, AFM
Surface properties	CD coupled with an enzyme-linked immunosorbent assay, time-of-flight secondary ion MS, ATR–FTIR, modified AFM, X-ray photoelectron spectroscopy
Protein corona (thickness and density)[a]	DLS, FCS, TEM, size exclusion chromatography, differential centrifugal sedimentation
Protein corona (composition and quantify)[a]	Polyacrylamide GE, LC-MS/MS
Protein corona (conformation)[a]	CD, simulation
Protein corona (affinity)[a]	Size exclusion chromatography, SPR, ITC

[a]Courtesy of Rahman et al. (2013).

(sensible or thermochemical), the utilization of nanomaterials for latent heat storage materials has become attractive among materials scientists and engineers.

The penetration of nanoscience and nanotechnology in materials research has paved the way for the development of advanced thermal storage materials with enhanced thermophysical properties. At this juncture, the contribution of some specific research studies toward exploring potential thermal storage properties of PCMs incorporated with the nanomaterials have been reviewed and are presented in the following section.

8.3.1 Evaluation of thermal storage properties

The intrinsic fact behind infusion of nanostructured materials into pure PCM is to acquire better thermal storage properties of the PCM compared to its bulk state. The incorporation of nanomaterials into the base PCM can have significant effects pertaining to the following:

- Thermal conductivity of PCMs can be improved (especially organic PCMs).
- Swift freezing and melting characteristics can be achieved.
- Thermal stability of PCMs can be enhanced.
- Heat storage performance in terms of reduced freezing time can be accomplished.

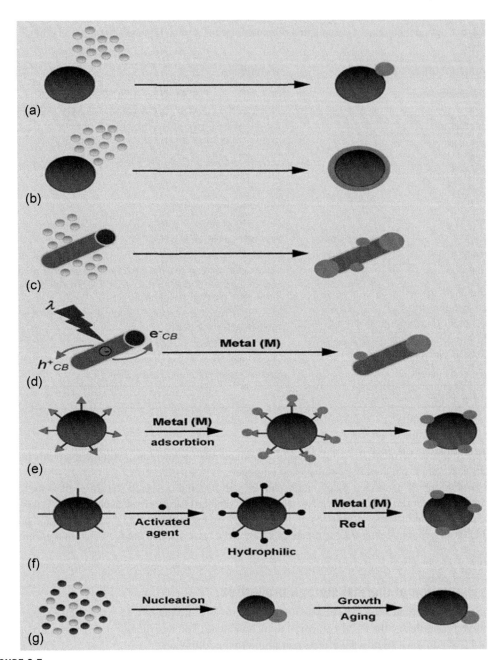

FIGURE 8.5

A general sketch of the reaction mechanisms for the formation of the hybrid nanostructures: (a–c) heterogeneous nucleation and growth of the secondary precursors on the preformed seeds; (d) secondary precursors adsorbed on the opposite-charged surfaced supports by photo-irradiation; (e) reduced precipitation of the secondary precursors on the hydrophilic-surfaced supports; (f) secondary precursor growth on the activated-surfaced supports; and (g) "one-pot" self-controlled nucleation growth [165].

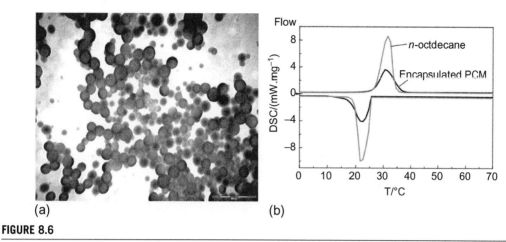

(a)   (b)

FIGURE 8.6

(a) TEM image of nanoencapsulated PCM (NEPCM) and (b) DSC curves of *n*-octadecane and the NEPCM [167].

From this perspective, the research study performed on the PCM incorporated with copper nanoparticles has enabled the PCM to exhibit better thermal storage performance, which can be suitable for real-time latent thermal energy storage (LTES) applications [166]. The copper nanoparticles added to the PCM in molar concentrations of 0.1 and 0.2 have yielded reduction in overall freezing time of the PCM. This in turn facilitates consuming less energy per unit mass of freezing the PCM.

Preparation of the nanoencapsulated PCM (NEPCM) with a polystyrene shell and *n*-octadecane as the core PCM material by means of the in situ ultrasonic-based polymerization method has been reported [167,168]. The transmission electron microscopy (TEM) image and the differential scanning calorimetry (DSC) graphs pertaining to the NEPCM are shown in Fig. 8.6. The size of the as-prepared NEPCM is found to be in the range of 100 to 123 nm, which exhibits a latent heat of fusion of about 124.4 kJ/kg. The polymerization factors involved during the synthesis enabled the NEPCM to have good thermal stability, enhanced specific heat capacity of 11.61 kJ/kg K, and viscous effects as well.

The enhancement of thermal storage properties including the thermal conductivity, transport phenomena, and heat interactions of palmitic acid PCM entangled with the multiwalled pristine carbon nanotubes (MWCNTs) has been investigated [202]. The pristine MWCNT as prepared through the mechano-chemical method while incorporated into the PCM contributes to an enhancement of thermal conductivity by 30% more than previously reported values. The increased thermal conductivity of this PCM allows it to be considered for LTES applications.

The enhancement of thermal conductivity of the aqueous solution of barium chloride has been achieved through the infusion of titanium dioxide (TiO_2) nanoparticles [196]. For 1.13% of concentration of the nanoparticles, the thermal conductivity of the PCM has been enhanced by 15.65%. This in turn contributes to the swift freezing and melting processes (reduction in time), enhanced nucleation effects, and heat transfer interactions of the PCM.

The NEPCM being prepared using *n*-tetradecane oil and urea-formaldehyde resin exhibited good freezing and melting characteristics [168]. The 100 nm nanoparticles are well encapsulated in the PCM, and the addition of recorcin (cross-link agent) up to 5% has enhanced the PCM concentration by 61.8%. Thus, the heat interactions of the NEPCM have been improved further. The latent heat capacity of this PCM ranged between 100 and 130 kJ/kg with the phase transition temperature in the range suitable for

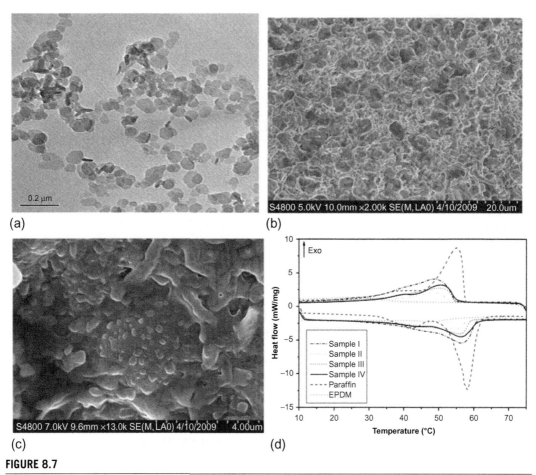

FIGURE 8.7

TEM photo of self-synthesized nano-MH (a) and the cross section SEM images (b) and (c) of flame retardant form-stable PCM blend (Sample IV) at different magnifications, and (d) DSC curves of paraffin, original EPDM polymer, and form-stable PCM blends [170].

TES applications. Similar research studies have also been reported in synthesizing the nanoparticles and incorporating them into the PCM for achieving improved thermal storage properties [169–172]. The TEM images and DSC curves of the fire retardant form stable PCM blend are shown in Fig. 8.7.

In addition to the NEPCMs, research studies dealing with the incorporation of nanofibers have also been increasingly attractive in recent years [173]. The schematic diagram of the cubic thermal containment unit (TCU) and the thermal energy storage characteristics during freezing and melting cycles are illustrated in Fig. 8.8. The inclusion of the graphite nanofibers into the PCM for enhancing thermal conductivity and achieving faster solidification and melting rates have been investigated using the TCU. The TCU has an aspect ratio of 0.5 and 2 with a total volumetric capacity of $1260\,cm^3$ and yielded considerable reduction in solidification time of the PCM.

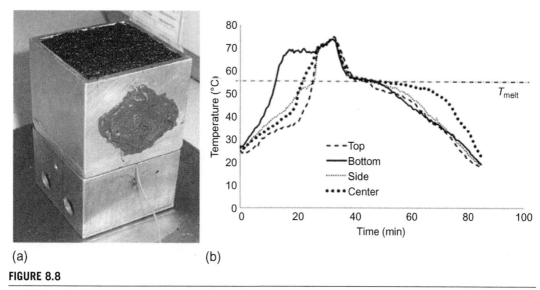

(a)   (b)

FIGURE 8.8

(a) 5.08 cm cubic thermal containment unit and heated base unit, and (b) temperature distribution during melt and solidification for the 5.08 cm side length cube with pure PCM during a heat load of 4 W/cm² [173].

Based on the figure of merit analysis, it is found that the presence of graphite nanofibers improves thermal diffusion effects within the PCM. However, initiation time for melting of the PCM is largely delayed due to the addition of graphite nanofibers, and this crucial parameter must be considered for investigating thermal performance of the PCM.

The thermal energy storage characteristics of organic PCM (1-octadecanol) have been analyzed with the incorporation of low concentration graphene nanoparticles [174]. The addition of graphene nanoparticles of 2% by weight results in increased thermal conductivity of the PCM by 63% with a minor reduction in the phase change enthalpy by 8.7%. By infusing 4% equivalent by weight of the graphene nanoparticles, thermal conductivity of the organic PCM is improved by 140%, but the reduction in latent heat potential is slightly on the higher side, which is estimated at 15%.

In a recent research study, the surface functionalized silver nanoparticles (AgNPs) were embedded into an organic ester PCM (ethyl trans-cinnamate, EC), which exhibited improved thermal conductivity up to 67%, congruent phase change temperature (6.93 °C), good latent heat capacity (89.47 kJ/kg), reduction in supercooling degree (1.96 °C), and reduction in freezing (30.8%) and melting times (11.3%) compared to pure PCM. For specific concentrations of the AgNP embedded into pure PCM, the latent heat capacity of the AgNP PCM decreases with respect to an increase in its viscosity [175].

Many research works pertaining to the incorporation of nanostructured materials in the form of fillers and enhancers have been performed to attain improved thermophysical properties of PCMs [176–181]. The morphology, phase change characteristics, and thermal stability of the bio-based PCM are depicted in Fig. 8.9. The nanomaterials that are used as fillers and nucleants actually enhance the thermal conductivity, freezing and melting rates, thermal stability, and thermal reliability of PCMs on a long-term basis. The summary of the various research outcomes on PCMs incorporated with nanostructures for enhancement of thermal conductivity are presented in Tables 8.6 through 8.8.

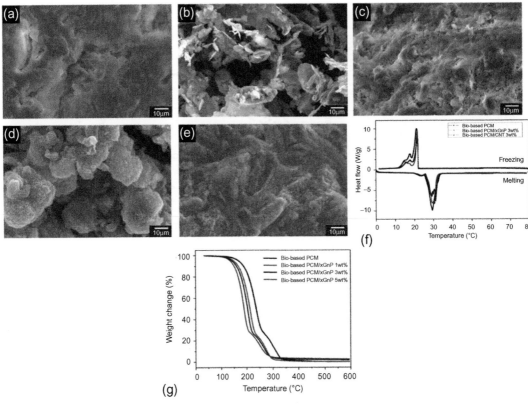

FIGURE 8.9

Scanning electron microphotographs (x1000) of (a) bio-based PCM, (b) xGnP, (c) bio-based PCM/xGnP 5wt%, (d) CNT, (e) bio-based PCM/CNT 5wt%, (f) DSC graphs of bio-based PCM and bio-based PCM composites, and (g) TGA curves of bio-based PCM/xGnP composites [181].

Table 8.6 Summary of utilized PCM and nano-structured thermal conductivity enhancers [166]

| Authors (Year) | PCM Materials and Properties | Nano-Structured Materials | | |
		Materials and Properties	Dimension, etc.	Fractions of Enhancers
[182]	Molten salts, k: 0.4-4 W/mK	Stainless steel, iron (Fe), aluminum (Al), and copper (Cu) particles, k: 15, 60, 204, and 386 W/mK	N/A	0–100 vol%[a]
[183]	Molten salts, k: 0.4-4 W/mK	Stainless steel and Cu particles, k: 50 and 380 W/mK	N/A	0–100 vol%

Table 8.6 Summary of utilized PCM and nano-structured thermal conductivity enhancers—cont'd

Authors (Year)	PCM Materials and Properties	Nano-Structured Materials		
		Materials and Properties	Dimension, etc.	Fractions of Enhancers
[184]	Paraffin wax, T_m: ~67 °C, k: ~0.24 W/mK, α: 1.61×10^{-7} m²/s	Carbon nanofibers (CNF), ρ: ~1600 kg/m³	Outer diameter:~100 nm, length:~20 μm	1, 2, 3, and 4 wt%
[185]	Water, ρ: 997.1 kg/m³, C_p: 4179 J/kg K, k: 0.6 W/mK, μ: 8.9×10^{-4} kg/ms, L: ~335,000 J/kg	Cu nanoparticles, ρ: 8954 kg/m³, C_p: 383 J/kg K, k: 400 W/mK	10 nm	10 and 20 vol%
[186,187]	PAC[b], EG solutions (with 50 vol% water), T_f: −35.6 °C	Single-walled carbon nanotubes (SWCNT), alumina (Al_2O_3), MgO nanoparticles	N/A	0.05, 0.1, and 0.2 wt%
[188]	1-Tetradecanol ($C_{14}H_{30}O$), T_m: ~38 °C, L:~230,000 J/kg	Silver (Ag) nanoparticles	500 nm	2–94 wt%
[189]	Deionized (DI) water	Al_2O_3, titanium oxide (TiO_2) nanoparticles	~8 nm (Al_2O_3); ~3.8 nm (TiO_2)	1.54–6.19 wt% (Al_2O_3); 2.616–7.85 wt% (TiO_2)
[190]	Paraffin wax, T_m: 56 °C, C_p: 2100 J/kg K, k: 0.25 W/mK, L: ~234,000 J/kg	Graphite nanofibers (GNF) of three types: ribbon, platelet, and herringbone	Diameter: 4–10 nm, length: ~1 μm	0.25, 0.5, 1, and 5 wt%
[191]	$C_{14}H_{30}O$ and $C_{14}H_{30}O$/PANI[c], T_m: ~35 °C, L: 221,250 J/kg and 119,140 J/kg	Multiwalled carbon nanotubes (MWCNT)	Outer diameter: 10–30 nm, length: 5–15 μm	0.5, 1, 2.5, and 5 wt%
[192]	Paraffin wax (shell wax 100), L: 156,300 J/kg	SWCNT, MWCNT, and CNF	Diameter: 1, 10, and 100 nm	0.1, 0.4, 0.7, and 1 vol%
[193]	Palmitic acid (PA), purity 98%, T_m: 62.5–64 °C, ρ: 853 kg/m³, k: 0.24 W/mK, L: 207,800 J/kg	MWCNT, surface-oxidized by a mixed acid with 1:3 of concentrated nitric and sulfuric acids	Diameter: 30 nm, length: 50 μm, specific surface area: 60 m²/g	0.5, 1, 2, and 5 wt%
[194]	n-Docosane, T_m: 53-57 °C, k: 0.26 W/mK, L: 157,300 J/kg	Exfoliated graphite nanoplatelets (xGnPs)	Diameter: 15 μm, thickness: <10 nm, surface area: 30 m²/g	1, 2, 3, 5, and 7 wt%
[195]	PA, T_m: 59.48 °C, T_f: 58.781 °C, k: 0.318 W/mK, L: ~201,000 J/kg	Long and short pristine MWCNT; surface-oxidized MWCNT (by two acids)	Outer diameter: 10–30 nm, length: 5–15 μm (long), 1~2 μm (short)	0.099-4.76 wt% (without surfactants), and 0.095-4.5 wt% (with surfactants)

Continued

Table 8.6 Summary of utilized PCM and nano-structured thermal conductivity enhancers—cont'd

Authors (Year)	PCM Materials and Properties	Nano-Structured Materials		
		Materials and Properties	Dimension, etc.	Fractions of Enhancers
[196]	Saturated barium chloride (BaCl$_2$) aqueous solution, T_f: −8 °C, pH: 8	TiO$_2$ nanoparticles	20 nm	0.167, 0.283, 0.565, and 1.13 vol%
[197]	Distilled water, k: 0.6008 W/mK	Al$_2$O$_3$ nanoparticles	20 nm	0.05, 0.1, and 0.2 wt%
[198,199]	n-Octadecane (C$_{18}$H$_{38}$), T_m: 26.5 °C, T_f: 25.1 °C, L: ~243,100 J/kg	Al$_2$O$_3$ nanoparticles, ρ: 3600 kg/m	33 nm (159.6 and 196.0 nm in suspensions)	5 and 10 wt%
[200]	Paraffin wax, T_m: 53 °C, L: ~165,300 J/kg	MWCNT, treated by ball milling	Diameter: 30 nm, length: 50 μm, specific surface area: 60 m²/g	0.2, 0.5, 1, and 2 wt%
[201]	C$_{14}$H$_{30}$O, k: 0.32 W/mK, L: ~220,000 J/kg	Ag nanowires	N/A	9.09–62.73 wt%
[202]	PA, purity 98%, T_m: 62.4 °C, ρ: 853 kg/m³, k: 0.22 W/mK (solid), 0.16 W/mK (liquid), L: 208,000 J/kg	MWCNT, treated by mechano-chemical reaction with potassium hydroxide/ ball milling	Diameter: 30 nm, length: 50 μm, specific surface area: 60 m²/g	0.2, 0.5, and 1 wt%
[203]	Paraffin wax, T_m: ~48.1 °C, L: 142, 200 J/kg	γ-Al$_2$O$_3$ nanoparticles ρ: 3900 kg/m³	20 nm	1, 2, and 5 wt%
[204]	PA, purity 98%, T_m: 62.5-64 °C, k: 0.223 W/mK (solid), 0.154 W/mK (liquid)	MWCNT, treated by surface oxidation, mechano-chemical reaction, ball milling, and grafting following acid oxidation	Diameter: 30 nm, length: 50 μm, specific surface area: 60 m²/g	0.2, 0.5, and 1 wt%
[205]	Paraffin wax, purity 99.99%, T_m: 58-60 °C	Cu nanoparticles, purity: 99.9%	Diameter: 25 nm, specific surface area: 30–50 m²/g	0.1, 0.5, 1, and 2 wt%
[206]	Paraffin, T_m: 58-60 °C, k: 0.2699 W/mK (solid), 0.1687 W/mK (liquid), L: 204,000 J/kg	Cu, Al, and C/Cu nanoparticles, purity: 99.9%	Diameter: 25 nm	0.1 wt% for all three kinds of nanoparticles, 0.1, 0.5, 1, and 2 wt% for Cu nanoparticles
[207]	DI water	MWCNT	Diameters: 10–30, 40–60, and 60–100 nm, length: 5–15 μm	0.1 wt%

Table 8.6 Summary of utilized PCM and nano-structured thermal conductivity enhancers—cont'd

Authors (Year)	PCM Materials and Properties	Nano-Structured Materials		
		Materials and Properties	Dimension, etc.	Fractions of Enhancers
[208]	Paraffin and soy wax, T_m: 52-54 °C	CNF, MWCNT (purity: 95%)	Outer diameter: 200 nm (CNF); diameter: 30 nm, length: 50 μm, specific surface area: 60 m²/g (MWCNT)	1, 2, 5, and 10 wt%
[209]	n-Docosane, T_m: 53-57 °C	Exfoliated graphite nanoplatelets (xGnPs); xGnP with ball milling treatment	Diameter: 15 μm, thickness: 10 nm, surface area: 20–40 m²/g, and diameter: 1 μm, surface area: 100–130 m²/g after ball milling	1, 2, 4, 6, 8, and 10 wt%
[210]	1-Octadecanol, T_m: ~66 °C, ρ: 812 kg/ m³, k: 0.38 W/mK, L: ~250,000 J/kg	Graphene flakes	N/A	0.2, 0.5, 1, 2, and 4 wt%
[211]	Cyclohexane, T_m: 6.51 °C, ρ: 850 kg/m³ (solid), 779 kg/m³ (liquid), L: 32,557 J/kg	Copper oxide (CuO) nanoparticles	Diameter: −-15 nm	1, 2, and 4 wt%
[212]	Eicosane, T_m: ~37 °C	Copper oxide (CuO) nanoparticles	Diameter: 5–15 nm	1, 2, 5, and 10 wt%

[a]Vol% and wt% stand for volumetric and mass fractions, respectively.
[b]PAC and EG stand for prediluted antifreeze coolant and ethylene glycol, respectively.
[c]PANI denotes polyaniline.

Apart from the inclusion of metal nanoparticles and carbon nanostructures, the interest shown toward incorporating metal oxide nanostructured materials into the PCM has been gaining momentum recently. In this regard, the metal oxide nanostructures including zirconium oxide, alumina (or aluminum oxide), and so on have been utilized as thermal conductivity enhancement materials.

For a thorough understanding of the freezing and melting characteristics on successful incorporation of such nanostructured materials into the PCM, investigation related to analytical modeling and experimental validation has been reported [169,203,213]. These studies reveal the fact that, by adding nanoparticles in different mass proportions, the reduction in the freezing and melting rates of PCMs are significant.

Although salt hydrate PCMs possess higher thermal conductivity compared to organic fatty acid or paraffin-based PCMs, the former may destabilize over a number of repeated thermal cycles. That is, the dissociation of the PCM takes place over several cycles leading to relatively low thermal storage abilities on a long-term basis. Thus, they are preferred much less compared to organic PCMs for LTES applications.

Table 8.7 Summary of preparation and characterization methods and instruments for studies of colloidal suspensions utilized as nano-enhanced phase change materials (NePCM) [166]

Authors (year)	Preparation		Thermal Conductivity[b]	Characterization		Study of heat Transfer
	Methods	Dispersion and Stabilization[a]		Other Measurements and Instruments	Stability Concerns	
[182]	N/A	N/A	Maxwell's equation	N/A	N/A	Analytical solutions
[183]	N/A	N/A	N/A	N/A	N/A	Theoretical analysis
[184]	Two-step	N/A	Calculated using measured α and C_p	DSC[c], laser flash technique	N/A	TC[d] readings for a thaw/freeze test; analytical model for predicting thermal conductivity
[185]	N/A	N/A	Maxwell's equation	N/A	N/A	CFD modeling
[186,187]	Two-step	Sonication surfactant: SDBS[e]	N/A	pH, freezing point (ASTM D1177), microscope	Visually observed in transparent beakers	N/A
[188]	One-step	N/A	N/A	DSC, FTIR[f], TEM, TGA, XRD	Two heating cycles	TC readings
[189]	Two-step	Sonication, magnetic force agitation	N/A	DSC, TEM, TGA, XRD	Severe aggregation of titanium oxide	N/A
[190]	Two-step	Sonication	Steady-state guided hot plate method, S-1T	DSC, TEM	Significant precipitation of GNF observed on several melting/freezing cycles	TC readings for melting experiments in a cubic cavity
[191]	Two-step	Sonication	Hot disk thermal constants analyzer (TPS[g]), S-1T	DSC, SEM[h], TGA	N/A	N/A
[192]	Two-step	Ultrasonication	N/A	DSC	N/A	Theoretical model by intermolecular attraction and numerical simulation
[193]	Two-step	Sonication	THW[i], SL-mT	DSC, SEM	Observed by SEM images after thawing/freezing 80 times	N/A

[194]	Two-step	Ultrasonication	Steady-state guarded heat flow meter method, S-1T	DSC, Femtostat/Potentiostat, SEM, TGA	N/A	N/A
[195]	Two-step	Sonication surfactants: CTAB, SDBS	Hot disk thermal constants analyzer (TPS), S-1T	DSC, FTIR	N/A	N/A
[196]	Two-step	Supersonic oscillator	THW, L-mT	TEM	N/A	Freeze/thaw cycling tests using a cold storage/supply system
[197]	Two-step	Ultrasonication surfactant: SDBS	THW, L-1T	Light scattering, zeta potential	Observed through particle size distribution and zeta potential	TC readings and infrared imaging for freeze/thaw tests
[198, 199]	Two-step	Ultrasonication (with unspecified surfactant SINO-POL20)	THW, L-mT	DSC, hydrometer, laser diffraction, viscometer	N/A	Melting experiments in a rectangular cavity
[200]	Two-step	Sonication	THW, SL-mT	DSC, SEM	Being subjected to 701 °C for 96h	N/A
[201]	Two-step	Sonication (with anhydrous ethanol)	Hot disk thermal constants analyzer (TPS), S-1T	DSC, FTIR, SEM, TGA, XRD	N/A	N/A
[202]	Two-step	Sonication	THW, SL-mT	DSC, FTIR, SEM	N/A	N/A
[203]	Two-step	Sonication (with oleylamine)	THW, SL-mT	DSC, FTIR, SEM, TEM, XRD	N/A	N/A
[204]	Two-step	Sonication	THW, SL-mT	FTIR, SEM, TEM, XRD	N/A	N/A
[205]	Two-step	Ultrasonication surfactant: gum arabic	N/A	DSC, TEM	Visually observed	TC readings for a thaw/freeze test
[206]	Two-step	Ultrasonication surfactants: gum arabic, Span-80, CTAB, SDBS, and Hitenol BC-10	THW, SL (no temperature information)	DSC, FTIR	Visually observed, thermal stability test for up to 100 cycles by DSC	TC readings for a thaw/freeze test

Continued

Table 8.7 Summary of preparation and characterization methods and instruments for studies of colloidal suspensions utilized as nano-enhanced phase change materials (NePCM)—cont'd

| Authors (year) | Preparation | | Characterization | | | |
---	Methods	Dispersion and Stabilization[a]	Thermal Conductivity[b]	Other Measurements and Instruments	Stability Concerns	Study of Heat Transfer
[207]	Two-step	Ultrasonication surfactant: SDS[k]	N/A	SEM	N/A	TC readings for freeze tests
[208]	Two-step	Ultrasonication	THW, S-1T	DSC, SEM	N/A	TC readings for heating tests of solids only
[209]	Two-step	Ultrasonication	Steady-state guarded heat flow meter method, S-1T	DSC, Femtostat/Potentiostat, SEM, TGA	N/A	N/A
[210]	Two-step	Sonication	Steady-state method, S-1T	DSC, SEM, TEM	N/A	N/A
[211]	Two-step	Surfactant: SOA[l]	Hot disk thermal constants analyzer (TPS), SL–mT	N/A	N/A	TC readings for freezing tests
[212]	Two-step	Surfactant: SOA	Hot disk thermal constants analyzer (TPS), SL–mT	N/A	N/A	N/A

[a] This column is only applicable to the two-step method.
[b] Under thermal conductivity column, SL–xT stands for S¼ solid phase, L¼ liquid phase, x¼1 for single and m for multiple, T¼ temperature measurements.
[c] DSC stands for differential scanning calorimetry.
[d] TC denotes thermocouples.
[e] SDBS denotes sodium dodecylbenzene sulfonate.
[f] FTIR, TEM, TGA, and XRD stand for Fourier transform infrared spectroscopy, transmission electron microscopy, thermal gravimetric analysis, and X-ray diffraction, respectively.
[g] TPS denotes transient plane source technique.
[h] SEM denotes scanning electron microscope.
[i] THW stands for transient hot wire method.
[j] CTAB denotes cetyltrimethyl ammonium bromide.
[k] SDS denotes sodium dodecyl sulfate.
[l] SOA denotes sodium oleate acid.

Table 8.8 Summary of the nanostructures utilized for thermal conductivity enhancement of phase change materials [166]

Supplier		Dimensions, etc.[a]	Authors (year)
A. Carbon-Based Nanostructures			
Carbon nanofibers (CNF)	Applied Sciences, Cedarville, OH Unknown	$D = 100\,nm$, $L = 20\,\mu m$	[184]
	Pyrograf Products, Inc., Cedarville, OH	$D = 100\,nm$	[192]
		$D = 200\,nm$	[208]
Graphite nanofibers	In-house	$D = 4\text{-}10\,nm$, $L = 1\,\mu m$	[190]
Graphite nanoplatelets	Asbury Graphite Mills, Inc., Asbury, NJ	$D = 15\,\mu m$, to 10 nm, $SSA = 30\,m^2/g$	[194,209]
	Asbury Graphite Mills, Inc., Asbury, NJ	$D = 1\,\mu m$, $t < 10\,nm$, $SSA = 100\text{-}130\,m^2/g$	[209]
Graphene flakes	In-house	N/A	[210]
B. Carbon Nanotubes			
Single-walled nanotubes (SWNT)	Carbon Nanotechnologies Inc. Houston, TX	N/A,	[186,187]
	Unknown	$D = 1\,nm$	[192]
Multiwalled nanotubes (MWNT)	Shenzhen Nanotech Port Co., Shenzhen, China	$D = 10\text{-}30\,nm$, $L = 5\text{-}15\,\mu m$	[191]
		$D = 10\,nm$	[192]
	Unknown	$D = 30\,nm$, $L = 50\,\mu m$, $SSA = 60\,m^2/g$	[193,200,202, 204,208]
	Chengdu Organic Chemicals Co., Ltd., China	$D = 10\text{-}30\,nm$, $L = 5\text{-}15$ & $1\text{-}2\,\mu m$	[195]
	Shenzhen Nanotech Port Co., Shenzhen, China	$D = 10\text{-}30$, $40\text{-}60$, $60\text{-}100\,nm$,	[207]
	Shenzhen Nanotech Port Co. Shenzhen, China	$L = 5\text{-}15\,\mu m$	
C. Nanoparticles			
Alumina (Al_2O_3)	Sigma-Aldrich, St. Louis, MO	N/A	[187]
Magnesium oxide (MgO)	Sigma-Aldrich, St. Louis, MO	N/A	[187]
Silver (Ag)	In-house	$D = 500\,nm$	[188]
Alumina (Al_2O_3)	Unknown	$D = 8\,nm$	[189]
Titanium oxide (TiO_2)	Unknown	$D = 3.8\,nm$	[189]
Titanium oxide (TiO_2)	Unknown	$D = 20\,nm$	[196]
Alumina (Al_2O_3)	Alfa Aesar, Ward Hill, MA	$D = 20\,nm$	[197]
Alumina (Al_2O_3)	Nanotech, Kanto Chemical Co., Inc., Japan	$D = 33\,nm$	[198,199]
Alumina (Al_2O_3)	Hangzhou Jingtian Nanotech Co., Ltd, China	$D = 20\,nm$	[203]

Continued

Table 8.8 Summary of the nanostructures utilized for thermal conductivity enhancement of phase change materials—cont'd

Supplier		Dimensions, etc.[a]	Authors (year)
Copper (Cu)	Shenzhen Junye Nano Material Ltd., China	$D=25\,nm$, $SSA=30\text{–}50\,m^2/g$	[205]
Al	Shenzhen Junye Nano Material Ltd., China	$D=25\,nm$	[206]
C/Cu	Shenzhen Junye Nano Material Ltd., China	$D=25\,nm$	[206]
Copper (Cu)	Shenzhen Junye Nano Material Ltd., China	$D=25\,nm$	[206]
Copper oxide (CuO)	In-house	$D=5\text{–}15\,nm$	[211,212]
D. Nanowires			
Silver (Ag) nanowires	Shenzhen Nanotech Port Co., Shenzhen, China	$D=10\text{–}30\,nm$, $L=5\text{–}15\,\mu m$	[201]

[a] In this column, D, L, SSA, and t stand for diameter, length, specific surface area, and thickness, respectively.

This does not mean that salt hydrate PCMs are incapable of TES applications, but their utilization in LTES systems may have certain technical limitations. On the other hand, salt hydrates are more pronounced in thermochemical energy storage, where they can serve as a good heat storage material depending on how the chemical process is activated.

Similar to salt hydrate PCMs, organic PCMs do have certain discrepancies in their thermal storage capabilities as was discussed in Chapter 5. For some organic PCMs, the condition of supercooling (drop in the temperature with respect to the phase change temperature during freezing) can arise during the phase transition processes. The incorporation of nanomaterials into such PCMs has been identified as an effective approach to counteract the supercooling degree.

Furthermore, the thermally conductive nanoparticles present in the PCM matrix can create a more effective network of thermal interfaces within the PCM. This in turn facilitates faster nucleation and growth kinetics of the stable nucleus (or ice crystals) in the PCM, while subjected to the freezing process. Similarly, during the discharging process swift melting of the ice crystal can take place through phonon heat interactions in the PCM due to the presence of nanoparticles.

Besides, the thermal stability and thermal reliability of PCMs are considered vital factors in determining their thermal storage performance over several repeated thermal cycles. Thermal stability of the PCM refers to the criterion of being characteristically stable or resistant to heat when the PCM is subjected to elevated temperature.

Likewise, thermal reliability of the PCM signifies its nondeteriorated thermal storage performance when undergoing several thousand repeated freezing and melting cycles. By embedding the nanostructured materials into the pure PCM, although there may be some chance for reduction in its latent heat potential, the positive lift shown on acquiring improved thermal storage properties of the PCM is appreciable.

In recent times, dedicated research studies have been performed on embedding hybrid nanocomposite (HyNC) materials into organic ester PCMs to enable them to be considered as potential candidates for cooling applications [214–218].

The organic ester PCMs utilized in these studies include EC, dimethyl adipate (DMA), and methyl cinnamate (MC). These PCMs are derived from the esterification process of naturally occurring cinnamic acid and adipic acid with the corresponding alcohol (ethyl or methyl) for producing the required organic esters as PCMs.

These organic ester compounds find their applications ranging from cosmetic to noncosmetic products. The attributes related to these organic compounds to serve as PCMs have not been explored in the past. Through the active works performed by the research group [214–218], the potential thermal storage properties of these organic ester PCMs embedded with different nanomaterials have been explored. The schematic representation of the preparation and thermophysical characterization of Ag-TiO$_2$ HyNC embedded PCMs (EC and DMA) are shown in Fig. 8.10.

From this perspective, the organic ester PCM embedded with the surface functionalized silver-titania (Ag-TiO$_2$) HyNC exhibited improved thermal conductivity up to 52%, congruent phase change temperature (6.8 °C), high latent heat capacity (90.81 kJ/kg), reduction in supercooling degree (1.82 °C), reduction in freezing (23.9%) and melting times (8.5%), thermal stability (191 °C), and chemical stability, compared to the pure PCM. For the same mass loading of the HyNC considered, the reduction in latent heat potential during freezing and melting, respectively, was ascertained by the increased viscosity of the HyNPCM from 0.35 to 3.89%.

The organic dibasic ester PCM embedded with the Ag-TiO$_2$ HyNC has exhibited improved thermal conductivity up to 58.4%, congruent phase change temperature (~9.6 °C), high latent heat potential (~160.22 kJ/kg), reduction in the supercooling degree (1.71 °C), reduction in freezing (29.9%) and melting times (9.17%), and thermal stability (~270 °C). The upgraded thermal properties have signified the role of HyNC present in the HiNPCM. The reduction in latent heat potential was anticipated by the increased viscous characteristics (2.53%) of the HiNPCM.

The organic ester PCM embedded with the copper-titania (Cu-TiO$_2$) HyNC exhibited improved thermal conductivity (0.347 W/m K), congruent phase change temperature (35.32 °C), high latent heat enthalpy (freezing: 109.05 kJ/kg, melting: 109.14 kJ/kg), and considerable reduction in freezing (21.2%) and melting times (29.2%) by virtue of the incorporation of HyNC.

The reason for the improved thermal conductivity of these HyNC PCMs is attributed to the nano-convection/Brownian motion and well-dispersed aggregation of the nanoparticles at lower and higher concentration of the HyNC, respectively. Furthermore, the surface functionalized and crystalline HyNC facilitates creating a densely packed network of thermal interfaces within PCMs for achieving enhanced phonon heat interactions during phase change processes.

Based on the inferences obtained from these studies, it is suggested that PCMs embedded with nanomaterials exhibit improved thermal properties and potential thermal energy storage characteristics suitable for them to be considered for LTES applications.

8.4 MERITS AND CHALLENGES

From the generalized viewpoint of the utilization potential of nanomaterials to real-world applications, some meritorious aspects and challenging factors are involved that include the following:

- Merits
  - The thermal storage characteristics of the latent heat storage materials can be enhanced through advanced materials being produced at nanoscale level.

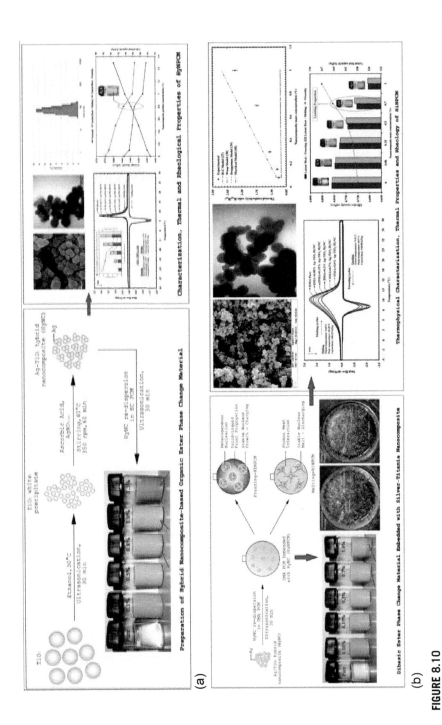

FIGURE 8.10

Preparation and thermophysical characterization of Ag-TiO$_2$ HyNC embedded PCMs (a) ethyl trans-cinnamate (EC), and (b) dimethyl adipate (DMA) [214, 218].

- Optimization of efficient production, absorption, and storage of energy can help to reduce energy consumption through nanotechnology.
- Nanotechnology facilitates the manufacturing of materials of desired physicochemical and thermophysical properties on large scale.
- It contributes to reducing environmental impact by means of the value chain of the energy sector.
- Costs related to energy generation, energy distribution, and energy storage can be reduced without consuming a large quantity of materials for the same expected outcomes.
- Nanoscience and nanotechnology can make energy dependency transitions from the fossil fuels to the renewable energy integration through advanced and energy efficient materials.
- Challenges
  - The synthesized nanomaterials or its by-products, if not treated properly before they are discharged or disposed, can have a significant impact on the environment.
  - Some of the chemical constituents used for the preparation of nanomaterials can be highly hazardous in nature and not easy to handle.
  - Nanomaterials embedded PCMs do possess a large and forefront challenge in terms of agglomeration or aggregation of particles, so that PCMs' suspension stability and thermal energy storage performance can be greatly reduced as well.
  - Waste of materials both at nanoscale level and bulk state can occur due to less proficient knowledge on the preparation of nanomaterials embedded PCMs.
  - Nanomaterials that enter the human system can have adverse effects on the human cycle; this calls for utmost care and safety while using them.
  - More standardization has to be used to characterize the exact enhancement of thermal properties of heat storage materials embedded with nanomaterials.

8.5 CONCISE REMARKS

Successful research on nanoscience and nanotechnology have contributed to the utilization of advanced materials for thermal energy storage. The utilization of nanomaterials for enhancing thermal storage properties of PCMs was discussed, and the key outcomes from a variety of research studies were presented. The nanomaterials dedicated for thermal energy storage can be prepared either through the top-down or bottom-up approach.

In most situations, the bottom-up approach, especially the chemical reduction method, has been largely preferred for the production of nanostructured materials. This is due to the fact that morphology (size and shape), yield efficiency, nucleation, and growth kinetics of nanoparticles can be effectively controlled through chemical reduction methods.

The incorporation of nanomaterials into pure PCM has shown remarkable enhancement of thermophysical properties over that exhibited in its purest form. The main objective of embedding nanomaterials into pure PCM is to achieve improved thermal conductivity, thermal stability, and swift freezing and melting rates of the PCM. The infused nanoparticles can help create a densely packed network of thermal interfaces within the PCM matrix and thereby augment thermal conductivity and heat transfer rate of the PCM.

Similarly, the condition of supercooling can also occur in some organic PCMs, wherein incorporating nanomaterials can reduce the supercooling degree significantly. The presence of thermally conductive nanomaterials in the PCM can induce phonon-like heat transfer between PCM molecules and nanoparticles. By this, the formation of stable nucleus and growth kinetics of the PCM crystals can be enhanced faster during freezing processes.

Likewise, during thawing processes the phonon-like heat transfer enables ice-like crystals to melt at a faster rate. More interestingly, it is suggested that enhancement of thermal conductivity can be expected due to the nanoconvection/Brownian motion occurring at lower concentration of the embedded nanomaterials.

On the other hand, at higher concentration of the nanomaterials, the thermal conductivity enhancement can be due to induced aggregation of the well-dispersed nanoparticles within the PCM. Having the positive aspects of the nanomaterials on one side, the challenges related to their preparation, utilization, discharge, and safety are equally important to be considered for real-time applications.

In total, it is suggested that PCMs embedded with nanomaterials exhibit improved thermal energy storage properties, which allow them to be considered as suitable candidates for future LTES applications.

References

[1] Parameshwaran R, Kalaiselvam S, Harikrishnan S, et al. Sustainable thermal energy storage technologies for buildings: a review. Renew Sust Energy Rev 2012;16:2394–433.

[2] Zäch M, Hägglund C, Chakarov D, Kasemo B. Nanoscience and nanotechnology for advanced energy systems. Curr Opinion Sol State Mater Sci 2006;10:132–43.

[3] Manikam VR, Cheong KY, Razak KA. Chemical reduction methods for synthesizing Ag and Al nanoparticles and their respective nanoalloys. Mater Sci Eng B 2011;176:187–203.

[4] Al-Mubaddel FS, Haider S, Al-Masry WA, et al. Engineered nanostructures: a review of their synthesis, characterization and toxic hazard considerations. Arab J Chem 2012; http://dx.doi.org/10.1016/j.arabjc.2012.09.010.

[5] Royal Society and Royal Academy of Engineering. Nanoscience and nanotechnologies: opportunities and uncertainties. London: The Royal Society; 2004.

[6] Yon JM, Jamie RL. Manufactured nanoparticles: an overview of their chemistry, interactions and potential environmental implications. Sci Total Environ 2008;400:396–414.

[7] Gleiter H. Nanostructured materials: basic concepts and microstructure. Acta Mater 2000;48:1–29.

[8] Pokropivny VV, Skorokhod VV. Classification of nanostructures by dimensionality and concept of surface forms engineering in nanomaterial science. Mater Sci Eng C 2007;27:990–3.

[9] Skorokhod V, Ragulya A, Uvarova I. Physico-chemical kinetics in nanostructured systems. Kyiv: Academperiodica; 2001 p. 180.

[10] Alonso JC, Diamant R, Castillo P, Acosta-García MC, Batina N, Haro-Poniatowski E. Thin films of silver nanoparticles deposited in vacuum by pulsed laser ablation using a YAG:Nd laser. Appl Surf Sci 2009;255:4933–7.

[11] Ferrando R, Jellinek J, Johnston RL. Nanoalloys: from theory to applications of alloy clusters and nanoparticles. Chem Rev 2008;108:845–910.

[12] Hu MZ, Easterly CE. A novel thermal electrochemical synthesis method for production of stable colloids of "naked" metal (Ag) nanocrystals. Mater Sci Eng C 2009;29:726–36.

[13] Khaydarov RA, Khaydarov RR, Gapurova O, Estrin Y, Scheper T. Electrochemical method for the synthesis of silver nanoparticles. J Nanopart Res 2009;11:1193–200.

[14] Jian Z, Xian Z, Yongchang W. Electrochemical synthesis and fluorescence spectrum properties of silver nanospheres. Microelec Eng 2005;77:58–62.

[15] Starowicz M, Stypuła B, Banas J. Electrochemical synthesis of silver nanoparticles. Electrochem Commun 2006;8:227–30.

[16] Ritson DR. The sonoelectrochemical synthesis of silver nanoparticles their applications to SERS. PhD dissertation, University of Liverpool; 2008.

[17] Liu S, Huang W, Chen S, Avivi S, Gedanken A. Synthesis of X-ray amorphous silver nanoparticles by the pulse sonoelectrochemical method. J Non-Cryst Solids 2001;283:231–6.

[18] Salkar RA, Jeevanandam P, Aruna T, Koltypin Y, Gedanken A. The sonochemical preparation of amorphous silver nanoparticles. J Mater Chem 1999;9:1333–5.

[19] Tunc I, Guvenc HO, Sezen H, Suzer S, Correa-Duarte MA, Liz-Marzán LM. Optical response of Ag-Au bimetallic nanoparticles to electron storage in aqueous medium. Nanosci Nanotechnol 2008;8:3003–7.

[20] Guzman MG, Dille J, Godet S. Synthesis of silver nanoparticles by chemical reduction method and their antibacterial activity. World Acad Sci Eng Technol 2008;43:357–64.

[21] Zhang W, Qiao X, Chen J. Synthesis of silver nanoparticles—Effects of concerned parameters in water/oil microemulsion. J Mater Sci Eng B 2007;142:1–15.

[22] Niemeyer CM. Nanoparticles, proteins, and nucleic acids: biotechnology meets materials science. Angew Chem Int Ed 2001;40:4128–58.

[23] Parameshwaran R, Kalaiselvam S, Jayavel R. Green synthesis of silver nanoparticles using Beta vulgaris: role of process conditions on size distribution and surface structure. Mater Chem Phys 2013;140:135–47.

[24] Lin P-C, Lin S, Wang PC, Sridhar R. Techniques for physicochemical characterization of nanomaterials. Biotechnol Adv 2013, http://dx.doi.org/10.1016/j.biotechadv.2013.11.006.

[25] Brar SK, Verma M. Measurement of nanoparticles by light-scattering techniques. TrAC Trends Anal Chem 2011;30:4–17.

[26] Domingos RF, Baalousha MA, Ju-Nam Y, Reid MM, Tufenkji N, Lead JR, et al. Characterizing manufactured nanoparticles in the environment: multimethod determination of particle sizes. Environ Sci Technol 2009;43:7277–84.

[27] Filipe V, Hawe A, Jiskoot W. Critical evaluation of nanoparticle tracking analysis (NTA) by nanosight for the measurement of nanoparticles and protein aggregates. Pharm Res 2010;27:796–810.

[28] Murdock RC, Braydich-Stolle L, Schrand AM, Schlager JJ, Hussain SM. Characterization of nanomaterial dispersion in solution prior to in vitro exposure using dynamic light scattering technique. Toxicol Sci 2008;101:239–53.

[29] Pan G-H, Barras A, Boussekey L, Qu X, Addad A, Boukherroub R. Preparation and characterization of decyl-terminated silicon nanoparticles encapsulated in lipid nanocapsules. Langmuir 2013;29:12688–96.

[30] Sapsford KE, Tyner KM, Dair BJ, Deschamps JR, Medintz IL. Analyzing nanomaterial bioconjugates: a review of current and emerging purification and characterization techniques. Anal Chem 2011;83:4453–88.

[31] Schacher F, Betthausen E, Walther A, Schmalz H, Pergushov DV, Müller AHE. Interpolyelectrolyte complexes of dynamic multicompartment micelles. ACS Nano 2009;3:2095–102.

[32] Wagner AJ, Bleckmann CA, Murdock RC, Schrand AM, Schlager JJ, Hussain SM. Cellular interaction of different forms of aluminum nanoparticles in rat alveolar macrophages. J Phys Chem B 2007;111:7353–9.

[33] Zhao T, Chen K, Gu H. Investigations on the interactions of proteins with polyampholyte-coated magnetite nanoparticles. J Phys Chem B 2013;117:14129–35, http://dx.doi.org/10.1021/jp407157n.

[34] Boukari H, Sackett DL. Fluorescence correlation spectroscopy and its application to the characterization of molecular properties and interactions. Methods Cell Biol 2008;84:659–78.

[35] Jing B, Zhu Y. Disruption of supported lipid bilayers by semihydrophobic nanoparticles. J Am Chem Soc 2011;133:10983–9.

[36] Nienhaus GU, Maffre P, Nienhaus K. Studying the protein corona on nanoparticles by FCS. In: Sergey YT, editor. Methods in enzymology. New York: Academic Press; 2013. p. 115–37.

[37] Choi J, Reipa V, Hitchins VM, Goering PL, Malinauskas RA. Physicochemical characterization and in vitro hemolysis evaluation of silver nanoparticles. Toxicol Sci 2011;123:133–43.

[38] Clogston J, Patri A. Zeta potential measurement. In: McNeil SE, editor. Characterization of nanoparticles intended for drug delivery. New York: Humana Press; 2011. p. 63–70.

[39] Khatun Z, Nurunnabi M, Cho KJ, Lee Y-K. Oral delivery of near-infrared quantum dot loaded micelles for noninvasive biomedical imaging. ACS Appl Mater Interfaces 2012;4:3880–7.

[40] Weiner BB, Tscharnuter WW, Fairhurst D. Zeta potential: a new approach. New York: Brookhaven Instruments Corporation; 1993.

[41] Xu R. Progress in nanoparticles characterization: sizing and zeta potential measurement. Particuology 2008;6:112–5.

[42] Kumar CS. Raman spectroscopy for nanomaterials characterization. Berlin: Springer Verlag; 2012.

[43] Popovic ZV, Dohcevic-Mitrovic Z, Scepanovic M, Grujic-Brojcin M, Askrabic S. Raman scattering on nanomaterials and nanostructures. Ann Phys 2011;523:62–74.

[44] Chang H-W, Hsu P-C, Tsai Y-C. Ag/carbon nanotubes for surface-enhanced Raman scattering. In: Kumar CSR, editor. Raman spectroscopy for nanomaterials characterization. Berlin/Heidelberg: Springer; 2012. p. 119–35.

[45] Kattumenu R, Lee C, Bliznyuk V, Singamaneni S. Micro-Raman spectroscopy of nanostructures. In: Kumar CSR, editor. Raman spectroscopy for nanomaterials characterization. Berlin/Heidelberg: Springer; 2012. p. 417–44.

[46] Kneipp J, Kneipp H, Wittig B, Kneipp K. Novel optical nanosensors for probing and imaging live cells. Nanomed Nanotechnol Biol Med 2010;6:214–26.

[47] Kumar J, Thomas KG. Surface-enhanced Raman spectroscopy: investigations at the nanorod edges and dimer junctions. J Phys Chem Lett 2011;2:610–5.

[48] Mannelli I, Marco MP. Recent advances in analytical and bioanalysis applications of noble metal nanorods. Anal Bioanal Chem 2010;398:2451–69.

[49] Braun GB, Lee SJ, Laurence T, Fera N, Fabris L, Bazan GC, et al. Generalized approach to SERS-active nanomaterials via controlled nanoparticle linking, polymer encapsulation, and small-molecule infusion. J Phys Chem C 2009;113:13622–9.

[50] Lin Z-H, Chang H-T. Preparation of gold–tellurium hybrid nanomaterials for surface-enhanced Raman spectroscopy. Langmuir 2007;24:365–7.

[51] Lucas M, Riedo E. Invited review article: combining scanning probe microscopy with optical spectroscopy for applications in biology and materials science. Rev Sci Instrum 2012;83:061101.

[52] Sinjab F, Lekprasert B, Woolley RAJ, Roberts CJ, Tendler SJB, Notingher I. Near-field Raman spectroscopy of biological nanomaterials by in situ laser-induced synthesis of tip-enhanced Raman spectroscopy tips. Opt Lett 2012;37:2256–8.

[53] Xiao M, Nyagilo J, Arora V, Kulkarni P, Xu D, Sun X, et al. Gold nanotags for combined multi-colored Raman spectroscopy and x-ray computed tomography. Nanotechnology 2010;21:035101.

[54] Cuche A, Masenelli B, Ledoux G, Amans D, Dujardin C, Sonnefraud Y, et al. Fluorescent oxide nanoparticles adapted to active tips for near- field optics. Nanotechnology 2009;20:015603.

[55] Kohli R, Mittal KL. Developments in surface contamination and cleaning-detection, characterization, and analysis of contaminants. Norwich, NY: William Andrew; 2011.

[56] W-f Lin, Li J-R, Liu G-Y. Near-field scanning optical microscopy enables direct observation of moiré effects at the nanometer scale. ACS Nano 2012;6:9141–9.

[57] Park HK, Lim YT, Kim JK, Park HG, Chung BH. Nanoscopic observation of a gold nanoparticle-conjugated protein using near- field scanning optical microscopy. Ultramicroscopy 2008;108:1115–9.

[58] Vancso GJ, Hillborg H, Schönherr H. Chemical composition of polymer surfaces imaged by atomic force microscopy and complementary approaches. In: Polymer analysis polymer theory. Adv Polym Sci, vol. 182. Berlin/Heidelberg: Springer; 2005. p. 55–129.

[59] Caminade A-M, Laurent R, Majoral J-P. Characterization of dendrimers. Adv Drug Deliv Rev 2005;57:2130–46.

[60] Ghosh PS, Han G, Erdogan B, Rosado O, Krovi SA, Rotello VM. Nanoparticles featuring amino acid-functionalized side chains as DNA receptors. Chem Biol Drug Des 2007;70:13–8.

[61] Huang R, Carney RP, Stellacci F, Lau BLT. Protein–nanoparticle interactions: the effects of surface compositional and structural heterogeneity are scale dependent. Nanoscale 2013;5:6928–35.

[62] Jiang X, Jiang J, Jin Y, Wang E, Dong S. Effect of colloidal gold size on the conformational changes of adsorbed cytochromec: probing by circular dichroism, UV–visible, and infrared spectroscopy. Biomacromolecules 2004;6:46–53.

[63] Knoppe S, Dharmaratne AC, Schreiner E, Dass A, Bürgi T. Ligand exchange reactions on Au38 and Au40 clusters: a combined circular dichroism and mass spectrometry study. J Am Chem Soc 2010;132:16783–9.

[64] Kobayashi N, Muranaka A, Mack J. Circular dichroism and magnetic circular dichroism spectroscopy for organic chemists. London: Royal Society of Chemistry; 2011, ISBN: 1847558690.

[65] Liu H, Webster TJ. Nanomedicine for implants: a review of studies and necessary experimental tools. Biomaterials 2007;28:354–69.

[66] Ranjbar B, Gill P. Circular dichroism techniques: biomolecular and nanostructural analyses—a review. Chem Biol Drug Des 2009;74:101–20.

[67] Ratnikova TA, Nedumpully Govindan P, Salonen E, Ke PC. In vitro polymerization of microtubules with a fullerene derivative. ACS Nano 2011;5:6306–14.

[68] Shang L, Wang Y, Jiang J, Dong S. pH-Dependent protein conformational changes in albumin: gold nanoparticle bioconjugates: a spectroscopic study. Langmuir 2007;23:2714–21.

[69] Gmoshinski IV, Khotimchenko SA, Popov VO, Dzantiev BB, Zherdev AV, Demin VF, et al. Nanomaterials and nanotechnologies: methods of analysis and control. Russ Chem Rev 2013;82:48.

[70] Lavigne J-P, Espinal P, Dunyach-Remy C, Messad N, Pantel A, Sotto A. Mass spectrometry: a revolution in clinical microbiology? Clin Chem Lab Med 2013;51:257–70.

[71] Tang Z, Xu B, Wu B, Germann MW, Wang G. Synthesis and structural determination of multidentate 2,3-dithiol-stabilized Au clusters. J Am Chem Soc 2010;132:3367–74.

[72] Tiede K, Boxall ABA, Tear SP, Lewis J, David H, Hassellöv M. Detection and characterization of engineered nanoparticles in food and the environment. Food Addit Contam Part A 2008;25:795–821.

[73] Gun'ko V, Blitz J, Zarko V, Turov V, Pakhlov F, Oranska O, et al. Structural and adsorption characteristics and catalytic activity of titania and titania-containing nanomaterials. J Colloid Interface Sci 2009;330:125–37.

[74] Kane SR, Ashby PD, Pruitt LA. ATR–FTIR as a thickness measurement technique for hydrated polymer-on-polymer coatings. J Biomed Mater Res B Appl Biomater 2009;91:613–20.

[75] Kazarian SG, Chan KL. Applications of ATR–FTIR spectroscopic imaging to biomedical samples. Biochim Biophys Acta 2006;1758:858–67.

[76] Zak AK, Majid W, Darroudi M, Yousefi R. Synthesis and characterization of ZnO nanoparticles prepared in gelatin media. Mater Lett 2011;65:70–3.

[77] Zhao Y, Qiu X, Burda C. The effects of sintering on the photocatalytic activity of N-doped TiO2 nanoparticles. Chem Mater 2008;20:2629–36.

[78] Bernier M-C, Besse M, Vayssade M, Morandat S, El Kirat K. Titanium dioxide nanoparticles disturb the fi bronectin-mediated adhesion and spreading of pre-osteoblastic cells. Langmuir 2012;28:13660–7.

[79] Boguslavsky Y, Fadida T, Talyosef Y, Lellouche J-P. Controlling the wettability properties of polyester fibers using grafted functional nanomaterials. J Mater Chem 2011;21:10304–10.

[80] Bootz A, Vogel V, Schubert D, Kreuter J. Comparison of scanning electron microscopy, dynamic light scattering and analytical ultracentrifugation for the sizing of poly(butyl cyanoacrylate) nanoparticles. Eur J Pharm Biopharm 2004;57:369–75.

[81] Hall JB, Dobrovolskaia MA, Patri AK, McNeil SE. Characterization of nanoparticles for therapeutics. Nanomedicine (London) 2007;2:789–803.

[82] Jin H, Wang N, Xu L, Hou S. Synthesis and conductivity of cerium oxide nanoparticles. Mater Lett 2010;64:1254–6.

[83] Johal MS. Understanding nanomaterials. Boca Raton, FL: CRC Press; 2011.

[84] Ratner BD, Hoffman AS, Schoen FJ, Lemons JE. Biomaterials science: an introduction to materials in medicine. New York: Academic Press; 2004.

[85] Dominguez-Medina S, McDonough S, Swanglap P, Landes CF, Link S. In situ measurement of bovine serum albumin interaction with gold nanospheres. Langmuir 2012;28:9131–9.

[86] Wang ZL. Transmission electron microscopy and spectroscopy of nanoparticles. In: Characterization of nanophase materials. Wiley-VCH Verlag; 2001. p. 37–80.

[87] Williams DB, Carter CB. The transmission electron microscope. In: Transmission electron microscopy. New York: Springer; 2009. p. 3–22.

[88] Fleming CJ, Liu YX, Deng Z, Liu GY. Deformation and hyper fine structures of dendrimers investigated by scanning tunneling microscopy. J Phys Chem A 2009;113:4168–74.

[89] Kocum C, Cimen EK, Piskin E. Imaging of poly(NIPA-co-MAH)-HIgG conjugate with scanning tunneling microscopy. J Biomater Sci Polym Ed 2004;15:1513–20.

[90] Nakaya M, Kuwahara Y, Aono M, Nakayama T. Nanoscale control of reversible chemical reaction between fullerene C60 molecules using scanning tunneling microscope. J Nanosci Nanotechnol 2011;11:2829–35.

[91] Ong QK, Reguera J, Silva PJ, Moglianetti M, Harkness K, Longobardi M, et al. High-resolution scanning tunneling microscopy characterization of mixed monolayer protected gold nanoparticles. ACS Nano 2013;7:8529–39.

[92] Overgaag K, Liljeroth P, Grandidier B, Vanmaekelbergh D. Scanning tunneling spectroscopy of individual PbSe quantum dots and molecular aggregates stabilized in an inert nanocrystal matrix. ACS Nano 2008;2:600–6.

[93] Wang H, Chu PK. Surface characterization of biomaterials. Chapter 4, In: Amit B, Susmita B, editors. Characterization of biomaterials. Oxford: Academic Press; 2013. p. 105–74.

[94] Mavrocordatos D, Pronk W, Boiler M. Analysis of environmental particles by atomic force microscopy, scanning and transmission electron microscopy. Water Sci Technol 2004;50:9–18.

[95] Parot P, Dufrêne YF, Hinterdorfer P, Le Grimellec C, Navajas D, Pellequer J-L, et al. Past, present and future of atomic force microscopy in life sciences and medicine. J Mol Recognit 2007;20:418–31.

[96] Yang PH, Sun XS, Chiu JF, Sun HZ, He QY. Transferrin-mediated gold nanoparticle cellular uptake. Bioconjug Chem 2005;16:494–6.

[97] Lundqvist M, Sethson I, Jonsson B-H. Transient interaction with nanoparticles "freezes" a protein in an ensemble of metastable near-native conformations†. Biochemistry 2005;44:10093–9.

[98] Mullen DG, Fang M, Desai A, Baker JR, Orr BG, Banaszak Holl MM. A quantitative assessment of nanoparticle–ligand distributions: implications for targeted drug and imaging delivery in dendrimer conjugates. ACS Nano 2010;4:657–70.

[99] Tomalia DA, Huang B, Swanson DR, Brothers HM, Klimash JW. Structure control within poly(amidoamine) dendrimers: size, shape and regio-chemical mimicry of globular proteins. Tetrahedron 2003;59:3799–813.

[100] Valentini M, Vaccaro A, Rehor A, Napoli A, Hubbell JA, Tirelli N. Diffusion NMR spectroscopy for the characterization of the size and interactions of colloidal matter: the case of vesicles and nanoparticles. J Am Chem Soc 2004;126:2142–7.

[101] Mirau PA, Naik RR, Gehring P. Structure of peptides on metal oxide surfaces probed by NMR. J Am Chem Soc 2011;133:18243–8.

[102] Zanchet D, Hall BD, Ugarte D. X-ray characterization of nanoparticles. In: Characterization of nanophase materials. Wiley-VCH Verlag; 2001. p. 13–36.

[103] Zhou C, Liu Z, Du X, Mitchell DR, Mai YW, Yan Y, et al. Hollow nitrogen-containing core/shell fibrous carbon nanomaterials as support to platinum nanocatalysts and their TEM tomography study. Nanoscale Res Lett 2012;7:165.

[104] Doniach S. Changes in biomolecular conformation seen by small angle X-ray scattering. Chem Rev 2001;101:1763–78.

[105] Hummer DR, Heaney PJ, Post JE. In situ observations of particle size evolution during the hydrothermal crystallization of TiO2: a time-resolved synchrotron SAXS and WAXS study. J Cryst Growth 2012;344:51–8.

[106] Grosso D, Ribot F, Boissiere C, Sanchez C. Molecular and supramolecular dynamics of hybrid organic–inorganic interfaces for the rational construction of advanced hybrid nanomaterials. Chem Soc Rev 2011;40:829–48.

[107] Rao CNR, Biswas K. Characterization of nanomaterials by physical methods. Annu Rev Anal Chem 2009;435–62 [Palo Alto: Annual Reviews].

[108] Biju V, Mundayoor S, Omkumar RV, Anas A, Ishikawa M. Bioconjugated quantum dots for cancer research: present status, prospects and remaining issues. Biotechnol Adv 2010;28:199–213.

[109] Bootz A, Vogel V, Schubert D, Kreuter J. Comparison of scanning electron microscopy, dynamic light scattering and analytical ultracentrifugation for the sizing of poly(butyl cyanoacrylate) nanoparticles. Eur J Pharm Biopharm 2004;57:369–75.

[110] Braun GB, Lee SJ, Laurence T, Fera N, Fabris L, Bazan GC, et al. Generalized approach to SERS-active nanomaterials via controlled nanoparticle linking, polymer encapsulation, and small-molecule infusion. J Phys Chem C 2009;113:13622–9.

[111] Caminade A-M, Laurent R, Majoral J-P. Characterization of dendrimers. Adv Drug Deliv Rev 2005;57:2130–46.

[112] Domingos RF, Baalousha MA, Ju-Nam Y, Reid MM, Tufenkji N, Lead JR, et al. Characterizing manufactured nanoparticles in the environment: multimethod determination of particle sizes. Environ Sci Technol 2009;43:7277–84.

[113] Hall JB, Dobrovolskaia MA, Patri AK, McNeil SE. Characterization of nanoparticles for therapeutics. Nanomedicine (London) 2007;2:789–803.

[114] Hurst SJ, Lytton-Jean AKR, Mirkin CA. Maximizing DNA loading on a range of gold nanoparticle sizes. Anal Chem 2006;78:8313–8.

[115] Jiang X, Jiang J, Jin Y, Wang E, Dong S. Effect of colloidal gold size on the conformational changes of adsorbed cytochrome c: probing by circular dichroism, UV–visible, and infrared spectroscopy. Biomacromolecules 2004;6:46–53.

[116] Mavrocordatos D, Pronk W, Boiler M. Analysis of environmental particles by atomic force microscopy, scanning and transmission electron microscopy. Water Sci Technol 2004;50:9–18.

[117] Murdock RC, Braydich-Stolle L, Schrand AM, Schlager JJ, Hussain SM. Characterization of nanomaterial dispersion in solution prior to in vitro exposure using dynamic light scattering technique. Toxicol Sci 2008;101:239–53.

[118] Nienhaus GU, Maffre P, Nienhaus K. Studying the protein corona on nanoparticles by FCS. In: Sergey YT, editor. Methods in enzymology. Academic Press; 2013. p. 115–37.

[119] Jiang X, Qu W, Pan D, Ren Y, Williford JM, Cui H, et al. Plasmid-templated shape control of condensed DNA-block copolymer nanoparticles. Adv Mater 2013;25:227–32.

[120] Powers KW, Brown SC, Krishna VB, Wasdo SC, Moudgil BM, Roberts SM. Research strategies for safety evaluation of nanomaterials. Part VI. Characterization of nanoscale particles for toxicological evaluation. Toxicol Sci 2006;90:296–303.

[121] Rao CNR, Biswas K. Characterization of nanomaterials by physical methods. Annu Rev Anal Chem 2009;435–62 [Palo Alto: Annual Reviews].

[122] Sapsford KE, Tyner KM, Dair BJ, Deschamps JR, Medintz IL. Analyzing nanomaterial bio-conjugates: a review of current and emerging purification and characterization techniques. Anal Chem 2011;83:4453–88.

[123] Schacher F, Betthausen E, Walther A, Schmalz H, Pergushov DV, Müller AHE. Interpolyelectrolyte complexes of dynamic multicompartment micelles. ACS Nano 2009;3:2095–102.

[124] Valentini M, Vaccaro A, Rehor A, Napoli A, Hubbell JA, Tirelli N. Diffusion NMR spectroscopy for the characterization of the size and interactions of colloidal matter: the case of vesicles and nanoparticles. J Am Chem Soc 2004;126:2142–7.

[125] Wang H, Chu PK. Surface characterization of biomaterials. Chapter 4, In: Amit B, Susmita B, editors. Characterization of biomaterials. Oxford: Academic Press; 2013. p. 105–74.

[126] Zanchet D, Hall BD, Ugarte D. X-ray characterization of nanoparticles. In: Characterization of nanophase materials. Wiley-VCH Verlag; 2001. p. 13–36.

[127] Choi J, Reipa V, Hitchins VM, Goering PL, Malinauskas RA. Physicochemical characteriza- tion and in vitro hemolysis evaluation of silver nanoparticles. Toxicol Sci 2011;123:133–43.

[128] Sapsford KE, Tyner KM, Dair BJ, Deschamps JR, Medintz IL. Analyzing nanomaterial bio-conjugates: a review of current and emerging purification and characterization techniques. Anal Chem 2011;83:4453–88.

[129] Xu R. Progress in nanoparticles characterization: sizing and zeta potential measurement. Particuology 2008;6:112–5.

[130] Rao CNR, Biswas K. Characterization of nanomaterials by physical methods. Annu RevAnal Chem 2009;435–62 [Palo Alto: Annual Reviews].

[131] Bothun GD. Hydrophobic silver nanoparticles trapped in lipid bilayers: size distribution, bilayer phase behavior, and optical properties. J Nanobiotechnol 2008;6:13.

[132] Gmoshinski IV, Khotimchenko SA, Popov VO, Dzantiev BB, Zherdev AV, Demin VF, et al. Nanomaterials and nanotechnologies: methods of analysis and control. Russ Chem Rev 2013;82:48.

[133] Grosso D, Ribot F, Boissiere C, Sanchez C. Molecular and supramolecular dynamics of hybrid organic–inorganic interfaces for the rational construction of advanced hybrid nanomaterials. Chem Soc Rev 2011;40:829–48.

[134] Gun'ko V, Blitz J, Zarko V, Turov V, Pakhlov E, Oranska O, et al. Structural and adsorptioncharacteristics and catalytic activity of titania and titania-containing nanomaterials. J Colloid Interface Sci 2009;330:125–37.

[135] Mirau PA, Naik RR, Gehring P. Structure of peptides on metal oxide surfaces probed by NMR. J Am Chem Soc 2011;133:18243–8.

[136] Mullen DG, Fang M, Desai A, Baker JR, Orr BG, Banaszak Holl MM. A quantitative assessment of nanoparticle–ligand distributions: implications for targeted drug and imaging delivery in dendrimer conjugates. ACS Nano 2010;4:657–70.

[137] Popovic ZV, Dohcevic-Mitrovic Z, Scepanovic M, Grujic-Brojcin M, Askrabic S. Raman scattering on nanomaterials and nanostructures. Ann Phys 2011;523:62–74.

[138] Tomalia DA, Huang B, Swanson DR, Brothers HM, Klimash JW. Structure control within poly(amidoamine) dendrimers: size, shape and regio-chemical mimicry of globular proteins. Tetrahedron 2003;59:3799–813.

[139] Gmoshinski IV, Khotimchenko SA, Popov VO, Dzantiev BB, Zherdev AV, Demin VF, et al. Nanomaterials and nanotechnologies: methods of analysis and control. Russ Chem Rev 2013;82:48.

[140] Sohaebuddin S, Thevenot P, Baker D, Eaton J, Tang L. Nanomaterial cytotoxicity is composition, size, and cell type dependent. Part Fibre Toxicol 2010;7:22.

[141] das Neves J, Sarmento B, Amiji MM, Bahia MF. Development and validation of a rapidreversed-phase HPLC method for the determination of the non-nucleoside reverse transcriptase inhibitor dapivirine from polymeric nanoparticles. J Pharm Biomed Anal 2010;52:167–72.

[142] Gugulothu D, Patravale V. Stability-indicating HPLC method for arteether and application to nanoparticles of arteether. J Chromatogr Sci 2013, http://dx.doi.org//10.1093/chromsci/bmt125. (in press).

[143] Khatun Z, Nurunnabi M, Cho KJ, Lee Y-K. Oral delivery of near-infrared quantum dot loaded micelles for noninvasive biomedical imaging. ACS Appl Mater Interfaces 2012;4:3880–7.

[144] Patri A, Dobrovolskaia M, Stern S, McNeil S, Amiji M. Preclinical characterization of engineered nanoparticles intended for cancer therapeutics. In: Nanotechnology for cancer therapy. CRC Press; 2006. p. 105–38.

[145] Bernier M-C, Besse M, Vayssade M, Morandat S, El Kirat K. Titanium dioxide nanoparticles disturb the fibronectin-mediated adhesion and spreading of pre-osteoblastic cells. Langmuir 2012;28:13660–7.

[146] Baer DR. Application of surface analysis methods to nanomaterials: summary of ISO/TC201 technical report: ISO 14187:2011 – surface chemical analysis – characterization of nanomaterials. Surf Interface Anal 2012;44:1305–8.

[147] Fujie T, Park JY, Murata A, Estillore NC, Tria MCR, Takeoka S, et al. Hydrodynamic transformation of a freestanding polymer nanosheet induced by a thermoresponsive surface. ACS Appl Mater Interfaces 2009;1:1404–13. .Guay-Bégin A-A, Chevallier P, Faucher L, Turgeon S, Fortin M-A. Surface modification of gadolinium oxide thin films and nanoparticles using poly(ethylene glycol)-phosphate. Langmuir 2011;28:774–82.

[148] Yang L, Watts DJ. Particle surface characteristics may play an important role in phytotoxicity of alumina nanoparticles. Toxicol Lett 2005;158:122–32.

[149] Milani S, Baldelli Bombelli F, Pitek AS, Dawson KA, Rädler J. Reversible versus irreversible binding of transferrin to polystyrene nanoparticles: soft and hard corona. ACS Nano 2012;6:2532–41.

[150] Nienhaus GU, Maffre P, Nienhaus K. Studying the protein corona on nanoparticles by FCS. In: Sergey YT, editor. Methods in enzymology. Academic Press; 2013. p. 115–37.

[151] Akhter S, Ahmad I, Ahmad MZ, Ramazani F, Singh A, Rahman Z, et al. Nanomedicines as cancer therapeutics: current status. Curr Cancer Drug Targets 2013;13:362–78.

[152] Röcker C, Pötzl M, Zhang F, Parak WJ, Nienhaus GU. A quantitative fluorescence study of protein monolayer formation on colloidal nanoparticles. Nat Nanotechnol 2009;4:577–80.

[153] Walczyk D, Bombelli FB, Monopoli MP, Lynch I, Dawson KA. What the cell "sees" inbionanoscience. J Am Chem Soc 2010;132:5761–8.

[154] Cedervall T, Lynch I, Lindman S, Berggård T, Thulin E, Nilsson H, et al. Understanding the nanoparticle–protein corona using methods to quantify exchange rates and affinities of proteins for nanoparticles. Proc Natl Acad Sci 2007;104:2050–5.

[155] Kapralov AA, Feng WH, Amoscato AA, Yanamala N, Balasubramanian K, Winnica DE, et al. Adsorption of surfactant lipids by single-walled carbon nanotubes in mouse lung upon pharyngeal aspiration. ACS Nano 2012;6:4147–56.

[156] Mahmoudi M, Lynch I, Ejtehadi MR, Monopoli MP, Bombelli FB, Laurent S. Protein–nano-particle interactions: opportunities and challenges. Chem Rev 2011;111:5610–37.

[157] Sacchetti C, Motamedchaboki K, Magrini A, Palmieri G, Mattei M, Bernardini S, et al. Sur-face polyethylene glycol conformation influences the protein corona of polyethylene glycol-modified single-walled carbon nanotubes: potential implications on biological performance. ACS Nano 2013;7:1974–89.

[158] Gebauer JS, Malissek M, Simon S, Knauer SK, Maskos M, Stauber RH, et al. Impact of the nanoparticle–protein corona on colloidal stability and protein structure. Langmuir 2012;28:9673–9.

[159] Laera S, Ceccone G, Rossi F, Gilliland D, Hussain R, Siligardi G, et al. Measuring protein structure and stability of protein–nanoparticle systems with synchrotron radiation circular dichroism. Nano Lett 2011;11:4480–4.

[160] Casals E, Pfaller T, Duschl A, Oostingh GJ, Puntes V. Time evolution of the nanoparticle protein corona. ACS Nano 2010;4:3623–32.

[161] Liu W, Rose J, Plantevin S, Auffan M, Bottero J-Y, Vidaud C. Protein corona formation for nanomaterials and proteins of a similar size: hard or soft corona? Nanoscale 2013;5:1658–68.

[162] Tassa C, Duffner JL, Lewis TA, Weissleder R, Schreiber SL, Koehler AN, et al. Binding affinity and kinetic analysis of targeted small molecule-modified nanoparticles. Bioconjug Chem 2009;21:14–9.

[163] Zhao T, Chen K, Gu H. Investigations on the interactions of proteins with polyampholyte-coated magnetite nanoparticles. J Phys Chem B 2013;117:14129–35, http://dx.doi.org/10.1021/jp407157n; .

[164] Rahman M, Laurent S, Tawil N, Yahia LH, Mahmoudi M. Analytical methods for corona evaluations. In: Protein–nanoparticle interactions. Berlin/Heidelberg: Springer; 2013. p. 65–82.

[165] Nguyen T-D. Portraits of colloidal hybrid nanostructures: controlled synthesis and potential applications. Colloid Surf B 2013;103:326–44.

[166] Khodadadi JM, Fan L, Babaei H. Thermal conductivity enhancement of nanostructure-based colloidal suspensions utilized as phase change materials for thermal energy storage: a review. Renew Sustain Energy Rev 2013;24:418–44.

[167] Fang Y, Kuang S, Gao X, Zhang Z. Preparation and characterization of novel nanoencapsulated phase change materials. Energy Conv Manage 2008;49:3704–7.

[168] Fang Y, Kuang S, Gao X, Zhang Z. Preparation of nanoencapsulated phase change material as latent functionally thermal fluid. J Phys D 2009;42:1–8.

[169] Ai D, Su L, Gao Z, Deng C, Dai X. Study of ZrO_2 nanopowders based stearic acid phase change materials. Particuology 2010;8:394–7.

[170] Song G, Ma S, Tang G, Yin Z, Wang X. Preparation and characterization of flame retardant form-stable phase change materials composed by EPDM, paraffin and nano magnesium hydroxide. Energy 2010;35:2179–83.

[171] Li MG, Zhang Y, Xu YH, Zhang D. Effect of different amounts of surfactant on characteristics of nanoencapsulated phase-change materials. Poly Bull 2011;67:541–52.

[172] Sebti SS, Khalilarya SH, Mirzaee I, Hosseinizadeh SF, Kashani S, Abdollahzadeh M. A numerical investigation of solidification in horizontal concentric annuli filed with nano-enhanced phase change material (NEPCM). World Appl Sci J 2011;13:9–15.

[173] Sanusi O, Warzoha R, Fleischer AS. Energy storage and solidification of paraffin phase change material embedded with graphite nanofibers. Int J Heat Mass Trans 2011;54:4429–36.

[174] Yavari F, Fard HR, Pashayi K, Rafiee MA, Zamiri A, Yu Z, Ozisik R, Tasciuc TB, Koratkar N. Enhanced thermal conductivity in a nanostructured phase change composite due to low concentration graphene additives. J Phys Chem C 2011;115:8753–8.

[175] Parameshwaran R, Jayavel R, Kalaiselvam S. Study on thermal properties of organic ester phase-change material embedded with silver nanoparticles. J Therm Anal Calorim 2013;114:845–58.

[176] Torgal FP, Jalali S. Eco-efficient construction and building materials. 1st ed. London: Springer; 2011.

[177] Torgal FP. Eco-efficient construction and building materials research under the EU Framework Programme Horizon 2020. Cons Build Mater 2014;51:151–62.

[178] He Y, Jin Y, Chen H, Ding Y, Cang D, Lu H. Heat transfer and flow behaviour of aqueous suspensions of TiO_2 nanoparticles (nanofluids) flowing upward through a vertical pipe'. Int J Heat Mass Transfer 2007;50:2272–81.

[179] Meng X, Zhang H, Sun L, Xu F, Jiao Q, Zhao Z, Zhang J, Zhou H, Sawada Y, Liu Y. Preparation and thermal properties of fatty acids/CNTs composite as shape-stabilized phase change materials. J Therm Anal Calorim 2013;111:377–84.

[180] Pielichowska K, Pielichowski K. Crystallization behaviour of PEO with carbon-based nanonucleants for thermal energy storage. Thermochim Acta 2010;510:173–84.

[181] Yu S, Jeong S-G, Chung O, Kim S. Bio-based PCM/carbon nanomaterials composites with enhanced thermal conductivity. Sol Energy Mater Sol C 2014;120:549–54.

[182] Siegel R. Solidification of low conductivity material containing dispersed high conductivity particles. Int J Heat Mass Transfer 1977;20:1087–9.

[183] Seeniraj RV, Velraj R, Narasimhan NL. Heat transfer enhancement study of a LHTS unit containing dispersed high conductivity particles. J Sol Energy Eng 2002;124:243–9.

[184] Elgafy A, Lafdi K. Effect of carbon nanofiber additives on thermal behavior of phase change materials. Carbon 2005;43:3067–74.

[185] Khodadadi JM, Hosseinizadeh SF. Nanoparticle-enhanced phase change materials (NEPCM) with great potential for improved thermal energy storage. Int J Heat Mass Transfer 2007;34:534–43.

[186] Hong H, Wensel J, Peterson S, Roy W. Efficiently lowering the freezing point in heat transfer coolants using carbon nanotubes. J Thermophys Heat Transfer 2007;21:446–8.

[187] Hong H, Zheng Y, Roy W. Nanomaterials for efficiently lowering the freezing point of anti-freeze coolants. J Nanosci Nanotechnol 2007;7:1–5.

[188] Zeng JL, Sun LX, Xu F, Tan ZC, Zhang ZH, Zhang J, et al. Study of a PCM based energy storage system containing Ag nanoparticles. J Therm Anal Calorim 2007;87:369–73.

[189] Xie H, Wan J, Chen L. Effects on the phase transformation temperature of nanofluids by the nanoparticles. J Mater Sci Technol 2008;25:742–4.

[190] Weinstein RD, Kopec TC, Fleischer AS, D'Addio E, Bessel CA. The experimental exploration of embedding phase change materials with graphite nanofibers for the thermal management of electronics. J Heat Transfer 2008;130(042405):8.

[191] Zeng JL, Liu YY, Cao ZX, Zhang J, Zhang ZH, Sun XL, et al. Thermal conductivityenhancement of MWNTS on the PANI/tetradecanol form-stable PCM. J Therm Anal Calorim 2008;91:443–6.

[192] Shaikh S, Lafdi K, Hallinan K. Carbon nanoadditives to enhance latent energy storage of phase change materials. J Appl Phys 2008;103(094302):6.

[193] Wang J, Xie H, Xin Z. Thermal properties of heat storage composites containing multiwalled carbon nanotubes. J Appl Phys 2008;104(113537):5.

[194] Kim S, Drzal LT. High latent heat storage and high thermal conductive phase change materials using exfoliated graphite nanoplatelets. Sol Energ Mat Sol C 2009;93:136–42.

[195] Zeng JL, Cao Z, Yang DW, Xu F, Sun LX, Zhang XF, et al. Effects of MWNTS on phase change enthalpy and thermal conductivity of a solid–liquid organic PCM. J Therm Anal Calorim 2009;95:507–12.

[196] Liu Y-D, Zhou Y-G, Tong M-W, Zhou X-S. Experimental study of thermal conductivity and phase change performance of nanofluids PCMs. Microfluid Nanofluidics 2009;7:579–84.

[197] Wu S, Zhu D, Li X, Li H, Lei J. Thermal energy storage behavior of Al_2O_3–H_2O nanofluids. Thermochimica Acta 2009;483:73–7.

[198] Ho CJ, Gao JY. Preparation and thermophysical properties of nanoparticle-inparaffin emulsion as phase change material. Int Commun Heat Mass Transfer 2009;36:467–70.

[199] Gao JY. An experimental study on melting heat transfer behavior of a phasechange-material containing Al_2O_3 nanoparticles in a vertical rectangular enclosure. MS thesis, Taiwan: National Cheng Kung University; 2008, 86 pages (available online at ⟨http://ethesys.lib.ncku.edu.tw/ETD-db/ETD-search/view_etd?URN= etd-0825108-153106⟩).

[200] Wang J, Xie H, Xin Z. Thermal properties of paraffin based composites containing multi-walled carbon nanotubes. Thermochim Acta 2009;488:39–42.

[201] Zeng JL, Cao Z, Yang DW, Sun LX, Zhang L. Thermal conductivity enhancement of Ag nanowires on an organic phase change material. J Therm Anal Calorim 2010;101:385–9.

[202] Wang J, Xie H, Xin Z, Li Y, Chen L. Enhancing thermal conductivity of palmitic acid based phase change materials with carbon nanotubes as fillers. Sol Energ 2010;84:339–44.

[203] Wang J, Xie H, Li Y, Xin Z. PW based phase change nanocomposites containing γ-Al2O3. J Therm Anal Calorim 2010;102:709–13.

[204] Wang J, Xie H, Xin Z, Li Y. Increasing the thermal conductivity of palmitic acid by the addition of carbon nanotubes. Carbon 2010;48:3979–86.

[205] Wang N, Yang S, Zhu D, Ju X. Preparation and heat transfer behavior of paraffin based composites containing nano-copper particles. In: Proceedings of the seventh international conference on multiphase flow, Tampa, FL; 2010, 4 pages.

[206] Wu S, Zhu D, Zhang X, Huang J. Preparation and melting/freezing characteristics of Cu/paraffin nanofluid as phase-change material (PCM). Energ Fuel 2010;24:1894–8.

[207] Mo S, Chen Y, Yang J, Luo X. Experimental study on solidification behavior of carbon nanotube nanofluid. Adv Mat Res 2011;171–172:333–6.

[208] Cui Y, Liu C, Hu S, Yu X. The experimental exploration of carbon nanofiber and carbon nanotube additives on thermal behavior of phase change materials. Sol Energ Mat Sol C 2011;95:1208–12.

[209] Xiang J, Drzal LT. Investigation of exfoliated graphite nanoplatelets (xGnP) in improving thermal conductivity of paraffin wax-based phase change material. Sol Energ Mat Sol C 2011;95:1811–8.

[210] Yavari F, Raeisi Fard H, Pashayi K, Rafiee MA, Zamiri A, Yu Z, et al. Enhanced thermal conductivity in a nanostructured phase change composite due to low concentration graphene additives. J Phys Chem C 2011;115:8753–8.

[211] Fan L, Khodadadi JM. An experimental investigation of enhanced thermal conductivity and expedited unidirectional freezing of cyclohexane-based nanoparticle suspensions utilized as nano-enhanced phase change materials (NePCM). Int J Therm Sci 2012;62:120–6.

[212] Fan L, Khodadadi JM. Temperature-dependent thermal conductivity of eicosane-based phase change materials with copper oxide nanoparticles. In: International symposium on thermal and materials nanoscience and nanotechnology, Antalya, Turkey; 2011, 8 pages.

[213] Kalaiselvam S, Parameshwaran R, Harikrishnan S. Analytical and experimental investigations of nanoparticles embedded phase change materials for cooling application in modern buildings. Renew Energy 2012;39:375–87.

[214] Parameshwaran R, Deepak K, Saravanan R, Kalaiselvam S. Preparation, thermal and rheological properties of hybrid nanocomposite phase change material for thermal energy storage. Appl Energ 2014;115:320–30.

[215] Parameshwaran R, Kalaiselvam S. Energy conservative air conditioning system using silver nano-based PCM thermal storage for modern buildings. Energ Buildings 2014;69:202–12.

[216] Parameshwaran R, Kalaiselvam S. Energy efficient hybrid nanocomposite-based cool thermal storage air conditioning system for sustainable buildings. Energy 2013;59:194–214.

[217] Parameshwaran R, Kalaiselvam S. Effect of aggregation on thermal conductivity and heat transfer in hybrid nanocomposite phase change colloidal suspensions. Appl Phys Lett 2013;103(193113):1–7.

[218] Parameshwaran R, Dhamodharan P, Kalaiselvam S. Study on thermal storage properties of hybrid nanocomposite-dibasic ester as phase change material. Thermochim Acta 2013;573:106–20.

Sustainable Thermal Energy Storage

9.1 INTRODUCTION

Energy is well recognized as the key catalyst for all human activities and plays a vital role in the economic and societal developments of a country. Energy, which is being produced, has to be supplied to the end user or demand sectors with minimum exergy losses. Since the energy crisis in early the 1970s and 1980s, concerns and challenges related to the extensive consumption of energy resources, greenhouse gas (GHG) emissions, and climate change have grown. It is of prime importance to bridge the gap between energy supply and energy demand, which would yield better economic and societal prospects.

In this context, thermal energy storage (TES) integration seems to be one of the best options for establishing a balance between energy supply and energy demand. The peak load shaving capabilities gained through off-peak charging of TES can benefit attainment of cost-energy savings without compromising energy efficiency. The integration of TES with conventional cooling/heating systems, especially in buildings, is more attractive in the sense that the oversizing of the system due to design safety factors can be minimized or eliminated. Furthermore, the amalgamation of renewable energy with TES can still augment the cost-energy savings potential, which helps facilitate the development of energy-efficient and sustainable buildings.

9.2 SUSTAINABLE THERMAL STORAGE SYSTEMS

9.2.1 Low energy thermal storage

The concept of low energy is a viable means for establishing the energy-efficient operation of systems and components toward achieving energy conservative potential in the long run. The equipment or working systems designed with the low energy concept can be expected to consume less energy while doing the same work.

In other words, low energy systems can perform efficiently by utilizing the available energy with minimum exergy losses. From the perspective of sustainable energy, low energy systems play a major role in acquiring energy savings potential without compromising energy efficiency.

It is pertinent to note that among energy consuming sectors, one of the major sectors is buildings, which accounts for one-quarter to one-third of overall energy generation worldwide [1]. The distinction between the quality of energy usage in conventional and low energy buildings is shown in Fig. 9.1. Although, several energy conservation schemes are available for reducing building energy needs, the systems developed depend one way or another on primary energy sources. In this context, the application of TES concepts in buildings can help reduce energy consumption as well as provide a positive lift toward conserving energy for a sustainable future.

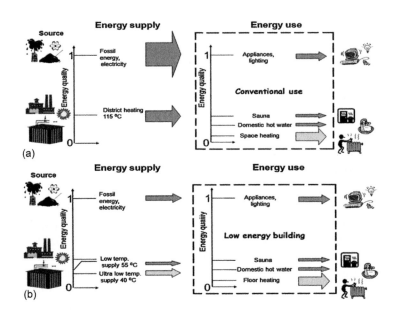

FIGURE 9.1

Schematic of energy quality of (a) fossil energy supply and its energy use at the conventional use of a building and (b) fossil and low temperature energy supply and its energy use at a low temperature building [2–5].

Low energy thermal storage is an integral concept for reducing energy consumption (especially in buildings) by bridging the gap between energy supply and energy demand. This concept has developed over the years, and due to technological developments and availability of advanced materials, TES capacities and energy redistribution capabilities have improved substantially.

The most commonly preferred types of low energy thermal storage include free cooling and building fabric storage concepts. As these concepts have been discussed in earlier chapters, for brevity a summary of research studies pertaining to free cooling using phase change material (PCM) thermal storage and hollow core technologies in buildings are presented in Tables 9.1 and 9.2, respectively.

9.2.2 Low carbon thermal storage

The term *low carbon* generally refers to the minimum output of GHG emissions into the environment. This terminology, which is well known in the energy sector in modern times, is expressed in different forms by different authors. To the best of knowledge of the authors of this book, no specific definition has been framed for revealing the meaning of low carbon. However, this terminology has a wide meaning depending on where it is used, and it has a vital role to play in the present energy scenario as well as for the development of a sustainable future.

For instance, the low carbon technologies (LCTs) literally pave the way for achieving relatively lower concentration of GHG emissions (especially CO_2 emissions) into the environment compared to conventional fossil fuel-based technologies. In a similar way, low carbon thermal storage expresses the same meaning, where the integration of TES with the existing cooling/heating system can help reduce GHG emissions considerably.

Table 9.1 Summary of Recent Theoretical and Experimental Studies Based on Free Cooling of Buildings [6]

| Reference | Type of Study | | For Climatic Condition | PCM Properties | | Heat Exchanger | Important Results/Findings |
	Theoretical Study	Experimental Study		Type	Melting Point (°C)		
[7]	Experimental work based on the outcomes of [10]		Lab scaled experiments	Inorganic PCM and organic PCM	25.0-27.0	Vertical duct type. Aluminum pouches and aluminum panels were used as PCM containers	• Aluminum panel encapsulation of PCM is more suitable for free cooling purpose. • Efforts should be made to design efficient heat exchangers instead of enhancing the PCM thermal conductivity. • Thermal conductivity enhancement of PCM puts an additional increase in the cost of the PCM.
[8]	✓	✓	UK	$Na_2SO_4 \cdot 10H_2O$ (salt hydrate)	22.0	Heat pipes embedded in PCM storage unit	• High temperature difference between PCM melting point and charging air will be beneficial to freeze and melt PCM in the required time period, otherwise high air flow rates will be needed to solidify the PCM completely in the required time span.
[9]	✓	✓	China	Fatty acid	24.5-25.5	PCM packed bed storage	• Night ventilation technique coupled with PCM storage can increase the comfort level of the buildings during daytime as 300 W cold was discharged from PCM to the living room. COP of the night ventilation with PCM packed bed storage (NVP) was found 80.

Continued

Table 9.1 Summary of Recent Theoretical and Experimental Studies Based on Free Cooling of Buildings—cont'd

| Reference | Type of Study | | For Climatic Condition | PCM Properties | | Heat Exchanger | Important Results/Findings |
	Theoretical Study	Experimental Study		Type	Melting Point (°C)		
[10]		✓	Laboratory experiment	RT25 paraffin	25.0	Flat plate heat exchanger	• Thickness of PCM slabs plays a vital during solidification process during night time.
[11]	✓	✓	Lab scale experiments	RT20 paraffin	22.0	PCM in finned rectangular container	• For small spaces the cooling load is lower so air flow rate will also be lower. • For large spaces the cooling load will be higher, and higher air flow rate will be needed. • Connecting cold storages in parallel will fulfill the required cooling load.
[12]	✓	✓	Japan	Granule PCM (GR)	22.5–25.0	PCM packed bed storage	• Climatic data should be considered for the selection of Phase change temperature of PCM. • Ventilation load can be reduced from 46% to 62%, using PCM storage unit in different cities of Japan.
[13]	✓	✓	Japan	Flocculated microcapsules PCM (paraffin)	20.0–23.0	PCM packed bed storage	• Stored cold during in PCM nighttime can be used to achieve a cooling load reduction of 92% during following daytime.

Continued

Table 9.1 Summary of Recent Theoretical and Experimental Studies Based on Free Cooling of Buildings—cont'd

Reference	Type of Study		For Climatic Condition	PCM Properties		Heat Exchanger	Important Results/Findings
	Theoretical Study	Experimental Study		Type	Melting Point (°C)		
[14]	✓	✓	European climate	RT20 paraffin	20.0	Packed bed model (cylindrical storage unit filled with PCM spheres)	• PCM melting temperature should be equal to the average temperature of the hottest summer month. • Air flow rate during charging of PCM should be at least three times higher than discharging air flow rate.
[15]	✓	✓	European climate	RT20–26 paraffin	20.0–26.0	Packed bed model (cylindrical storage unit filled with PCM spheres)	• For efficient performance the PCM melting temperature should be in the range of 72°C from the operating temperature.
[16]	✓	✓	European climate	RT20 paraffin	20.0	Packed bed model (cylindrical storage unit filled with PCM spheres)	• 6.4 kg of PCM per square meter of the floor area was found optimum for the selected location. • Comfort temperature of the building was kept between 25 and 26°C.
[17]	✓		Dry and hot climatic conditions	SP27	27–29	Flat plate heat exchanger	• When melting point of the storage materia is equal to the comfort temperature of the hottest summer month, performance of the storage unit in terms of cooling capacity is maximized for the whole summer season. • The performance of the storage unit is more sensitive to phase change temperatures of the PCM as compared to air flow rates.

Table 9.2 Summary of Recent Theoretical and Experimental Studies Based on Free Cooling of Buildings [18]

Year	Authors	Technologies or Models	Pros and Cons	Applications	Remarks
1970s	[19]	Black box model	Dealed active hollow core slab as a complete "black box" unit, interaction within slab was not modeled	Hollow core slab could be modeled	This model is used in a simulation
1985	[20]	Finite element model	Ignored the heat transfer in the concrete mass in the direction of the air flow, and concrete section was split up in smaller section with adiabatic boundaries	Calculate the complicated temperature in the slab	This model is two-dimensional and steady
1985	[21]	Finite element model	Simplified the above model [20]	Calculate the outlet air temperature of the active core slab	This model is one-dimensional and steady
1988	[22]	Finite difference model	Simplified as two parallel plates with air passing between them, and a constant heat transfer coefficient along the air path was assumed	Study the thermal performance of an active hollow core concrete slab	This model is two-dimensional and steady
1996	[23]	RC model	Considered the thermal resistance and the heat capacity, and simplified the slab as two parallel plates with air passing between them	Calculate the indoor space temperature	This model is dynamic and simplified thermal network model
1996	[24]	Mathematical model	Mass transfer through a three-layer dynamic building element has been calculated	Enable the designer to explore the potential of possible "breathing" envelope constructions	This model is one-dimensional and steady
1997	[25]	CFD model	This package was based on "PHOENICS" equation solver to solve the dynamic problem	Evaluate the heat transfer behaviors of an active hollow core slab	This model is dynamic
1997	[26]	CFD model	Same as [25]	Investigate thermal behaviors of the active hollow core slab at different boundary conditions	This model is dynamic
1997	[27]	A multinode model	Incorporated the multinode model into ESP-r for easy and practical application of the active hollow core slab	Simulate an active hollow core slab of any length, with any flow rate in a full building model	This is a simplified model

Continued

Table 9.2 Summary of Recent Theoretical and Experimental Studies Based on Free Cooling of Buildings—cont'd

Year	Authors	Technologies or Models	Pros and Cons	Applications	Remarks
1998	[28]	RC model	Considered the thermal resistance and the heat capacity, and the heat flow from one surface to the other has not been included in the model It has been validated by the experiment measurement	Predict the mass temperature in the slab	This mode is a dynamic and simplified thermal network model
2000	[29]	Finite element model	The heat conduction in concrete mass in the direction of the air flow was negligible; the heat conduction between the cores was negligible; the model has been validated by the experiment	Investigate the thermal performance of ventilated hollow core slab	This model is one-dimensional, simplified, and steady
2001	[30]	Numerical simulation technique	Did not consider the heat transfer in the slab mass along the slab core and the air temperature decrease along the air passage in the core	Describe temperature, heat flux, or convection with the local ambient environment	This model is dynamic and two-dimensional
2002	[31]	Finite difference model	Ignored the influence of adjacent cores, assumed the temperatures to be constant in the upper and lower room spaces, and air flows through a continuous straight tube; the model was validated	Study on thermal performance of the active hollow core slab	This model is dynamic and two-dimensional
2002	[32]	Finite element method	The flow in the air gap was assumed steady, and the equations for conservation of mass and energy were given in terms of specific enthalpy	Calculate the surface temperature and air temperature for each step length along the air flow	This model is simplified one-dimensional steady model
2003	[33]	Mathematical model	Analyzed the heat transfer process of air and slab, and heat transfer coefficients were assumed to be constant along the length of the duct	Suitable for design applications by using ventilated facades, the heat quantity took away by the airflow has been calculated	This model is simplified one-dimensional steady model
2005	[34]	Thermal network model	This model was validated based on detailed in situ measurements	Describe the heat transfer in the active hollow core slab	This model is simplified dynamic thermal model

Continued

Table 9.2 Summary of Recent Theoretical and Experimental Studies Based on Free Cooling of Buildings—cont'd

Year	Authors	Technologies or Models	Pros and Cons	Applications	Remarks
2007	[35]	Finite difference model	The slab was expressed in typical differential Formulation; the ventilation system and the room were modeled with the finite difference	Investigate the thermal performance of active hollow core slabs	This model is dynamic
2011	[36]	CFD techniques	Considered solar radiation in the air gap of the facade	Analyze temperature and heat flux distributions, and energy consumption	The models are three-dimensional, steady-state/quasi-steady-state
2012	[37]	Finite element method	Considered the convection and radiation, negligible the heat transfer in the slab mass along the gap	Describe the heat transfer process and the internal temperature distribution characteristics	This model is dynamic one-dimensional
2013	[38]	A computational model	The developed model showed agreement with an existing model	Investigate its implementation in EnergyPlus	Dynamic simulation

From the sustainability point of view, the integration of TES facilitates accomplishing energy redistribution needs from on-peak to off-peak load periods, thereby reducing the dependency on primary energy sources and CO_2 emissions as well. The summary of the CO_2 savings potential, the local/location factors, and cost considerations of some major LCTs [39] are listed in Table 9.3.

The description of solar thermal-based systems, wind-TES, ground coupled (source) heat pumps (GSHPs), and the combined heat and power (CHP) system are discussed in forthcoming sections. LCTs referring to cooling using underground water and a district cooling/heating facility have already been demonstrated in other chapters.

Because solar photovoltaics and biomass systems are not within the scope of this book, descriptions pertaining to these systems have not been included. From Table 9.3, it can be clearly observed that the prescribed LCTs have proven to be beneficial in terms of reducing GHG emissions into the environment.

Solar thermal energy-based active and passive systems are known for their potential energy savings opportunities and GHG emissions reduction. Solar energy, which is renewable but intermittent, can contribute to acquiring peak load shifting of energy demand through the integration of a TES system. The principal element of a TES system is that the heat storage materials have the capability to absorb and release thermal energy depending on the energy demand.

The heat from solar radiation can be effectively captured and stored using the collector and heat storage materials. The stored heat energy can then be reutilized or discharged whenever the demand arises. For instance, in building space heating applications, the heat energy trapped from solar radiation can be stored using sensible or latent heat storage (LHS) materials. The stored heat energy can be

Table 9.3 Summary of Potential Low Carbon Technologies

Low Carbon Technology	Potential Savings of CO_2	Local/Location Factors	Cost Consideration
Solar thermal energy-based systems	High	High	High-to-medium
Solar photovoltaics (PV)	Medium	Medium	Medium-to-low
Wind power systems	Medium	Medium	Medium
Ground coupled (source) heat pumps	High	Medium	Medium
Biomass boiler	Medium-to-low	Medium	Medium-to-low
Cooling using underground water	Based on type of building/ infrastructure	Based on type of building/ infrastructure	Medium
District cooling and heating facility	Medium	High-to-medium	Medium
Combined heat and power			
Biomass-based	Medium	Medium-to-low	High-to-medium
Gas-based	Medium	Medium	Medium

High: 😊, High-to-medium: 🙂, Medium: 😐, Medium-to-low: 😕, Low: 😞.

discharged from the heat storage materials and fed to heating elements located in indoor spaces. The transfer of thermal energy from and to the heat storage materials can be effectively achieved by means of heat transfer fluid (either air or water). TES capacity, charging and discharging rates, and stability of the heat storage materials depend on their thermophysical properties, type of configuration, packing density, and so on. The operating strategies of low and medium temperature solar TES systems and materials have already been covered in earlier chapters.

In high temperature solar thermal applications, concentrating solar power (CSP) plants utilizing TES systems are particularly attractive because of their energy generation capacity and highly dispatchable abilities with conventional systems. Modern scientific and technological advancements in materials research have paved the way for accomplishing integration of TES systems with CSP plants. This not only improves the energy efficiency of power plants, but also helps to maintain environmental sustainability. The storage of heat energy through TES can help drive the heat engine at a later period when power demand arises.

The prime requirements for integration of TES systems for CSP plants are prescribed in Table 9.4. For effective functioning of a CSP plant integrated with a TES system, the important three levels of criteria to be taken into consideration include the following:

- Plant level
- Component level
- System level

The *plant level* criterion involves the design considerations and the overall operating strategies of CSP plants, the integration of the TES system into the CSP plant, and its compatibility with the other utilities of the power plant. In the *component level* strategy, the design of the TES system starting from selection to performance evaluation of the heat storage materials is considered. The design phase contains the parametric analysis of heat storage materials in terms of heat storage capacity, heat transfer rate between the heat transfer fluid (HTF) and the heat storage material, heat transfer enhancement, and so on.

Table 9.4 Key Requirements for Developing TES Systems for Concentrating Solar Power (CSP) Plants [40]

1	High energy density capability of storage material
2	Efficient heat transfer between the storage material and HTF provided by properly designed heat exchange equipment.
3	Fast response to load changes in the discharge mode
4	Low chemical activity of storage material and HTF toward the materials of construction
5	Good chemical stability of storage material/HTF and temperature reversibility in a large number of thermal charge/discharge cycles comparable to a life span of the power plant, 30 years
6	High thermal efficiency and low parasitic electric power for the system
7	Low potential contamination of the environment caused by an accidental spill of large amounts of chemicals used in the TES system
8	Low cost of storage material, taking into account the embodied energy (carbon)
9	Ease of operation and low operational and maintenance costs
10	Feasibility of scaling up TES designs to provide at least 10 full load operation hours for large-scale solar power plants of 50 MW electrical generation capacity and larger

The *system level* design considers the amalgamation of essential components such as heat exchangers, storage tanks, control modules for regulating the charging and discharging processes, and pumps for HTF circulation. This is an essential criterion that signifies the improvement of plant efficiency, reducing any system level losses due to process conditions, and the costs associated with the system itself.

In a similar way, the three major elements of a CSP plant include the solar field, TES, and the power block. The three-level criteria and the essential components of a CSP plant integrated with TES are depicted in Figs. 9.2 and 9.3, respectively. The description, status, and major characteristics of different CSP technologies are summarized in Table 9.5. Generically, the operational performance of the CSP

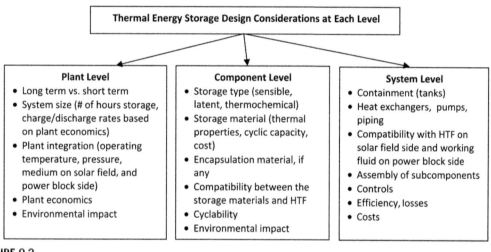

FIGURE 9.2

Thermal energy storage design considerations at each level [40].

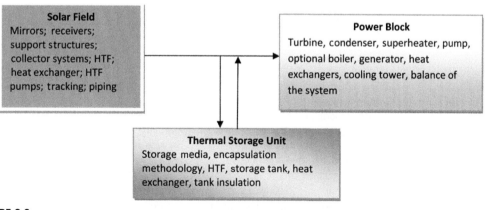

FIGURE 9.3

Main parts of a concentrating solar power (CSP) plant and its components [40].

Table 9.5 Description and Status of CSP Technologies [40,41]

	Parabolic Trough	Solar Tower	Linear Fresnel	Dish-Stirling
Maturity of technology	Commercially proven	Pilot plants, commercial projects under construction	Pilot projects	Demonstration projects
Key technology providers	Abengoa Solar, Sener Group, Acciona, Siemens, NextEra, ACS, SAMCA, etc.	Abengoa Solar, BrightSource Energy, eSolar, SolarReserve, Torresol, SunBorne Energy	Novatec Solar, Areva	
Technology development risk	Low	Medium	Medium	Medium
Operating temperature of solar field (°C)	290–550	250–650	250–390	550–750
Plant peak efficiency (%)	14–20	23–35[a]	~18	~30
Annual solar-to-electricity efficiency (net) (%)	11–16	7–20	13	12–25
Annual capacity factor (%)	25–28 (no TES) 29–43 (7 h TES)	55 (10 h TES)	22–24	25–28
Collector concentration	70–80 suns	>1000 suns	>60 suns (depends on secondary reflector)	>1300 suns
Receiver/absorber	Absorber attached to collector, moves with collector, complex design	External surface or cavity receiver, fixed	Fixed absorber, no evacuation, secondary reflector	Absorber attached to collector, moves with collector
Storage system	Indirect 2-tank molten salt at 380°C ($\Delta T = 100$°C) or direct 2-tank molten salt at 550°C ($\Delta T = 300$°C)	Direct 2-tank molten salt at 550°C ($\Delta T = 300$°C)	Short-term pressurized steam storage (<10 min)	No storage, chemical storage under development
Grid stability	Medium to high (TES or hybridization)	High (large TES)	Medium (backup firing possible)	Low
Cycle	Superheated steam Rankine	Superheated steam Rankine	Saturated steam Rankine	Stirling
Steam conditions (°C/bar)	380-540/100	540/100-160	260/50	n.a.
Water requirement (m³/MWh)	3 (wet cooling) 0.3 (dry cooling)	2e3 (wet cooling) 0.25 (dry cooling)	3 (wet cooling) 0.2 (dry cooling)	0.05e0.1 (mirror washing)

Table 9.5 Description and Status of CSP Technologies—cont'd

	Parabolic Trough	Solar Tower	Linear Fresnel	Dish-Stirling
Suitability for air cooling	Low to good	Good	Low	Best
Storage with molten salt	Commercially available	Commercially available	Possible, but not proven	Possible, but not proven

aUpper Limit for Solar Tower with Combined Cycle Turbine

plant depends on several factors, of which the crucial factors are considered to be the solar field collection efficiency and the power cycle efficiency, which can be defined by

$$\text{Solar field collection efficiency} = \frac{\text{Useful thermal energy received or collected}}{\text{Incident solar energy}} \tag{9.1}$$

$$\text{Power cycle efficiency} = \frac{\text{Net power output}}{\text{Heat input}} \tag{9.2}$$

The state-of-the-art molten salt two tank TES system being amalgamated into a parabolic trough power plant is schematically depicted in Fig. 9.4. The detailed description on the operational strategy of this system can be referred from [42]. A computer simulation-based study has been performed on the operational performance, cost, and economic factors of this system integrated with a TES system [43].

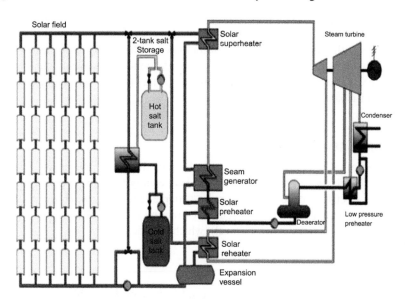

FIGURE 9.4

Schematic diagram of a parabolic trough solar power plant with a two-tank molten salt storage [40,42].

The annual performance of the $50\,MW_e$ capacity of the CSP combined with the 6h of TES is analyzed and compared to outcomes of the same system without TES. The results of the study infer the following:

- The annual solar-to-electric efficiency has improved by 13.2%, which can account for the much reduced requirement for dumping energy during high to very high insolation periods and lower turbine start-up losses for buffering during intermittent periods. However, the steam cycle efficiency is expected to be marginally decreased to 37.5% from 37.9%, which is due to the lower steam temperature obtained while TES is in operation.
- Due to the higher capacity factor, the levelized cost of energy is decreased by 10%.
- There is a marginal increase in the thermal losses at the receiver/collector, which is because of the high return temperature of the HTF entering into the solar field.
- Apart from thermal losses at the receiver, inherent losses are occurring at the TES system during charging or discharging processes, as well as the energy loss that takes place when storage is full and the power plant cannot accept more energy because it has already reached the maximum load condition.
- This integrated system requires a comparatively larger solar energy collection field area compared to the conventional system without TES.
- The system generates more energy at affordable cost of electricity, but increases the capital cost.
- Because of TES integration and higher annual energy generation, the factors related to electric parasitics and off-line parasitic consumption is minimized.
- The start-up instances of the turbine can become smaller compared to total energy usage by utilizing the TES, because the turbine can be made operational for more periods (or hours) with fewer frequencies of starting.

Although the low carbon technology utilizing solar energy seems to be intermittent or variable in nature, proper harnessing of the heat energy using TES systems can considerably improve overall plant efficiency. Indeed, the design of CSP plants combined with TES systems has to be thoroughly evaluated on the parameters mentioned in Table 9.4, in addition to the conventional design approach. As a ready reference, Table 9.6 lists the key parameters of the TES system being integrated with a $600\,MWh_{th}$ system for a parabolic trough power plant.

9.2.3 Geothermal energy storage

The term *geothermal* often refers to the use of a high temperature energy resource, which is available from the depths of the earth. This is one way that heat energy from the depths of the earth can be harnessed to produce electricity or to fulfill direct heat source requirements. However, there is a basic distinction between geothermal energies utilized for extracting heat from the depths of the earth and from underground (below the earth's surface level), which includes

- High grade geothermal energy
- Low grade geothermal energy

In *high grade geothermal energy*, the heat produced by virtue of the earth's pressure can be utilized for converting water into steam, which in turn can be used to drive turbines to generate electrical power. On the other hand, *low grade geothermal energy* refers to a process through which heat trapped within the earth's crust can be used to offset the heating load demand, especially in dwellings. A number

Table 9.6 TES System Parameters for a 600 MWh$_{th}$ System for a Parabolic Trough Power Plant [40,44]

	Internally Insulated Carbon Steel Storage Tank	Stainless Steel Storage Tank
Diameter, m	22.4	22.4
H, m	11	11
Qtotal, kW	187	187
Weight of steel, tons	279	410
Thickness of the lateral insulation material, mm	125	250
Roof insulating thickness, mm	125	200
Foamglas® thickness, mm	40	200
No. of brick_foundation	2	2
No. of brick_vessel (radially mounted)	1	0
No. of brick bottom	5	0
TIC, M€	7.45	9.13
Cost of steel, M€	1.04	3.27
Cost of flexible liner, k€	482	0

of synonyms have been used to express low grade geothermal technology or systems, including the following:

- Ground source heat pump (GSHP) systems
- Ground-coupled heat pump (GCHP) systems
- Earth-coupled heat pump (ECHP) systems
- Earth heat exchange (EHEX) systems
- Geo-exchange (GHEX) systems
- Earth-coupled water source heat pump (ECWSHP) systems

Notwithstanding the aforementioned designations, the technology driven by low grade geothermal energy remains the same. The basic functioning of a low grade geothermal energy technology is that, during summer, the cold energy trapped within the earth's crust is transferred to the dwelling for offsetting the space cooling load demand.

Likewise, during winter, heat energy from underground is transferred to the building for meeting the space heating load demand. The underground (beneath the earth's surface) remains at a relatively constant temperature throughout the year. That is, low grade geothermal technology takes advantage of this, where the underground serves as a heat source in winter and a heat sink in summer. The low grade geothermal system basically includes three major components or subsystems, which include

- Geothermal-earth connection
- Geothermal-heat pump
- Geothermal-heat distribution

The selection of the location or site is the most vital aspect of attaining maximum benefits from the utilization of geothermal energy for real-time applications. The relationship persisting between geological and heat transfer characteristics at different zones beneath the earth's surface is presented in Table 9.7.

Table 9.7 The Relationship Between the Lithology and the Heat Transfer Characteristics at Different Zones from Cap Rock Down to Zone of Intrusion [45]

Unit	Lithology	Heat Transfer
Cap rock	Thick unconsolidated silt, sand, gravel, and anhydrite-rich deposits	Heat flow by conduction
Slightly altered reservoir	Shale and sand	Enhanced conductivity resulting from presence of sand; still part of thermal cap
	Small sand units	
	Upper reservoir	Convection within sand units; shales separate region into isolated hydrologic systems
	Shales, siltstone, and sandstone cemented by calcite or silica	
	Major shale break	
Highly altered reservoir	Lower reservoir	Fractures allow more extensive convection patterns
	Reduced permeability results when altered by replacement of calcite with epidote; extensively fractured	
Zone of intrusion	Intrusion of small basaltic dikes and sills into sedimentary section; less than 20% intrusive bodies	Rate of heat release is a function of rate of intrusion

By taking into account the aforementioned characteristics, the selection and design of geothermal energy technologies can be devised for achieving improved energy efficiency and reduced CO_2 emissions without sacrificing sustainable prospects. From this perspective of TES integration, heat energy either trapped underground or from solar radiation can be effectively stored and can then be reutilized later.

In this context, a GCHP system combined with an ice storage system has been proposed for building cooling applications in recent years [47]. The floor area of the building accounts for about 184,000 m². The heat pump system can be operated in three different modes, which includes the (1) heating mode, (2) ice storage mode, and (3) cooling mode. A schematic representation of the proposed system is shown in Fig. 9.5. The operating strategy of this system can be presumed to be self-explanatory.

The operating cost of this system during summer and winter conditions has been estimated to be reduced by 42.7%–71.4% and 50%, respectively, compared to conventional heating and air-conditioning systems. The TES system in conjunction with the GCHP system enables peak load shifting strategy during summer, which can be ascertained for enhanced performance of the GCHP system. The key approaches of GCHP system integration with TES and other energy-efficient technologies are summarized in Table 9.8 for quick reference.

A study on the evaluation of the performance of GSHP combined with an LHS system for greenhouse heating application has also been reported [48]. The experimental system developed for analysis is schematically represented in Fig. 9.6. The experimental system is located in the Elazığ, East Anatolia region of Turkey. This system essentially consists of five major components, which include

- GSHP and ground heat exchanger (GHS)
- Latent heat storage (LHS) unit
- Heat storage material
- Experimental greenhouse
- Heat transfer unit with necessary data acquisition unit

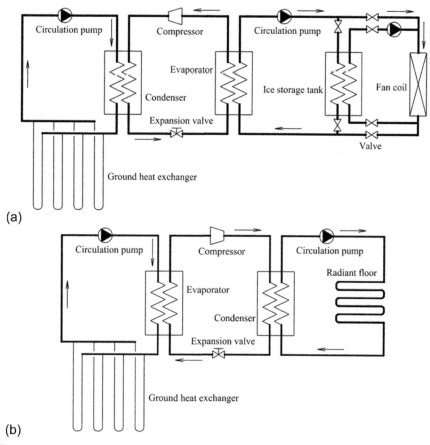

FIGURE 9.5

Flow diagram of the ground-coupled heat pump (GCHP) system integrated with an ice storage system. Operating modes in (a) summer and (b) winter [46].

Table 9.8 Main Integrated Approaches of GCHP Systems [46]

Integrated Approaches	Heating-Dominated Buildings	Cooling-Dominated Buildings
Integrated with solar energy	✓	
Integrated with cooling towers		✓
Integrated with thermal storage technologies	✓ (Heat storage)	✓ (Cold storage)
Integrated with conventional air-conditioning systems	✓	✓
Integrated with dehumidification systems		✓
Integrated with heat recovery technologies		✓

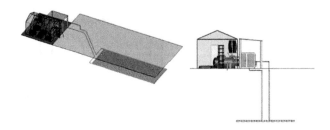

FIGURE 9.6

Experimental equipment of greenhouse heating system [48].

Table 9.9 The Main Components Specification and Characteristics of the GSHP System Studied [48]

Main Circuit	Element	Technical Specification
Ground coupling unit	Horizontal ground heat exchanger (HGHE) length of 246 m	Horizontal heat exchanger; pipe distance 0.3 m; pipe diameter 0.016 m; piping depth 2 m; material polyethylene, PX-b cross link
	Water–antifreeze solution circulating pump	Manufacturer: DAB A50/180×3 speed; speed step (2710, 2540, and 1715 rpm); power (160, 148, and 140 W); flow rate 1-12 m³/h; pressure head 8 m
	Expansion tank	50 l
Refrigerant circuit	Compressor	Type, hermetic reciprocating, manufacturer: Tecumseh; model TFH 5532 F volumetric flow rate 9.2 m³/h; speed 2900 rpm; the rated power of electric motor driving 2.5 HP (1.86 kW); refrigerant R-22; capacity 5.484 kW (at cooling/condensing temperatures of 0/46 °C)
	Heat exchanger	Manufacturer: Altıntaş Isı type—ID 23-01; capacity 10 kW heat transfer surface 0.85 m²
	Condenser for heating	Manufacturer: Azak Soğutma type—AS169 25 model; m2D capacity 11.63 kW; 25 m² surface area; 45 cm fan diameter
	Dryer	Manufacturer: DE-NA/233-083 Dry-101 connection 3/8 in.
	Observe glass	Manufacturer: Honeywell S21; connection 3/8 in.
Fan circuit	Fan of air cooled condenser	Manufacturer: Aldag type—SAS 228; diameter 380 mm; air volumetric flow rate 600 m³/h; power180 W
	Fan of discharging of PCM	Manufacturer: Bahçıvan motor—BDRKF 180; diameter 200 mm; volumetric flow rate 860 m³/h; power 85 W; speed 2350 d/d

The specifications of the key components and characteristics of the GSHP system considered in this study are given in Table 9.9. The layout of the ground heat exchanger (GHE) and the pictorial views of the greenhouse system are shown in Fig. 9.7. Detailed information on the operational strategy and modeling of this system can be referred from [48]. The daily variation of COP_{sys} of the GSHP and the total power consumption including the compressor, condenser fan, and circulation pump are shown in Fig. 9.8. For brevity, the major outcomes of this study are presented here:

- During the winter and cold seasons, the COP_{HP} of the GSHP is found to be comparatively higher than the COP_{HP} of the air-source heat pump.
- The heat transfer rate of the COP_{sys} of system increases with the increase of the mass flow rate of brine water.

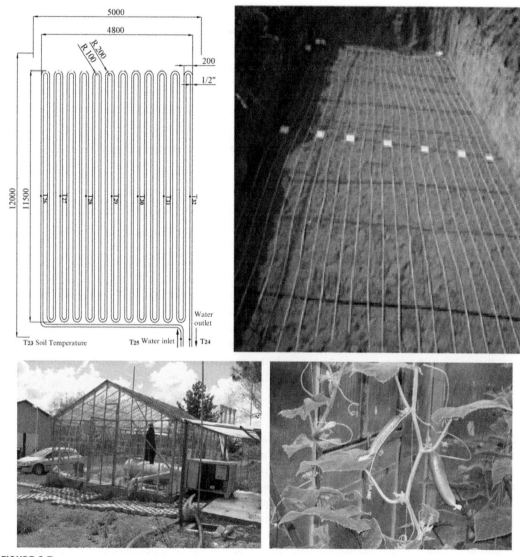

FIGURE 9.7

Layout of the GHE and the pictorial views of the greenhouse system [48].

- The incorporation of the chemical heat storage material and the charging and discharging characteristics are appreciable. The heat energy redistribution into the greenhouse was effectively achieved using the thermal storage system with good stability of the heat storage material being observed.
- As the groundwater temperature is rather higher than the environment, the rate of compression of the compressor is relatively lower, leading to less energy consumption from external sources.

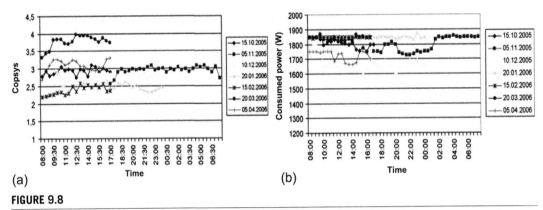

FIGURE 9.8

(a) Daily variation of COP$_{sys}$ of the ground source heat pump (GSHP) and (b) the total power consumption including the compressor, condenser fan, and circulation pump [48].

- Based on the temperature persisting in the greenhouse environment, the heat pump increases the temperature by 5–10 °C, as well as the chemical heat storage material, which increases the temperature by 1–3 °C on average. The maximum COP$_{HP}$ of the GSHP and the overall system are expected to achieve 4.3 and 3.8, respectively, during the month of March.
- In case of low ambient temperature, the inherent heat losses occurring from the greenhouse cannot be compensated using the univalent central heating system. Bivalent operation including another heating system is thus suggested for better operational performance of the system, if peak heating load could not be easily controlled.

9.2.4 Wind-thermal-cold energy storage

The concept of integrating wind energy, which is basically a renewable energy, with thermal and cold storage facilities has gained momentum in recent years. The objective behind the development of wind-thermal-cold energy storage systems is to take advantage of excess energy being produced by wind farms. This type of system would also help to avoid grid congestion and wind curtailment issues. Furthermore, the wind-thermal-cold energy storage system can benefit in more ways to utilize the excess energy produced, thereby eliminating the establishment of new fossil fuel-based plants.

In a recent research work, it is interesting to note that the concept of renewable energy integration, especially wind energy with a thermal and cold storage facility, has been demonstrated [49]. This study proposed the idea of utilizing excess energy (electricity) produced by wind farms to be stored in residential cold stores to yield the maximum benefit to the electrical network, utility, or cold store owner.

The excess electricity can be supplied to cold store chiller plants (or the refrigeration unit) during off-peak periods or in low electricity tariff periods. The products in the cold stores can then be further cooled to a very low temperature, which enables storing the cold energy. During on-peak periods, the refrigerating units can be turned off, and the stored cold energy can be reutilized for maintaining the initial storage temperature of the products. Thus, the redistribution of cold energy has been made possible through integration of wind energy with cold storage.

In this study, a simple linear programming model has been developed to demonstrate the usefulness of curtailed energy redistribution in a cost-effective way to local end users and the independent power producer as well. In a similar way, the surplus electricity being generated from wind farms can be utilized for space heating and thermosiphon heating applications in dwellings. The outcome of this study is summarized in Table 9.10 for immediate reference.

9.2.5 Hybrid TES

As the name signifies, this system combines the essential features of solar thermal energy and chemical looping combustion (CLC) strategies for providing the diurnal TES requirements for concentrated solar thermal energy. This system fulfills the base load power generation for solar energy at an affordable cost through utilization of sensible and chemical energy storage principles.

CLC is a system that integrates sensible and chemical energy storage elements for the desired purpose. The schematic representation of a hybrid TES system is illustrated in Fig. 9.9. The operating principle of the hybrid TES system can be gleaned from [51] for more information. The essential operating conditions, model validations pertaining to the fuel reactor and the air reactor of the hybrid TES system, are summarized in Tables 9.11 through 9.13.

The outcome of the hybrid TES system can be summarized as follows:

- The operating temperature is maintained constant at the outlet from the air reactor, albeit with variations occurring in the fuel reactor operating temperature by virtue of variations in solar thermal energy.
- The predictions obtained from the simulation reveal that 70% of the absorbed solar energy in the solar fuel reactor is stored in the particles.
- Due to the reradiation through the solar fuel reactor aperture, 14.5% of the total energy has been lost.
- Exergy efficiency has been found to be 7% higher than the reference CLC system without the solar input while averaged over the 24 h day.
- The oxygen carrier particles serve to store solar thermal energy in the form of sensible heat.
- The major restraint of this system is its lower solar fraction of about 6.5% while averaged over the 24 h day, but this could be addressed with other options leading to higher solar fraction.

9.2.6 CHP thermal storage

The pathway toward a sustainable future can also be achieved through the amalgamation of CHP and TES systems. A CHP system is one that can reduce dependency on fossil fuel-based sources and operational costs as well as GHG emissions into the environment by utilizing the waste heat recovery principle. According to this principle, the heat that is produced during various processes can be reutilized effectively for serving the heating load demand in buildings.

However, the prime restraint in using a CHP system can be understood from the mismatch between the amount of electrical energy generated and the heat energy being provided by the CHP system versus the amount of electrical and thermal energy required for the building as intended to be delivered by the CHP system. This energy gap can be effectively bridged through the integration of TES and CHP systems. A properly designed TES system would offer the peak load shifting strategy by which the CHP system operates energy efficiently on a long-term basis with minimum exergy losses.

One such combined cooling, heating, and power plant (CHCP) integrated TES system has been investigated for the evaluation of its thermodynamic and economic aspects pertaining to the optimal sizing of TES for a certain system. Three different buildings have been considered in the tertiary sector

Table 9.10 Power Output Measurements, ExCF, Wind Curtailment, and ACF [49,50]

WF1	2001	2002	2003	2004	2005	2006	2007	2008	2009
Net Electricity Produced Delivered to the System (WF1)									
(kWh)	30,477,639	20,786,341	30,965,827	29,702,400	29,294,652	28,254,624	25,817,520	25,587,563	28,141,526
ExCF (%)	35.14	23.97	35.71	34.25	33.78	32.58	29.77	29.50	32.45
Wind curtailment (kWh)	2,114,700	1,534,370	4,526,880	4,620,950	2,944,500	2,779,640	2,462,730	2,118,430	2,489,760
ACF (%)	37.58	25.74	40.93	39.58	37.17	35.79	32.61	31.95	35.32
Net Electricity Produced Delivered to the System (WF2)									
(kWh)	14,989,487	10,191,754	14,939,494	12,962,400	13,575,480	14,570,000	13,491,600	12,399,600	13,033,200
ExCF (%)	34.57	23.50	34.45	29.89	31.31	33.60	31.11	28.60	30.06
Wind curtailment (kWh)	1,052,700	670,850	1,964,530	1,294,440	1,440,010	1,887,270	1,552,760	1,321,820	1,229,250
ACF (%)	37.00	2505	38.98	32.88	34.63	37.95	34.69	31.64	32.89

WF, Wind farm; ExCF, Exergetic capacity factor; ACF, Accumulated capacity factor.

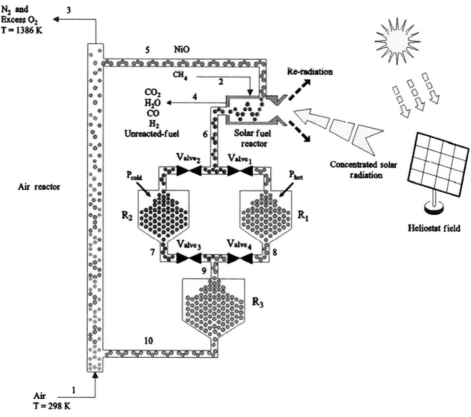

FIGURE 9.9

The configuration of the solar–chemical looping combustion (CLC) hybrid system assessed here to provide storage of solar thermal energy. Reservoirs R_1 and R_2 are used to store the hot and cold particles, respectively, produced with and without the addition of solar thermal energy. The hot and cold particles are then mixed and stored in R_3 to provide a constant temperature feed to the air reactor.

for analysis. The extended aggregated thermal demand method is adopted for case studies of the buildings considered that incorporate CHCP and TES systems. The schematic diagram of the combined system is depicted in Fig. 9.10.

The integration of a TES system with a CHCP system has actually delivered both thermodynamic and economic benefits, provided the TES system is properly sized. It is pertinent to note that the thermal energy contribution shows strong dependence on the operational performance of the CHCP system due to the interaction between energy production and demand profiles. There is an optimum size for the TES system to be thermodynamically beneficial, and above this level, further utilization of heat energy does not produce any significant effect. Likewise, an optimum volume of the storage system maximizes the economic benefits of integrating the TES system.

Table 9.11 Main Assumptions for the Solar–CLC Hybrid System [51]

Parameter	Value
Effective absorptance [52,53]	1
Effective emittance [52,53]	1
Optical efficiency of heliostat field (%) [54]	65
Mean flux concentration ratio	2000
Normalized fuel molar flow rate (mol/mol$_{CH4,in}$)	1
Air mass flow (mol/mol$_{CH4,in}$)	23.8
Input fuel pressure (atm)	1
Input air pressure (atm)	1
Input fuel temperature (°C)	25
Input air temperature (°C)	25
Normalized heliostat area to input fuel mass flow $\left(\dfrac{m^2}{mol_{Ch4,in}/s} \right)$	450
Fuel reactor pressure (atm)	1
Air reactor pressure (atm)	1
Oxygen added ratio	2

Table 9.12 Comparison of the Predictions of the Present Fuel Reactor Model with the Simulation Results Reported by Hong et al. [55,56] for the Purpose of Validation [51]

Input Parameter	Value		
Operating pressure (bar)	15.0		
Input CH$_4$ temperature (°C)	15.0		
Input CH$_4$ molar flow rate (mol/s)	1.0		
Input NiO temperature (°C)	530.0		
Input NiO molar flow rate (mol/s)	4.0		
Absorbed solar thermal energy (kW)	183.4		
Calculated Value	**Hong et al. [55,56]**	**Present Model**	**Difference (%)**
Fuel reactor temperature (°C)	530.0	562.0	6.0
ΔH_{530} (kJ/mol)[a]	158.0	150.0	5.0

[a]ΔH_{530} is the CH$_4$ and NiO reaction enthalpy change at 530 °C.

Generically, the integration of TES system with CHP (or CHCP) systems would yield benefits from the standpoint of reducing overall operational cost, thermodynamic efficiency, and economic aspects as well. Indeed, the oversizing of the TES system to meet the energy demand and redistribution requirements can sometimes affect the satisfactory performance of the overall system. Moreover, integration of the TES system may not fully replace the requirements of boilers or other heating components, thereby creating certain limitations on the reduction of operating costs of the system [58].

Table 9.13 Comparison of the Predictions of the Present Air Reactor Model with the Simulation Results Reported by Hong et al. [55,56] for the Purpose of Validation [51]

Input Parameter	Value		
Operating pressure (bar)	15.0		
Input air temperature (°C)	530.0		
Input air molar flow rate (mol/s)	37.8		
Input Ni temperature (°C)	530.0		
Input Ni molar flow rate (mol/s)	4.0		
Calculated Value	**Hong et al. [55,56]**	**Present Model**	**Difference (%)**
Fuel reactor temperature (°C)	1200.0	1195.0	0.4
$\Delta H'_{1200}$ (kJ/mol)[a]	−959.0	−934.0	2.7

[a]$\Delta H'_{1200}$ is the Ni and O_2 reaction enthalpy change at 1200 °C.

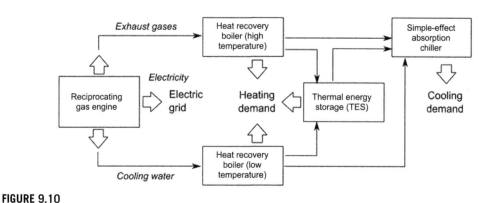

FIGURE 9.10

Schematic diagram of the cooling, heating, and power plant (CHCP) system [57].

Thus, adopting proper and viable design strategies for the implementation of TES systems for CHP applications can be beneficial on a long-term basis.

9.3 LEADERSHIP IN ENERGY AND ENVIRONMENTAL DESIGN (LEED) AND SUSTAINABILITY PROSPECTS

The term *sustainability* is the most citable topic of interest among the architects, engineers, and clients (owners) involved in building design worldwide. Bridging the gap between energy supply and energy demand is said to be the underlying factor for successfully achieving sustainable development. The most important factors substantiating sustainable development through sustainable energy are presented in Table 9.14.

Table 9.14 Sustainable Energy Performance Indicators [60,61]

No.	Criteria	Explanation
C1	Renewable energy	Encourage and recognize increasing levels of on-site renewable energy self-supply to reduce environmental and economic impacts associated with fossil fuel energy use
C2	Minimum energy performance	Establish the minimum level of energy efficiency for the proposed building and systems
C3	Fundamental commissioning of the building energy systems	Verify that the building's energy-related systems are installed, are calibrated, and perform according to the owner's project requirement, basis of design, and construction documents
C4	Enhanced commissioning	Begin the commissioning process early during the design process and execute additional activities after system performance verification is completed
C5	Measurement and verification	Provide for the ongoing accountability of building energy consumption over time
C6	Optimize energy performance	Achieve increasing levels of energy performance above the baseline in the prerequisite standard to reduce environmental and economic impacts associated with excessive energy use

It is a well-known fact that buildings are one of the high energy consuming sectors, amounting to approximately one-quarter to one-third of the total energy generated globally. The issues related to GHG emissions and climate change have presented continuous challenges and issues over the years. In this context, it is of prime importance to achieve energy efficiency at every step of the design process, which would contribute to the development of high performance sustainable buildings or green buildings.

The basic distinction between a green building and a sustainable building is the extent to which the proposed design can prove its performance satisfactorily toward maintaining the ecological balance. The renewable energy integration can enhance the possibility for achieving energy efficiency in buildings as well as environmental sustainability. The deliverables, merits, and challenges of utilizing renewable energy sources in buildings are provided in Tables 9.15 through 9.17.

Because building architecture is comprised of several intricate designs, the incorporation of energy-efficient materials and energy systems can also help to acquire energy efficiency in the long run. Based on a survey conducted with building professionals, a new ranking system has been developed for estimating sustainable energy performances [59]. The new rating system is based on two broad categories, namely, energy-efficiency criteria and material efficiency criteria, as shown in Figs. 9.11(a) and 9.11(b).

It is obvious that the parameters pertaining to energy consumption, durability, low embodied energy, avoiding environmentally toxic materials, and the use of recycled materials possess a high relative importance index. Providing sufficient considerations on these parameters can yield success toward the pathway of sustainable development in both buildings and the environment. The sphere of sustainability depicted in Fig. 9.12 signifies the interdependencies of various factors including technological,

Table 9.15 Versatile Types of Renewable Energy [60,61]	
Renewable Energy	**Description and Benefits**
Active solar energy	☐ Convert solar energy into another more useful form of energy
	☐ This would normally be a conversion to heat or electrical energy
	☐ Inside a building this energy would be used for heating, cooling, or offsetting other energy use or costs
	☐ The basic benefit is that controls can be used to maximize its effectiveness
	☐ Photovoltaic solar panels are in this group
Passive solar energy	☐ In passive solar building design, windows, walls, and floors are made to collect, store, and distribute solar energy in the form of heat in the winter and reject solar heat in the summer
	☐ The key to designing a passive solar building is to best take advantage of the local climate
	☐ Elements to be considered include window placement and glazing type, thermal insulation, thermal mass, and shading
Wind energy	☐ Wind power is the conversion of wind energy into a useful form of energy, such as using wind turbines to make electricity, windmills for mechanical power, wind pumps for water pumping or drainage
Geothermal energy	☐ Geothermal energy is thermal energy generated and stored in the earth
Fuel cell	☐ A fuel cell is a device that converts the chemical energy from a fuel into electricity through a chemical reaction with oxygen or another oxidizing agent. Hydrogen is the most common fuel used

socioeconomic, environmental, and cost-effective measures for the development of high performance sustainable buildings.

From this perspective, thermal storage systems offer huge potential for reducing GHG emissions and operational costs without sacrificing energy efficiency in buildings. The incorporation of TES systems in buildings can be more functional in the sense that TES systems not only facilitate experiencing peak load shifting and energy redistribution requirements, but they also enable utilizing the power (or electricity) that is available at a considerably lower cost, for charging of TES during nighttime. These aspects are attributable to TES integration with conventional cooling units in large infrastructural projects, wherein oversizing of the chiller plant on the basis of design safety factors can be minimized or eliminated [62].

Moreover, TES integration with chiller plants can benefit in terms of meeting about 20% of the design cooling load, which rules out the condition of uplifting chiller capacity to about 20% higher than the actual cooling load. This in turn contributes to accomplishing enhanced energy efficiency of the cooling plant, which fulfills the LEED credit requirements for sustainable buildings.

TES systems, on the other hand, integrated with heating systems in buildings can also provide a means for attaining cost-energy savings potential with reduced power consumption at heating systems. The heating energy redistribution requirements in buildings in terms of space heating and hot water demand can be effectively met by using TES systems. In the past decade, approximately 60 new TES installations have been performed in China, of which about 20 TES installations in Beijing are dedicated for heating applications in buildings [63].

Table 9.16 The Strengths of Renewable Energy Types [60,61]

Renewable Energy	Ranking		Feasible use in Urban Areas and Buildings	The Amount of Fossil Fuel Consumption Reduction	The Amount of Initial Construction Costs Enhancement	The Amount of Maintenance and Operation Costs Reduction
	AS	N				
Active solar energy	4.57	1.00		✓		✓
Passive solar energy	4.43	0.95	✓		✓	
Wind energy	2.79	0.43		✓		
Geothermal energy	1.79	0.11			✓	
Fuel cell	1.43	0.00	✓		✓	

Table 9.17 The Challenges Toward the Use of Renewable Energy [60,61]

Renewable Energy	Ranking		High initial Costs	Lack of Government Support	Lack of Public Awareness	Lack of Technical Technology	Lack of Proper and Required Equipment	Poor Planning Approach
	AS	N						
Passive solar energy	2.38	1.00	4	2	3	3	5	1
Wind energy	2.36	0.97	1	5	6	3	2	4
Active solar energy	2.31	0.89	1	4	3	6	5	2
Geothermal energy	1.75	0.15	1	5	2	4	3	6
Fuel cell	1.64	0.00	1	4	5	3	2	4

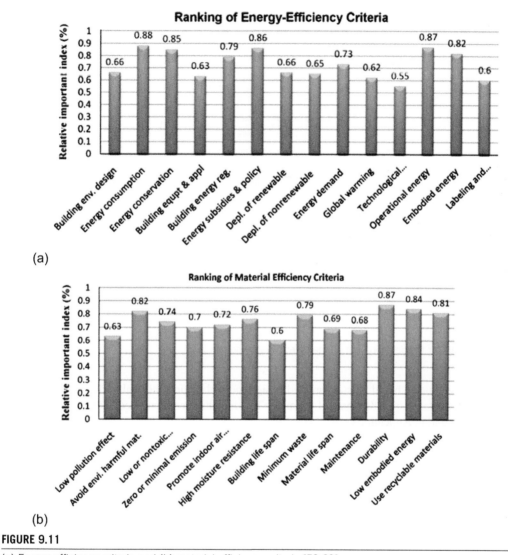

(a)

(b)

FIGURE 9.11

(a) Energy efficiency criteria and (b) material efficiency criteria [59,60].

More focus can be given to research pertaining to the development of new and advanced thermal storage materials, which actually enhance the operational performance of TES systems and yield cost-energy savings potential over time. The incorporation of eco-friendly heat storage materials into the TES system can add value in terms of acquiring more green credits for building cooling/heating applications [64]. The inherent operational characteristics of TES integrated building cooling/heating

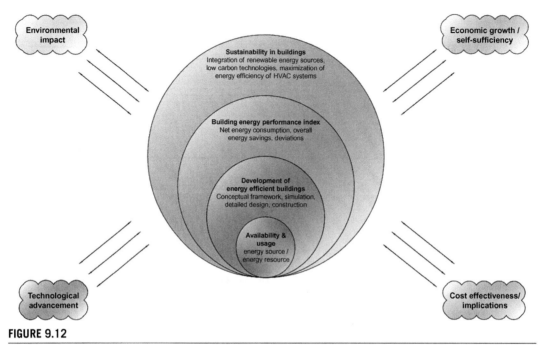

FIGURE 9.12

Sustainability chart for the development of energy efficient buildings [1].

systems can be evaluated based on standardized design methodologies, thermal performance analysis, and testing procedures [1,65–68]. Some of the major TES installations accredited with the LEED rating system can be found in Appendix V.

9.4 CONCISE REMARKS

In the view of sustainable design, thermal storage technology can be considered one of the most energy-efficient and best options for addressing energy redistribution requirements in modern buildings. The integration of TES with cooling/heating systems in buildings can provide peak load shifting and storage of thermal energy during off-peak periods at a relatively low cost energy strategy.

The most promising aspect of utilizing thermal storage technology is that it can effectively create a balance between energy supply and energy demand in buildings. The amalgamation of TES with a conventional cooling system can provide about 20% of the design cooling load in buildings. That is, due to the peak load shaving attributes of TES, the cooling load sharing capacity between the chiller plant and the TES system can be effectively redistributed.

In addition, renewable energy integration including solar, wind, geothermal, low energy, and CHP technologies with TES systems can enhance operational performance and cost effectiveness of the overall system. The energy efficiency and cost-energy savings potential of TES systems integrated with conventional and renewable energy systems can collectively contribute to reduction in GHG emissions and paves the way for the development of a sustainable future.

References

[1] Parameshwaran R, Kalaiselvam S, Harikrishnan S, et al. Sustainable thermal energy storage technologies for buildings: a review. Renew Sust Energ Rev 2012;16:2394–433.

[2] Hepbasli A. Low exergy (LowEx) heating and cooling systems for sustainable buildings and societies. Renew Sustain Energy Rev 2012;16:73–104.

[3] Schmidt D. Low exergy systems for high-performance buildings and communities. Energ Build 2009;41:331–6.

[4] Schmidt D. Benchmarking of "low-exergy" buildings. In: 1st international exergy, life cycle assessment, and sustainability workshop & symposium (ELCAS); 2009.

[5] Schmidt D, Sager C, Schurig M, Torio H, Kühl L. Abschlussbericht des Forschungsvorhabens: Projektverbund LowEx: Nutzung von regenerativen Energiequellen in Gebäuden durch den Einsatz von Niedrig-Exergiesystemen [Utilisation of Renewable Energy Sources in Buildings by using Low Exergy Systems]. The German Ministry of Economics and Technology (BMWi), Report No. ES-342 01/2009, Kassel, April 2009 [in German].

[6] Waqas A, Din ZU. Phase change material (PCM) storage for free cooling of buildings—a review. Renew Sustain Energy Rev 2013;18:607–25.

[7] Lazaro A, Dolado P, Marın M, Zalba B. PCM–air heat exchangers for free-cooling applications in buildings: experimental results of two real-scale prototypes. Energy Convers Manage 2009;50:439–43.

[8] Turnpenny J, Etheridge D, Reay D. Novel ventilation cooling system for reducing air conditioning in buildings. Part I: testing and theoretical modeling. Appl Therm Eng 2000;20:1019–37.

[9] Yanbing K, Yi J, Yinping Z. Modeling and experimental study on an innovative passive cooling system—NVP system. Energ Build 2003;35:417–25.

[10] Marin J, Zalba B, Cabeza F, Mehling H. Free-cooling of buildings with phase change materials. Int J Refrig 2004;27:839–49.

[11] Stritih U, Butala V. Energy saving in building with PCM cold storage. Int J Energ Res 2007;1532–44.

[12] Takeda S, Naganao K, Mochida T, Shimakura K. Development of a ventilation system utilizing thermal energy storage for granules containing phase change material. Sol Energ 2004;77:329–38.

[13] Naganao K. Development of the PCM floor supply air-conditioning system. Thermal Energy Storage for Sustainable Energy Consumption 2007;367–73.

[14] Medved S, Arkar C. Correlation between the local climate and the free-cooling potential of latent heat storage. Energ Build 2008;40:429–37.

[15] Arkar C, Vidrih B, Medved S. Efficiency of free cooling using latent heat storage integrated into the ventilation system of a low energy building. Int J Refrig 2007;30:134–43.

[16] Arkar C, Medved S. Free cooling of a building using PCM heat storage integrated into the ventilation system. Sol Energ 2007;81:1078–87.

[17] Waqas A, Kumar S. Utilization of latent heat storage unit for comfort ventilation of buildings in hot and dry climates. Int J Green Energy 2011;8:1–24.

[18] Xu X, Yu J, Wang S, Wang J. Research and application of active hollow core slabs in building systems for utilizing low energy sources. Appl Energ 2014;116:424–35.

[19] Bring A. BRIS-computer program developed by National Swedish building research for calculation of room climate and heating and cooling effects. National Swedish Building Research Summaries. 1974 p. S:23.

[20] Augenbroe GLM, Vedder HA. Accurate modeling of air-supplied heat storage in hollow core slabs. In: CLIMA 2000, world congress on heating, ventilation and air conditioning, August; 1985. p. 441–6.

[21] Ham ER. Report C-43 Experimental verification of the computational model for the energon floor slab. Building physics group, department of civil engineering, Delft University of Technology; 1985.

[22] Zmeureanu R, Fazio P. Thermal performance of a hollow core concrete floor system for passive cooling. Build Environ 1988;23(3):243–52.

[23] Holmes MJ, Wilson A. Assessment of the performance of ventilated floor thermal storage systems. ASHRAE Trans 1996;102(1):698–707.

[24] Taylor BJ, Cawthorne DA, Imbabi MS. Analytical investigation of the steady-state behavior of dynamic and diffusive building envelopes. Build Environ 1996;31(6):519–25.

[25] Winwood R, Benstead R, Edwards R, Letherman KM. Building fabric thermal storage: use of computational fluid-dynamics for modelling. Build Serv Eng Res Technol 1994;15(3):171–8.

[26] Winwood R, Benstead R, Edwards R. Advanced fabric energy storage II: computational fluiddynamics modelling. Build Serv Eng Res Technol 1997;18(1):7–16.

[27] Winwood R, Benstead R, Edwards R. Advanced fabric energy storage III: theoretical analysis and whole-building simulation. Build Serv Eng Res Technol 1997;18(1):17–24.

[28] Ren MJ, Wright JA. A ventilated slab thermal storage system model. Build Environ 1998;33(1):43–52.

[29] Augenbroe GLM, Vedder HA. Accurate modelling of air-supplied heat storage in hollow core slabs. In: CLIMA 2000, world congress on heating, ventilation and air conditioning, August; 1985. p. 441–6.

[30] Russell MB, Surendran PN. Influence of active heat sinks on fabric thermal storage in building mass. Appl Energy 2001;70(1):17–33.

[31] Barton P, Beggs CB, Sleigh PA. A theoretical study of the thermal performance of the TermoDeck hollow core slab system. Appl Therm Eng 2002;22(13):1485–99.

[32] Balocco C. A simple model to study ventilated facades energy performance. Energ Build 2002;34(5):469–75.

[33] Ciampi M, Leccese F, Tuoni G. Ventilated facades energy performance in summer cooling of buildings. Sol Energ 2003;75(6):491–502.

[34] Weber T, Jóhannesson G. An optimized RC-network for thermally activated building components. Build Environ 2005;40(1):1–14.

[35] Corgnati SP, Kindinis A. Thermal mass activation by hollow core slab coupled with night ventilation to reduce summer cooling loads. Build Environ 2007;42:3285–97.

[36] Sanjuan C, Suarez MJ, Gonzalez M, et al. Energy performance of an open-joint ventilated façade compared with a conventional sealed cavity facade. Sol Energ 2011;85(9):1851–63.

[37] Mavromatidis LE, Mohamed AB, Mankibi E, et al. Numerical estimation of air gaps influence on the insulating performance of multilayer thermal insulation. Build Environ 2012;49:227–37.

[38] Chae YT, Strand RK. Modeling ventilated slab systems using a hollow core slab: implementation in a whole building energy simulation program. Energ Build 2013;57:165–75.

[39] Renewable energy sources for buildings. CIBSE TM38: 2006. The Chartered Institution of Building Services Engineers.

[40] Kuravi S, Trahan J, Goswami DY, Rahman MM, Stefanakos EK. Thermal energy storage technologies and systems for concentrating solar power plants. Prog Energy Combus Sci 2013;39:285–319.

[41] International Renewable Energy Agency (IRENA). Renewable energy technologies: cost analysis series e concentrating solar power 2012.

[42] Herrmann U, Kelly B, Price H. Two-tank molten salt storage for parabolic trough solar power plants. Energy 2004;29:883–93.

[43] Price H, Kearney D. Reducing the cost of energy from parabolic trough solar power plants. In: International solar energy conference Hawaii Island, Hawaii; 2003.

[44] Gabbrielli R, Zamparelli C. Optimal design of a molten salt thermal storage tank for parabolic trough solar power plants. J Sol Energy Eng 2009;131:041001–10.

[45] Younker LW, Kasameyer PW, Tewhey JD. Geological, geophysical, and thermal characteristics of the Salton sea geothermal field, California. J Volcano Geotherm Res 1982;12:221–58.

[46] Zhai XQ, Qu M, Yu X, Yang Y, Wang RZ. A review for the applications and integrated approaches of ground-coupled heat pump systems. Renew Sustain Energy Rev 2011;15:3133–40.

[47] Wei W, Zhang J, Chen T, Zhang Y. A novel heating and cooling system combining a ground source heat pump with an ice storage (in Chinese). Fluid Mach 2007;35:72–5.

[48] Benli H, Durmuş A. Evaluation of ground-source heat pump combined latent heat storage system performance in greenhouse heating. Energ Build 2009;41:220–8.

[49] Xydis G. Wind energy to thermal and cold storage—a systems approach. Energ Build 2013;56:41–7.

[50] Strenecon SA, Personal contact. Available from: http://www.strenecon.gr/, 2011.

[51] Jafarian M, Arjomandi M, Nathan GJ. A hybrid solar and chemical looping combustion system for solar thermal energy storage. Appl Energy 2013;103:671–8.

[52] Steinfeld A, Larson C, Palumbo R, Foley Iii M. Thermodynamic analysis of the co-production of zinc and synthesis gas using solar process heat. Energy 1996;21:205–22.

[53] Steinfeld A, Kuhn P, Reller A, Palumbo R, Murray J, Tamaura Y. Solar-processed metals as clean energy carriers and water-splitters. Int J Hydrogen Energy 1998;23:767–74.

[54] Kribus A, Krupkin V, Yogev A, Spirkl W. Performance limits of heliostat fields. J Sol Energy Eng 1998;120:240–6.

[55] Hong H, Jin H. A novel solar thermal cycle with chemical looping combustion. Int J Green Energy 2005;2:397–407.

[56] Hong H, Jin H, Liu B. A novel solar-hybrid gas turbine combined cycle with inherent CO_2 separation using chemical-looping combustion by solar heat source. J Sol Energy Eng 2006;128:275–84.

[57] Martínez-Lera S, Ballester J, Martínez-Lera J. Analysis and sizing of thermal energy storage in combined heating, cooling and power plants for buildings. Appl Energy 2013;106:127–42.

[58] Smith AD, Mago PJ, Fumo N. Benefits of thermal energy storage option combined with CHP system for different commercial building types. Sustain Energy Tech Assess 2013;1:3–12.

[59] Mwasha A, Williams RG, Iwaro J. Modeling the performance of residential building envelope: The role of sustainable energy performance indicators. Energ Build 2011;43:2108–17.

[60] GhaffarianHoseini A, Dahlan ND, Berardi U, GhaffarianHoseini A, Makaremi N, GhaffarianHoseini M. Sustainable energy performances of green buildings: A review of current theories, implementations and challenges. Renew Sustain Energy Rev 2013;25:1–17.

[61] Qaemi M, Heravi G. Sustainable energy performance indicators of green building in developing countries. In: Construction Research Congress 2012 ©ASCE, US; 2012. p. 1961–70.

[62] MacCracken M. Thermal energy storage myths. ASHRAE Journal 2003;45:1–6.

[63] Wang Q, Zhao X. Thermal energy storage in China. ASHRAE Journal 2001;43:53–5.

[64] Ice storage as part of a LEED® building design. Trane Engineers Newsletter 2007;36:1–6.

[65] ASHRAE. Method of testing thermal storage devices with electrical input and thermal output based on thermal performance. ANSI/ASHRAE 94.2;1981:(RA06).

[66] ASHRAE. Method of testing active sensible thermal energy devices based on thermal performance. ANSI/ASHRAE 94.3;1986:(RA06).

[67] ASHRAE. Method of testing active latent-heat storage devices based on thermal performance. ANSI/ASHRAE 94.1;2002:(RA06).

[68] ASHRAE fundamentals handbook. ASHRAE Inc;2009.

Thermal Energy Storage Systems Design

10.1 INTRODUCTION

The most important part of recommending thermal energy storage (TES) systems, either sensible heat or latent heat, to be integrated with the cooling/heating systems in buildings essentially depends on their inherent design aspects. To enable thermal storage systems as fully functional, some crucial factors need to be considered during the design phase, which include the following:

- Thermal load (cooling/heating) profile of the building under consideration
- Type of TES system (e.g., full storage or partial storage)
- Selection of proper heat storage material
- Mode of operation (e.g., active or passive system)
- Availability periods of power supply
- Cooling/heating plant redundancy
- Space availability
- Capital and operating costs

In this context, the basic design of some sensible and latent TES systems with example calculations are demonstrated in forthcoming sections.

10.2 SENSIBLE HEAT STORAGE SYSTEMS

In the sensible heat storage system, the quantity of heat energy that can be stored in the material (solid or liquid) by virtue of raising the temperature can be expressed as

$$Q = mc_p \Delta T \qquad (10.1)$$

Equation (10.1) can be expressed in another form as

$$Q = mc_p \left(T_h - T_l \right) \qquad (10.2)$$

where m is the mass of the storage medium, T_h and T_l are the maximum and initial temperature of the material (K), and $(T_h - T_l)$ is referred to the temperature swing (K).

The volume (in m³) of the material that is required to store a given quantity of heat energy can be calculated by

$$V = \frac{m}{d} \qquad (10.3)$$

where, d is density of the material (kg/m³).

The volume of the material required can also be estimated using the relation given by

$$V = \frac{E}{dc_p \Delta T}$$ (10.4)

where, c_p is specific heat of the storage material (kJ/(kg K)).

For a typical office space subjected to a temperature swing of $2\,°C$, the quantity of energy that can be stored is about $0.250\,MJ/m^2$. For an industrial space undergoing a temperature swing of $3\,°C$, the amount of energy that can be stored is about $0.370\,MJ/m^2$. Thus, the quantity of sensible heat energy storage can be expressed as

$$E = 0.250 \times A \left(\text{for office space application}\right)$$ (10.5)

$$E = 0.370 \times A \left(\text{for industrial application}\right)$$ (10.6)

where, E is thermal energy being stored (kJ) and A is floor or building area (m²).

10.3 LATENT HEAT STORAGE SYSTEMS

From the perspective of latent heat storage systems, thermal energy can be stored or released by virtue of the phase transition process of a material occurring at or near isothermal conditions. The change in the state of the material from either liquid to solid (freezing) or solid to liquid (melting) is the key phenomenon for enabling the TES in phase change materials. The change in the latent heat enthalpy of the material during the freezing and melting processes determines the storage capacity of phase change materials (PCMs). This physical process can be suitably expressed in the form as given by

$$Q = m \left[\left(c_p \Delta T_{\text{sensible}} \right) + H + \left(c_p \Delta T \right)_{\text{latent}} \right]$$ (10.7)

where, H is latent heat enthalpy of the heat storage material (kJ/kg).

The simple design procedures for *ice thermal energy storage* (*ITES*) and chilled water-packed bed TES systems are presented in forthcoming sections:

10.3.1 Sizing of ITES system

Inputs for design of ITES system:

Requirement of cooling energy on daily basis—3000 kWh
Time of charging (freezing) of the ice storage—8 h
Time of discharging (melting) of the ice storage—10 h
Peak cooling load demand persisting in the building space—500 kW
A rotary screw type chiller (cooling plant) considered
Energy storage efficiency factor—0.95 (assumed)
Storage capacity—50%

Estimation of the ice storage capacity:

The energy generation requirement on a per day basis can be determined by

$$E_g = \text{Cooling energy requirement} \left(\text{per day}\right) * \text{storage capacity}$$ (10.8)

Thus, $E_g = 3000 \times (50/100) = 1500\,kWh$
The total energy storage capacity can be evaluated by

$$E_t = \frac{E_g}{\text{Energy storage efficiency factor}} \tag{10.9}$$

Thus, $E_T = 1500/0.95 \approx 1580\,kWh$
The charging (freezing) of ice storage using the chilled water medium can be determined by

$$\text{Ice storage charging} = \frac{E_T}{\text{Time of charging of the ice storage}} \tag{10.10}$$

That is, ice storage charging $= (1580/8) = 198\,kWh$
The chiller cooling capacity during the design day operation can be estimated by

$$\text{Chiller cooling capacity} = \frac{E_T}{\text{Time of discharging of the ice storage}} \tag{10.11}$$

That is, chiller cooling capacity $= (1580/10) = 158\,kWh$.
Thus, the chiller nominal capacity can be determined by

$$\text{Chiller nominal capacity} = \left(\begin{array}{c} \text{Peak cooling load demand persisting} \\ \text{in building} - \text{Chiller cooling capacity} \end{array} \right) \tag{10.12}$$

Therefore, chiller nominal capacity $= (500 - 158) = 342\,kWh$.
The chiller plant being selected for charging the ice storage (during nighttime) is capable of shifting the peak load demand in a building during daytime at 50% higher capacity. Hence, the design cooling capacity of the chiller can be estimated to by $\approx 300\,kWh$ (i.e., $198\,kW \times 1.5$).

10.3.2 Sizing of chilled water packed bed LTES system

The design of a chilled water-packed bed LTES system is proposed for a building located in a hot and humid climatic condition. The on-peak and off-peak cooling load requirements are represented by red and blue colors, respectively, in Table 10.1. The total cooling load of the building considered is estimated to be 4130 kW. The efficiency factors for the charging and discharging processes are assumed to be 0.9 and 0.8, respectively.
The nominal cooling capacity of the chiller plant can be estimated and is given by

$$\text{Nominal cooling capacity} = \frac{\text{Total cooling load}}{\begin{array}{c} (\text{Total charging hours} * \text{efficiency factor} + \\ \text{Total discharging hours} * \text{efficiency factor}) \end{array}} \tag{10.13}$$

Thus, nominal cooling capacity $= (4130/\{(14*0.9) + (10*0.8)\}) = 204\,kW$.
The cooling capacity of the chiller during direct cooling (on-peak load conditions) can be determined by

$$\left(\text{Cooling capacity of chiller}\right)_{\text{on-peak}} = 204 * 0.9 = 184\ kW$$

Table 10.1 Design of the chilled water-packed bed LTES system

Time (h)	Total Cooling Energy Demand (kW)	Load Satisfied by		Chiller+TES Charging (kW)	TES Balance (kW)
		Chiller (kW)	TES Tank (kW)		
0	100	164	100	164	353
1	100	164	100	164	416
2	100	164	100	164	480
3	101	164	101	164	542
4	103	164	103	164	603
5	101	164	101	164	666
6	107	164	107	164	722
7	112	164	112	164	774
8	120	164	120	164	818
9	225	184	41	0	777
10	245	184	61	0	716
11	270	184	86	0	630
12	300	184	116	0	514
13	320	184	136	0	378
14	300	184	116	0	262
15	270	184	86	0	176
16	255	184	71	0	105
17	245	184	61	0	45
18	229	184	44	0	0
19	113	164	113	164	50
20	110	164	110	164	104
21	105	164	105	164	162
22	101	164	101	164	225
23	100	164	100	164	289
Total	4130	1840	2290	2290	

Likewise, the cooling capacity of the chiller during charging of the LTES system (off-peak conditions) can be obtained by

$$(\text{Cooling capacity of chiller})_{\text{off-peak}} = 204 * 0.8 = 164 \text{ kW}$$

The cooling capacity shared by the LTES system (energy redistribution) during on-peak condition can be estimated as follows:

At the commencement of the off-peak load condition (typically at 19 h):

$$\text{Cooling capacity of LTES} = (\text{Cooling capacity of chiller at the specific hour of} \\ \text{part load condition} - \text{Total cooling load at that hour}) \\ + \text{Storage balance at the previous hour} \quad (10.14)$$

Thus, $(\text{cooling capacity of LTES})_{\text{commencement}} = (164 - 113) + 0 = 50 \text{ kW}.$

At the completion of the off-peak load condition (typically at 8 h):

Cooling capacity of LTES = (Total cooling load ((Nominal cooling capacity of the chiller

$$* \text{ peak load operating hours} * \text{efficiency factor})$$ (10.15)

$$+ \Sigma \text{ Cooling capacity of chiller during charging at part load conditions}))$$

Thus, $\left(\text{cooling capacity of LTES}\right)_{\text{completion}} = (4130 - ((204 * 10 * 0.9) + \Sigma 164 * 9))$

$$= 818 \text{ kW}$$

Cooling capacity shared by LTES = (Cooling capacity of LTES at the completion of part load

$$- (\text{Total cooling load}$$ (10.16)

$$- \text{Chiller cooling capacity during direct cooling}))$$

Thus, the cooling load shared by LTES during on-peak load conditions can be estimated by

$$\left(\text{Cooling load shared by LTES}\right)_{\text{on peak}} = \left(818 - \left(225 - 184\right)\right) = 777 \text{ kW}$$

Therefore, it can be observed from Table 10.1 that the LTES system is completely discharged at 18 h (completion of on-peak load demand period), which signifies that the LTES system is ready for the commencement of the next cycle of the charging process.

10.4 DESIGN EXAMPLES

10.4.1 Long-term thermal storage option

For a building cooling application, a chiller of 350 kW cooling capacity is considered to be fully operational for about 720 h. The criteria is to store a minimum annual cooling load, which is given by

$$\left(\text{Annual cooling load to be stored}\right)_{\text{min}} = 350 * 720 = 252,000 \text{ kWh}$$

For ice to be used as a latent heat storage medium, for every megajoule of cooling energy, about 3 liters of ice (equivalent to 3 kg) is required.

Thus, the volume of ice required for thermal storage can be determined by

$$\left(\text{Ice required}\right)_{\text{volume}} = 3 \text{ L / MJ} * 252,000 \text{ kWh} * 3.6 \text{ MJ/kWh}$$

$$= 2,721,600 \text{ liters} \approx 2722 \text{ m}^3$$

Assuming the depth of the storage tank to be 2 m,

$$\text{Area of the storage tank}, A = \left(2722\right)/2 = 1361 \text{ m}^2$$

Considering a cylindrical configuration of the storage tank,
Diameter of the storage tank = $(4A/\pi)^{0.5} = ((4*1361)/3.1415)^{0.5} \approx 42$ m.

Thus, based on the preceding calculations, it is seen that a very large diameter of the storage tank is actually required for meeting the seasonal cooling energy redistribution requirements in the building. Furthermore, this condition can also be treated as a major constraint in establishing long-term TES options for cooling applications in buildings, which involves additional cost factors.

10.4.2 Short-term thermal storage option

For the same cooling capacity of the chiller as discussed earlier, if the chiller is operated for about 10h, then the minimum annual cooling load is given by

$$(\text{Annual cooling load to be stored})_{min} = 350*10 = 3500 \, \text{kWh}$$

For ice to be used as a latent heat storage medium, for every megajoule of cooling energy, about 3 liters of ice (equivalent to 3 kg) is required.

Thus, the volume of ice required for thermal storage can be estimated by

$$\left(\text{Ice required}\right)_{volume} = 3 \, \text{L/MJ} * 3,500 \, \text{kWh} * 3.6 \, \text{MJ/kWh}$$
$$= 37,800 \, \text{liters} \approx 37.8 \, \text{m}^3$$

Assuming the depth of the storage tank to be 2 m,

$$\text{Area of the storage tank, } A = \left(37.8\right)/2 = 18.9 \, \text{m}^2$$

Considering a cylindrical configuration of the storage tank,

Diameter of the storage tank $= (4A/\pi)^{0.5} = ((4*18.9)/3.1415)^{0.5} \approx 5 \, \text{m}$

It is easily observed that the storage tank diameter is almost 8–9 times smaller than to that of the long-term storage option. Thus, the LTES systems can be integrated with building cooling systems, which can be advantageous in meeting the energy redistribution requirements on diurnal or short-term basis.

10.4.3 Short-term thermal storage option in piping systems

The chilled water circulating in the hydronic piping network, which connects the building side with the chiller plant, can be effectively used as a cool thermal storage medium for offsetting the peak load demand in buildings. Usually, the chilled water at a temperature of 8–9 °C is supplied to the cooling coil heat exchanger of the building air handling unit. Reducing the supply temperature of the chilled water by 4 or 5 °C prior to on-peak load conditions can help to reduce the load at the chiller during peaking periods of electricity. For instance, the volume of chilled water in the piping network is considered to be 4000 liters with the chilled water temperature differential remaining at 4 °C, and the refrigeration capacity of the chiller can be estimated by

$$\text{Refrigeration capacity} = \left(V*d*\Delta T*c_p\right)/3600 = (4000*1*4*4.2)/3600$$
$$= 18.7 \, \text{kWh}$$

This means that, 18.7 kWh of thermal energy is stored in the chilled water, which can be utilized an hour prior to the commencement of the on-peak load for experiencing a reduction in the load at the chiller side during electrical peak periods.

Assuming the electricity demand charges for this system to be $10/kW and the coefficient of performance of the chiller to be 3.5, the cost savings on annual basis can be obtained by

$$\text{Annual cost savings} = (18.7 * 10 * 12)/3.5 \approx \$642$$

Assuming the cost incurred for a timer and reset control for the system to be $1000, the payback period can be calculated by

$$(\text{Payback period})_{\text{simple}} = \$1000/\$642 \approx 1.6 \text{ years}$$

10.4.4 Heating thermal storage option with pressurized water systems

For a domestic hot water application, a pressurized storage system is considered to be integrated with the hot water circuit, in which the water present in the storage tank can be heated up to 135 °C prior to the occurrence of the peak electrical demand period. The hot water can be distributed at 90 °C during the peak period. Assuming 1 liter of water, the additional heat storage capacity of the water can be estimated based on Eq. (10.1) given by

$$Q = mc_p \Delta T = 1 * 4.2 * (135 - 90)$$

Therefore, $Q = 189$ kJ/kg $= 0.189$ MJ/kg or MJ/L

For a building with 2600 kW of electric water heating facility and with four pressurized storage tanks with a of volume 9500 liters installed, the storage energy required for the same temperature conditions of hot water as above can be evaluated by

$$\begin{aligned}
\text{Heating energy stored} &= \text{Volume} * \text{additional capacity of energy storage} \\
&= 4 * 9500 * 0.189 \\
&= 7182 \text{ MJ}
\end{aligned}$$

The electrical load reduction for 1 h can be determined by

$$\begin{aligned}
\text{Reduction in electrical load} &= (\text{Heating energy stored} * \text{Peak hour shift})/3.6 \\
&= (7182 * 1)/3.6 = 1995 \text{ kW}
\end{aligned}$$

where, the value of 3.6 is expressed in terms of MJ/kWh.

Assuming the electricity demand charge for this system to be $10/kW and for a heating season period of about 4.5 months, the cost savings on annual basis can be obtained by

$$\text{Annual cost savings} = (1995 * 10 * 4.5) \approx \$89,775/\text{year}$$

Assuming the cost incurred for the installation of the four pressurized storage tanks and the associated piping network to be $260,000, the payback period can be calculated by

$$(\text{Payback period})_{\text{simple}} = \$260,000/\$89,775 \approx 3 \text{ years}$$

10.4.5 TES option with waste heat recovery

An industrial batch process is considered in which about 20,000 liters of water at 90 °C are discharged to the drain. By installing a holding tank with a capacity 20,000 liters and a heat exchanger facility, the heat that is wasted from the discharged water can be effectively utilized for heating the incoming makeup water. Assuming the temperature of the water exiting the heat exchanger to be 30 °C, the heat energy being stored in the waste water tank during each drain cycle can be estimated using Eq. (10.1), given by

$$Q = mc_p \Delta T = \left(20000 * 4.2 * \left(90 - 30\right)\right)/1000$$

Therefore, $Q = 5023$ MJ

It is understood that roughly 80% of the heat energy can be transferred from the hot water to the incoming makeup water, whereas the remaining 20% can be accounted for inherent thermal losses occurring at the tank and piping elements. The waste heat that is recovered from the heat exchanger can help to reduce the preheating energy required for the gas-fired boiler to heat the incoming makeup water.

Assuming the costing charge for the gas to be $0.50/m³, the cost savings on an annual basis can be obtained by

$$\text{Annual cost savings} = \frac{QEf_eCfN}{Ef_b\left(\text{fuel heating value}\right)} \quad (10.18)$$

where, Ef_e is efficiency of heat exchanger (assumed to be 80%), Cf is fuel cost ($/m³), N is number of tank cycles/year (assumed to be 1000), and Ef_b is efficiency of boiler (assumed to be 70%).

Therefore, annual cost savings = (5023*0.8*0.5*1000)/(0.7*37.2) = $77,158.

Assuming the cost incurred for the heat exchanger, tank, and associated piping network to be $180,000, the payback period can be calculated by

$$\left(\text{Payback period}\right)_{\text{simple}} = \$180,000/\$77,158 \approx 2.3 \text{ years}$$

The capacity of the TES for this option depends on the volume of the liquid (water) being drained in each tank cycle as well as the temperature gradient between the entering and leaving water at the heat exchanger or heat pump. By properly maintaining the heat exchanger system from contamination or fouling effects, the reduction in the effectiveness of heat transfer can be prevented.

10.5 CONCISE REMARKS

The successful implementation of TES systems for cooling and heating applications primarily depends on the way they are designed to match the purpose. The crucial factors in determining the performance of the TES systems have to be carefully considered during the design phase. The inclusion of modest safety factors into the design of TES systems can help reduce oversizing or underestimation of chiller and storage systems. Flexibility in design methodology is most vital in the sense that it can accept additional storage capacities depending on the future load demand.

Further Reading

[1] ASHRAE. Thermal storage (Chapter 51). ASHRAE handbook—HVAC systems and equipment, 2012.

[2] Thermal storage. Energy management series 19 for industry commerce and institutions, Energy mines and Resources, Canada, https://oee.nrcan.gc.ca/sites/oee.nrcan.gc.ca/files/pdf/commercial/password/downloads/EMS_19_thermal_storage.pdf (accessed January 2014)

[3] Parameshwaran R, Kalaiselvam S. Energy efficient hybrid nanocomposite-based cool thermal storage air conditioning system for sustainable buildings. Energy 2013;59:194–214.

[4] Seaman A, Martin A, Sands J. HVAC thermal storage: practical application and performance issues; 2000, BSRIA Application Guide AG 11/2000.

[5] Parameshwaran R, Kalaiselvam S. Thermal energy storage technologies, nearly zero energy building refurbishment: a multidisciplinary approach. London: Springer Publications; 2013, pp. 483–536.

[6] Cool thermal energy storage: shifting cooling load off peak, energy design resources, 2009.

[7] Niehus TL. Designing a thermal energy storage program for electric utilities. In: Proceedings: ninth symposium on improving building systems in hot and humid climates, Arlington, TX, May 19-20; 1994. p. 285–8.

Review on the Modeling and Simulation of Thermal Energy Storage Systems

11.1 INTRODUCTION

In the scenario of ever-increasing energy demand and climate change issues, it is of prime importance to develop and integrate energy-efficient latent thermal energy storage (LTES) systems in building envelopes that will fulfill the energy redistribution requirements from on-peak to off-peak load periods. To achieve energy efficiency in the LTES systems, the materials capable of storing and releasing thermal energy depending on load conditions and that would possess good thermophysical properties must be properly selected. In addition, the phase change material (PCM) being selected for a specific application must also satisfy desirable TES performance at phase transition temperature. Prior to implementation of the PCM into real-time building applications the operational performance can be effectively analyzed by modeling and simulation methods. In this context, the major attributes of a variety of modeling and simulation approaches are discussed in the forthcoming sections.

11.2 ANALYTICAL/NUMERICAL MODELING AND SIMULATION

11.2.1 Latent thermal energy storage

The performance of LTES systems for real-time applications can be ascertained by conducting modeling and simulation procedures. The inherent ideology of carrying out performance analysis of LTES systems through analytical/numerical modeling and simulation is to achieve optimal design of the system with appropriate selection of the heat storage material. Many research studies pertaining to the modeling and simulation of LTES systems have been reported in recent years. The predictions obtained from research studies can be helpful in estimating overall thermal storage performance of the PCMs separately or integrated with real-time cooling/heating system applications.

The general formulation of the mathematical model of a simple freezing and melting processes comprises the most significant and fundamental phenomena known as the *classical Stefan problem*. The main aspect of the phase change problems is attributed to the formulation of moving boundary or the Stefan condition, in which the location of the solid-liquid interface in addition to the temperature field must be determined. For pure materials, the distinction between the solid and liquid phase is obvious, and the phase change process occurs at isothermal conditions [1–3].

In the case of conduction dominated heat transfer process, the generalized governing equation pertaining to the solid and liquid phases satisfying the Stefan condition can be expressed by [1]

For heat transfer in solid phase:

$$\rho c_s \frac{\partial T_s}{\partial t} = \frac{\partial}{\partial x}\left(k_s \frac{\partial T_s}{\partial x}\right)$$

(11.1)

For heat transfer in liquid phase:

$$\rho c_l \frac{\partial T_l}{\partial t} = \frac{\partial}{\partial x}\left(k_l \frac{\partial T_l}{\partial x}\right)$$

(11.2)

The heat balance or the Stefan condition imposed at the solid-liquid interface can be expressed by

$$\frac{\partial}{\partial x}\left(k_s \frac{\partial T_s}{\partial x}\right)n - \frac{\partial}{\partial x}\left(k_l \frac{\partial T_l}{\partial x}\right)n = \rho Lvn$$

(11.3)

By framing a set of assumptions to the phase change problems, the criticality in solving the moving boundary conditions during solid and liquid phase transformation processes can be reduced substantially and would lead to the so-called Stefan condition. The summary of the major physical factors involved in the phase transition processes and the assumptions leading to the Stefan condition are presented in Table 11.1.

Very few literatures are available to deal with the closed form for formulating phase change problems, and hence, approximate numerical solutions are generally considered for such a class of problems. The numerical methods intended for solving such a class of problems can be categorized into three main types, which include

- Fixed grid method
- Deforming grid method or front tracking scheme
- Hybrid method

In a *fixed grid method*, the interface boundary of the phase change problem associated with the fixed space grids can be tracked by using an auxiliary function. Albeit, this method is considered to be a weak solution, it is widely preferred for estimating the phase transition of the latent heat storage materials. The *deforming grid or the front tracking method* is a strong approach for the phase change problems, wherein the nodes of the grid are allowed to move along with the moving boundary layer (interface). By this, the deformation of the space grids occurs as the solution develops. The classical Stefan condition is explicitly used for tracking the interface position. The *hybrid method* is a combination of the fixed grid and the deforming grid methods, in which the features related to the fixed background grid and the local front tracking schemes are utilized to follow the movement of the interface boundary.

Simulation of a system refers to the imitation of the system being considered by a convenient approach or "stand-in" with which the operational performance of the system can be understood to a greater extent. The simulation of a system usually includes either a field trail method instead of an actual unmonitored process, or a laboratory experiment instead of the field trial or a mathematical model in place of a laboratory experiment. As general practice, the simulation of a system is performed through computational approach or computer codes and programs. The most important

---

Table 11.1 Summary of Major Physical Factors in Phase Change Processes and Assumptions for Stefan Problem [4]

Physical factors influencing phase change processes	Simplifying assumptions toward Stefan problem	Observations on the assumptions
Heat and mass transfer by virtue of conduction, convection, radiation with possible effects due to the gravitational, elastic, chemical, and electromagnetic forces	Isotropic heat transfer takes place by only conduction. Other effects are assumed to be negligible	Valid for pure materials, small container, and conditions with moderate temperature gradients
Storage and release of latent heat	Latent heat stored or released at the phase change temperature and remains constant	Reasonable and consistent assumption
Fluctuation in the phase transition temperature	Phase change temperature is a property of the material and known to be a fixed quantity	Reasonable and consistent assumption
Nucleation and supercooling effects	Assumed to be not present in the phase change material during its phase transformation processes	Reasonable and consistent assumption in many situations
Interfacial surface tension forces and curvature effects	The effects caused by surface tension forces and curvature can be assumed insignificant	Reasonable and consistent assumption
Thermophysical properties variation during phase change processes	For simplicity, the change in thermophysical properties in each phase is assumed constant ($c_s \neq c_L$ and $k_s \neq k_L$)	Reasonable and consistent assumption. The discontinuity across the interface is, however, allowed
Changes in density of the heat storage material	Assumed to be constant ($\rho_s \neq \rho_L$)	Unreasonable assumption, but considered to prevent the movement of the heat storage material

---

factors to be considered for carrying out the simulation using the computational method includes the following:

- Physical problem definition—the physical parameters or variables that decide the operational performance of the system must be identified and defined into the computational model with proper input information or data.
- Formulation and analysis—translate the physical problem into a precise mathematical problem to make it more sensible for analysis and arriving at the solution for the problem as well.
- Problem discretization—perform approximation of the formulated problem by replacing the continuous entities with discrete ones accordingly to develop a numerical algorithm for obtaining the solution for the problem.
- Construct and incorporate numerical algorithms in the form of computer codes for them to be analyzed. The resulting numerical solution can then be validated using the experimental measurements.

Table 11.2 Available Approaches to Model Phase Change Materials in Building Applications [2]

Approach	Accucacy	Time and CPU Cost
Enthalpy models	Good	Very high
TTM and VVM	Acceptable	Very high
RC models	Fair	High
CTF	Fair	Medium
Stefan models	Poor	Low

TTM, Temperature transforming model; VVM, Variable viscosity of the medium; RC, Resistant and capacitors; CTF, Conduction transfer function.

A variety of computational simulation and modeling techniques have recently been developed for analyzing the TES performance of PCMs dedicated to building cooling/heating applications. The summary of some major modeling approaches followed to model the phase transition behavior of the PCMs in building applications are presented in Table 11.2.

The method of combining the features pertaining to the enthalpy and the heat capacity models, in which a prediction step followed by a correction step [5], is introduced in Fig. 11.1(a). In this method, based on the guessed values, new nodal temperatures are predicted as indicated by point (2) on the graph.

The enthalpy is then determined based on the new temperature point subsequently. The predicted temperatures are corrected to be in line with the temperature-enthalpy curve as depicted by point (3) on the graph. This is the main aspect of the proposed method, and it is also referred to as *the quasi-enthalpy* method [6].

It is pertinent to note that the proposed method in [5] may not conserve energy at every time step as mentioned by [7]. Thus, a better conservative iterative approach has been proposed [8], in which the iteration between the predicted and corrected immediate values is performed until the convergence is accomplished. This method as illustrated graphically in Fig. 11.1(b) can be considered an alternative to the PCM simulation algorithm incorporated in ESP-r [9].

The method of temporal averaging has also been proposed in recent years [Morgan] as shown in Fig. 11.1(c), which can be expressed in the form of equation given by

$$C^A = \frac{\Delta h}{\Delta T} = \frac{h^n - h^{n-1}}{T^n - T^{n-1}} \tag{11.4}$$

This method is more prone to issues related to convergence and thus requires some precautionary steps to be taken before arriving at the final solution to the phase change problems. From this perspective, the major features, merits, and demerits and the possible solution schemes for the mathematical models dedicated for the latent heat evolution are summarized in Table 11.3.

It is noteworthy that the enthalpy method is found to be more attractive compared to other methods used for solving the phase change problems. This is due to the fact that the enthalpy model exhibits better computational efficiency, flexibility in selecting the schemes for the solutions, and modeling accuracy. Specifically, it can be advantageous if the corrective iterative scheme involving a fast and energy conservative approach and the noniterative scheme involving a quick but conservative approach at low time steps are realized.

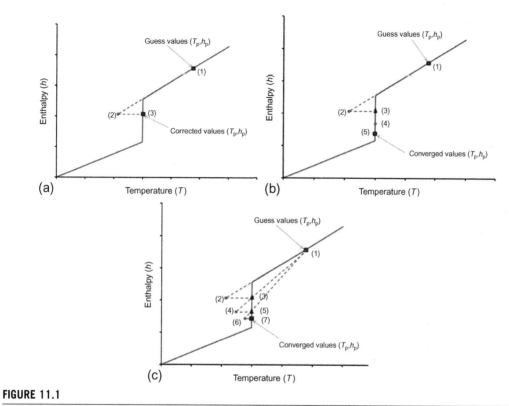

FIGURE 11.1

(a) Corrective noniterative scheme in the quasi-enthalpy method at a node during one time step, (b) corrective iterative scheme in the enthalpy method at a node during one time step, and (c) apparent heat capacity approximation at a node during one time step using iterative methods [1].

In the spectrum of modeling and simulation, there has been a continuous interest toward developing several models pertaining to building envelopes using simple, intermediate, and sophisticated methods. The summary of the modeling approaches for latent heat evolution in building envelopes is presented in Table 11.4. It can be observed from the comparison provided in Table 11.4 that the *simple models* offer simplified solutions, but they could not handle the complex heat transfer process involved in phase transition problems.

Intermediate models, on the other hand, are developed to study operational performance of the heat storage materials dedicated for specialized building envelopes. Like simple models, the intermediate models also cannot cope with the complex heat transfer process during phase change and advanced design alternatives. Thus, this type of model offers relatively fewer optimal solutions in reality.

Sophisticated models are much preferable to simple and intermediate models in the sense that flexibility in solving complex and multiphysics problems can be achieved using sophisticated models. However, their application toward modeling PCMs is not fully explored due to computational inefficiencies, detailed data inputs, very limited access to the source codes, the length of the model setup, and validations.

Table 11.3 Feature, Advantages, and Disadvantages of Mathematical Methods Used for Phase Change Problems [1]

Mathematical Model for Latent Heat Evolution	Main Feature	Advantages	Disadvantages	Possible Solution Schemes	References
Enthalpy method	Enthalpy accounts for sensible and latent heat	• Fast if proper scheme is selected • Deal with sharp as well as gradual phase change	• Difficult to handle supercooling problems • The temperature at a typical grid point may oscillate with time	Iterative scheme with nonlinear solvers (e.g., Newton's methods) Linearized-enthalpy: corrective iterative scheme Quasi-enthalpy: noniterative temperature correction scheme	[4,10] [7,8] [5]
Heat capacity method	Heat capacity accounts for both sensible and latent heat	• Intuitive because dealing with one dependent variable, "Temperature" • Easy to program • Suitable for gradual phase change	• Lack of computational efficiency • Small time step and fine grids are required for accuracy • Difficult in handling cases where the phase-change temperature range is small • Difficult to obtain convergence with this technique, and there is always a chance that the latent heat is underestimated • Not applicable for cases where phase change occurs at fixed temperature	Iterative scheme (e.g., Gauss-Seidel iterative scheme) if a proper heat capacity approximation is selected	[7,11–15]

Method	Description	Pros	Cons	Solution scheme	Ref.
Temperature transforming method	Heat capacity and source term are used to account for sensible and latent heat	• Deal with sharp and gradual phase change • Handle large time step and course grids	• Not a common method and therefore not tested to evaluate the pros and cons	Iterative scheme (e.g., Gauss-Seidel iterative scheme) after linearizing the source term	[16–21]
Heat source method	Latent heat is treated as a source term	• Intuitive due to separating the latent heat from sensible • Deal with sharp and gradual phase change	• Requires under-relaxation and therefore extra efforts is needed to determine the optimum relaxation factor • Lack of computational efficiency • Problems with round off errors if melting occurs over temperature range	Iterative scheme (e.g., Gauss-Seidel iterative scheme) after linearizing the source term	[22–25]

Table 11.4 Modeling Approaches for Latent Heat Evolution in Building Enclosure [1]

Complexity Level	Latent Heat Evolution's Approach	Building Enclosure Case Studied	Modeling Formulation	Solution Strategy	Validation	References
Simplified Models						
	Heat capacity method	Wall and roof	Steady-state analytical model			[26,27]
		Wallboard			Experimental	[28]
	Optimum nodes for heat capacity distribution using genetic algorithm	Wall	R-C Network		N/A	[29,30]
Intermediate Models						
	Enthalpy method	Wall	FVM: 1D	Newton's method	Experimental	[31]
		BIPV	FVM: 2D & 3D	Nonlinear solver	Experimental	[32–34]
		Wall	FVM: 1D	Iterative corrective scheme	Comparative	[9,35]
	Heat capacity method	Wall	FDM: 1D	G-S	Experimental	[36]
		Ceiling/roof	FDM: 1D	TDMA	Experimental	[37–39]
		Floor	FDM: 1D	G-S	Experimental	[40–43]
		Glazed-Windows	FDM: 1D	EM	Experimental	[44,45]
		Wallboard	FDM:2D		Analytical and experimental	[46]
		Wallboard	FDM: 3D	EM	Experimental	[47–50]
	Heat source method	Wall	FDM: 1D		Experimental	[51]
		Wall	FDM: 1D	TDMA	Experimental	[52]
		Wall	FDM: 1D	Iterative scheme	Analytical	[53]
		Wall	FDM: 2D	TDMA	Analytical, comparative and experimental	[54]

Sophisticated Numerical Packages

COMSOL	Heat capacity method and heat source method	Wall	FEM: 2D		Experimental	[55]
	Heat capacity method	Wall	FEM		Comparative	[56]
	Heat capacity method	Wall	FEM: 3D		Experimental	[57]
FLUENT	Heat source method	Wall	FEM: 3D	Simple algorithm	Experimental	[58]
	Heat source method	Wall	FEM: 2D		Experimental	[59]
HEATING	Heat capacity method	Wall and roof	FDM: 1D, 2D, and 3D	Point-successive over-relaxation iteration	Experimental	[60–62]

FVM, finite volume method; FDM, finite difference method; FEM, finite element method; G-S, Gauss-Seidel iterative method; TDMA, tridiagonal matrix algorithm; EM, explicit time stepping marching; R-C, resistance-capacitance.

In short, it is suggested to analyze phase change problems using these models and methods with small time steps, which in turn would enable achieving higher solution accuracy. However, for investigating the typical building thermal performance on a year-round basis, the simulation process using these models and methodologies may be relatively slower in nature. Moreover, the phenomena of supercooling and hysteresis, which are inherent characteristics of PCMs, are to be considered irrespective of the type of simulation analysis being performed, for accomplishing better predictions on the thermal storage performance of the PCMs.

The whole building simulation programs are equally important for analyzing thermal storage performance and the economic viability of PCMs for their integration into building structures. The comparison of various numerical methods for latent heat evolution in building simulation programs is presented in Table 11.5. Most of the models related to the PCMs being included in the whole building simulation programs are based on the heat capacity method. This necessitates reducing the simulation time step from 1 hour to smaller time steps in the order of minutes for achieving better accuracy.

The whole building simulation program intended for one-year analysis often exhibits solution instability, computational ineffectiveness, and nonconvergence of the solution to the PCM problems defined. Hence, to facilitate the whole building simulation programs, get rid of these constraints, be more functional and reliable, the efficient mathematical models, which are faster, more accurate, and more numerically stable under real-time steps are suggested to be implemented into the simulation programs.

11.3 CONFIGURATIONS-BASED MODEL COLLECTIONS

The thermal storage mechanism of PCMs can be greatly influenced by the type of geometric configuration of the storage containment. The heat transfer rate between the storage container and the PCM during phase transition is dependent on a number of factors, which include

- Size and shaper of the container
- Encapsulation methods
- Incorporation of heat transfer enhancement materials into the PCM
- Providing extended surfaces, etc.

Several configurations are available in practice for establishing the thermal storage performance of PCMs for different applications. The PCM configurations that are most commonly preferred for the containment of PCMs include the following:

- Rectangular configuration
- Cylindrical configuration
- Spherical configuration
- Finned configuration
- Packed-bed configuration
- Porous and fibrous materials configuration
- Slurry configuration

Table 11.5 Numerical Methods for Latent Heat Evolution in Building Simulation Programs [1]

Building Simulation	Module Identification	Numerical Formulation	Numerical Method Used for Latent Heat Evolution	Time Stepping Scheme	Limitations/ Constrains	Validation	References
EnergyPlus	CondFD	FDM: 1D	Heat capacity method	1. Implicit 2. Semi-implicit	• Time step <3 min • Small grids • Hysteresis in PCM is not modeled • Phase change at isothermal temperature is not modeled	Analytical, Comparative, and Experimental	[62–68]
TRNSYS	Modified "TYPE36"	FDM: 1D	Enthalpy method	Explicit	• Low time step No access to the code	Limited validation using experimental results for concrete	[69,70]
	"TYPE58"	FDM: 2D	Enthalpy method	Explicit	No access to the code	Experimental	[71]
	"TYPE204"	FDM: 3D	Heat capacity method	Select an appropriate factor for implicit, semi-implicit or explicit	Computationally inefficient	N/A	[72]
	"TYPE101"	FDM: 1D	Heat capacity method	Semi-implicit (Crank-Nicolson)	• A correction factor to account for cold bridges has to be used for model accuracy	Experimental	[73]

Continued

Table 11.5 Numerical Methods for Latent Heat Evolution in Building Simulation Programs—cont'd

Building Simulation	Module Identification	Numerical Formulation	Numerical Method Used for Latent Heat Evolution	Time Stepping Scheme	Limitations/ Constrains	Validation	References
	TRNSYS "Active Wall"	Equivalent heat transfer coefficients	Variable heat source function mimicking PCM behavior		• Real heat transfer physics in PCM is not modeled	Experimental	[74]
	"TYPE241"	FDM: 1D	Heat source method		No Published data	N/A	[75]
	"TYPE260"	FDM: 1D	Heat capacity method	Implicit	• Thermal properties including heat capacity are based on previous time step (i.e., explicit scheme)	Experimental	[76]
	Modified "TYPE101"	FDM: 1D	Heat capacity method	Implicit	• Developed for Internal partition wall	Experimental	[77]
	"TYPE1270"	Lumped method using heat balance	Quasi-heat source method		• Very simplified model • Internal layer within an envelope • Based on lumped heat balance (not a finite volume), low accuracy • For phase change at fixed temperature	N/A	[78]

ESP-r	SPMCMP53-SPMCMP56	FDM: 1D	Heat capacity and heat source methcd		• Low time step	N/A	[79,80]
BSim		FDM: 1D	Heat capacity methcd	Implicit	• Low time step to avoid instability	Experimental	[81]
RADCOOL		FDM: 1D	Heat capacity method	Implicit		Experimental	[82]
ESim		FDM: 1D	Heat capacity method	Explicit	• Explicit scheme requires low time step to avoid instability	Experimental	[83]

FVM, finite volume method; FDM, finite difference method.

Each configuration has its inherent influence on the heat transfer processes taking place between the PCM and the heat transfer source (or medium). Depending on the geometry of the containment, the TES or release capacity of the PCMs is varied. Also, depending on the application, the geometry of the PCM containment can be selected for achieving good thermal storage capability and heat transfer effectiveness.

It is well known that prior to large-scale implementation of the PCM TES systems, the investigation on their operational performance has to be carried out using the appropriate modeling and simulation tools. It is noteworthy that many interesting modeling and simulation-based research studies dealing with performance analysis of the PCM TES systems with different configurations of the containment have been reported in recent years. For brevity, some major research works that infer the significance of the modeling and simulation analyses being performed on different PCM configurations are reviewed and presented in Tables 11.6–11.12.

Table 11.6 Models for Rectangular Geometry [3]

Model	Material	Numerical Formulation	Comment	Validation
[84,85]		FD 2D	Surface integrated heat-transfer rates; boundary temperatures solidified fraction and interface position all as function of the time	
[86]	n-Octadecane	A 1D	Propagation and inclination of the interface and energy storage rate predicted	Experimental [87] Numerical [88,89]
[90–93]	n-Octadecane [51]	FG 2D	The melting process is chiefly governed by the magnitude of the Stefan number	Experimental [92,94] Numerical [95]
[96,97]		FD FG 2D	Multilayer PCM storage Parametric model formulated from numerical study	Numerical [98]
[99]	n-Octadecanol gallium, tin	FD 2D	For n-octadecanol, there is poor agreement with experimental results. Variation of viscosity with temperature, as heating losses in the wall or different initial supercooling are the proposed explanation for this discrepancy	Experimental [94,100–103]

Table 11.6 Models for Rectangular Geometry—cont'd

Model	Material	Numerical Formulation	Comment	Validation
[104]	Multiple PCM	FD 1D 2D	For 1D model, calculations have been made for the melt fraction and energy stored for conduction and convection, while for 2D model, calculations have been made for conduction only	Numerical [105,106]
[107]	$CaCl_2 \cdot 6H_2O$	A 2D	Thin flat containers and air is passed through gaps between them	On going, not published
[108]	$KF \cdot 4H_2O$	FE 1D A 2D FD CFD	Encapsulated PCM slab	Experimental [109]
[110]	Paraffin wax RII-56	FD 3D	Interpolating cubic spline function method is used for determining an effective specific heat	Yes
[111]	Paraffin wax	MM 1D	Charge and discharge of a PCM encapsulated slab	Yes
[112]	Lauric acid	MMLE* 2D	Laplace–Euler: new numerical method* Applicable to water and gallium	Experimental [113]
[114]	$CaCl_2\text{-}6H_2O$	FD 1D	Isothermal phase change of encapsulated PCM	Numerical [115] Yes Numerical [116]

11.4 MODELING AND SIMULATION ANALYSIS

11.4.1 Numerical solution and validation

It is appropriate to signify here that the development of numerical models and simulation analyses in one way greatly helps better understanding of the operational performance of the PCM TES systems. On the other hand, the validation of the numerical solutions using experimental measurements can be beneficial in terms of achieving optimum operational performance of PCM TES systems. However, in recent years, the focus has been on validating the numerical solution with respect to the numerical model being developed or with similarly published literatures data [3]. Hence, it is suggested to perform dedicated experimental procedures for validating numerical solutions using the test data and to tune the simulation problems accordingly toward the optimum solution.

Table 11.7 Models for Cylindrical Geometry [3]

Model	Material	Numerical Formulation	Comment	Validation
[117,118]	Paraffin wax RT 30	FG 2D	Melting+solidification	Yes
[119]	80.5% LiF+19.5% CaF2	FG 2D	Coaxial exchanger	No
[120]	Paraffin wax, $CaCl_2 \cdot 6H_2O$	A 2D	Second law	No
[121]		M 2D	Vertical cylinder with PCM at the bottom	Yes
[122]		MM 2D	Closed loop with a flat plate collector	Yes
[123]		MM 2D	Several MCPs with different melting temperatures	Yes
[124]	n-Eicosane	MM 2D	Multiblock FVM	Yes
[125]	$CaCl_2 \cdot 6H_2O$ paraffin,$Na_2SO_4 \cdot 10H_2O$ paraffin	FD FG 2D	Energy storage reservoir	Experimental [126]
[127]	n-Hexacosane	1D	Triplex concentric tube Temperature and thermal resistance iteration method	Yes

Table 11.8 Models for Spherical Geometry [3]

Model	Material	Numerical Formulation	Comment	Validation
[128]		MM 2D FD	Treatment of convection Two zones model	Numerical [129]
[130]	37.5% NH_4NO_3+62.5% $Mg(NO_3) \cdot 6H_2O$	A 1D	Transient position of interface, temperature distribution, melting fraction, energy released, and duration of complete solidification	No
[131]	n-Octadecane	A 2D	Contact melting on unfixed solid phase	Experimental [132] Numerical [133,134]
[135,136]	Ice		Capsule filled a 80% with an air cell on the top	Eames and Adref [137]
[138]		MM FD 1D	Numerical correlation relating the working fluid temperature to the time has been produced	Numerical [139–142]
[143]	Ice	MM FD 1D	Analysis of the impact of thermal conductivity of the shell material	Yes
[144]	Various PCM	A 2D	Solidification and melting of sphere with conduction, natural convection, and heat generation	Experimental [137]
[145]	Paraffin wax	FD	Convective environment outside capsule	Yes

Table 11.8 Models for Spherical Geometry—cont'd

Model	Material	Numerical Formulation	Comment	Validation
[146]		A 1D	Quasi-steady analysis	Numerical [137,147–151]
[152]		PG 1D	Correlations that express the dimensionless total solidification time of the PCM in terms of Stefan number, Biot number, and superheat parameter were derived	Numerical [150,153]
[154]	Parrafin wax RT27	MM 2D CFD	Partly filled capsule with open end	Yes + experimental [155,156]
[157]	Beewax	FG 2D	Constrained melting	Yes
[158]	*n*-Octadecane	MM CFD 2D	Constrained and unconstrained melting	Experimental [159]

Table 11.9 Models for Finned Surfaces [3]

Model	Material	Numerical Formulation	Comment	Validation
[160,161]		3D	Performance independent of geometry only surface area	Yes
[162]	*n*-Octadecane	FG 2D	An analytic 1D model is also produced	Yes
[163]		FG		
[164]		FD 2D FG	Validation of semianalytic result of [165]	Numerical [162] Kays and Crawford [166]
[167–169]	Paraffin RT 60 [167] LiF·MgF$_2$ [169]	FD 2D FG	Phase-change material kept inside a longitudinal internally finned vertical tube	Yes [167] [168] num. [163] [169] num. [170]
[171–174]	Paraffin	FE 2D	Analytical model validated	Yes
[175,176]	Sodium acetate trihydrate with graphite	A	Horizontal and vertical fins around circular vertical cylinder	Yes
[177]	Paraffin wax	MM 2D	Double rectangular enclosure where the top enclosure is filled with paraffin wax and the bottom is filled with water	Experimental [178]
[179]	Paraffin RT27	FD 2D FG	Slabs containing paraffin and metal fins made of aluminum	Yes
[180,181]	Ice [182] Paraffin [183]	FD 2D	Radially finned horizontal tube [180] Axially finned vertical tube [181]	Yes [181]

Continued

Table 11.9 Models for Finned Surfaces—cont'd

Model	Material	Numerical Formulation	Comment	Validation
[184]	Ice	FD 2D FG	Radially finned horizontal tube	Numerical [162] Zhang and Faghri [164]
[185]	*n*-Eicosane	FD 2D FG	Array of vertical fins	Numerical [186]
[182,183,187]	Paraffin wax RT25	2D 3D	Array of vertical fins	Numerical [182]
[173,188]	$CaCl_2 \cdot 6H_2O$	1D 3D	Also analytical 1D	Yes

Table 11.10 Models for Packed Bed [3]

Model	Material	Numerical Formulation	Comment	Validation
[189]		A 1D	A general model, great reference paper	Numerical [190–194]
[195]	Paraffin wax	FD 2D	Mass flow rate is determinant	Yes
[196–202]		PM 2D	Supercooling studied	Experimental [197,201,203]
[204]	Ice	1D	Supercooling is reduced by lower inlet heat transfer fluid (HTF) temperature	Yes
[205]	Paraffin wax RT20	PM FD FG 2D	Continuous solid phase model	Yes
[206]	Paraffin wax FNP-0090	FD FG 1D	Geometry analysis of capsules	Yes
[207]			Heat leaks through sidewalls	Experimental [208]
[209]	Ice	FD MM 1D	Effective heat conduction coefficient	Yes
[210]		1D	Parametric model	
[208,211]	Ice	1D	Solution by Laplace transform	Experimental [208]
[212]	$MgCl_2 \cdot 6H_2O$		Their own dimensionless number!	Yes
[213]	ZrO_2 Cu		First and second law efficiency, high temp	Yes
[214]	$KNO_3 \cdot NaNO_3$ NaCl, Pb Al-Si	1D	Six materials with various melting points	Yes
[215,216]		1D	Second law model	Yes
[217]		2D	First and second law efficiency. Fluent	Experimental [218]

Table 11.11 Models for Porous and Fibrous Materials [3]

Model	Material	Numerical Formulation	Comment	Validation
[219]	n-Octadecane	MM 1D	Second law efficiency calculated	Yes
[220,221]		FD 2D FG	Carbon foam impregnated with PCM	Khillarkar et al. [222]
[223,224]	n-Octadecane	FD 2D FG	Carbon-fiber brushes in PCM	Yes
[225]	n-Octadecane	FD 2D FG	Carbon-fiber chips in PCM	Yes
[226]	Paraffin wax	A 1D	Carbon nanofibers in paraffin wax	Lafdi and Matzek [226,227]
[228]	37:26:38 NH_4NO_3 $Mg(NO_3)_2 \cdot 6H_2O$ $MgCl_2 \cdot 6H_2O$	2D	Carbon fibers in PCM	Yes

Table 11.12 Models for Slurries [3]

Model	Material	Numerical Formulation	Comment	Validation
[229]		FD 1D	Circular ducts	Ahuja [230]
[231]		FD FG 1D	Circular tubes, constant heat flux	Experimental Goel et al. [232]
[233,234]		CFD	Effective heat capacity	Numerical Petukhov [235]
[236]	10% n-hexadecane in water	FD 3D FG	Turbulent heat transfer	Sparrow [237] Experimental Hartnett [238] Choi [239,240]
[191,241]		FD 2D FG	Circular tubes, constant heat flux	Experimental Goel et al. [232]
[242]		FD FG	Micro and nano particles	Experimental [232] Numerical [229,231]
[243,244]	30:5:65 Paraffin surfactant water	FD 2D FG	Natural convection, Newtonian, and non-Newtonian fluids	Yes Numerical Vahl Davis and Jones [245] Ozoe and Churchill [246]

Continued

Table 11.12 Models for Slurries—cont'd

Model	Material	Numerical Formulation	Comment	Validation
[247]	Octacosane	FD 2D FG	Mixed convection	Yes
[248]		FD 1D FG	Millimetric particles in oil	Numerical Bird et al. [249]
[250]		CFD 3D	3D single phase flow, FLUENT	Experimental Goel et al. [232]
[251]		FE 3D	Bulk fluid with adaptative specific heat	Experimental Goel et al. [232]
[252,253]		FD 2D FG	Analogy with thermal sources	Charunyakorn et al. [229] Alisetti and Roy [234]
[254]	Paraffin wax	2D	Neutrally buoyant ceramics	Experimental Jones et al. [124]

11.4.2 Materials selection and configuration

The most important aspects of the selection of TES materials and the configurations rely on the geometry and the thermophysical properties of the PCMs. The strong interdependencies between the convection effects on geometry and the size of the containment, as well as the viscous effects and thermal conductivity of the PCM, have to be carefully considered. Likewise, cost issues related to the embedding of the micro or nanomaterials into the pure PCM, viscosity increase, stratification effects, and thermal expansion during the phase transition processes and pumping power are also equally crucial from the perspective of modeling and simulation analyses [3].

11.4.3 Economic perspectives

The operational performance of PCM TES systems has to be evaluated from the economic point of view such that operational costs can be reduced considerably. The process of thermoeconomic optimization or exergoeconomic optimization can be performed on the LTES system as a whole, in addition to the modeling and simulation outcomes.

11.5 CONCISE REMARKS

The establishment of a variety of models ranging from simple to sophisticated enables the designer to correlate the significance of utilizing PCMs for LTES applications. These models can serve as a tool for the implementation of numerical method in the design of LTES systems. The numerical solutions being obtained if validated with the properly conducted experimental measurement values can facilitate the achievement of optimum or near optimum solutions for the phase change problems. The thermoeconomic or exergoeconomic evaluation systems are highly regarded to yield the best solutions for enhancing the functional aspects of PCM LTES systems on a long-term basis.

References

[1] AL-Saadi SN, Zhai ZJ. Modeling phase change materials embedded in building enclosure: a review. Renew Sustain Energy Rev 2013;21:659–73.

[2] Mirzaei PA, Haghighat F. Modeling of phase change materials for applications in whole building simulation. Renew Sustain Energy Rev 2012;16:5355–62.

[3] Dutil Y, Rousse DR, Salah NB, Lassue S, Zalewski L. A review on phase-change materials: mathematical modeling and simulations. Renew Sustain Energy Rev 2011;15:112–30.

[4] Alexiades V, Solomon AD. Mathematical modeling of melting and freezing processes. Hemispheres Publishing; 1993.

[5] Pham QT. A fast, unconditionally stable finite-difference scheme for heat conduction with phase change. Int J Heat Mass Transfer 1985;28:2079–84.

[6] Pham QT. Modelling heat and mass transfer in frozen foods: a review. Int J Refrig 2006;29:876–88.

[7] Voller VR. An overview of numerical methods for solving phase change problems. In: Minkowycz WJ, Sparrow EM, editors. Advances in numerical heat transfer. Taylor & Francis; 1997. p. 341–80.

[8] Swaminathan CR, Voller VR. On the enthalpy method. Int J Numer Meth Heat Fluid Flow 1993;3:233–44.

[9] Sadasivam S, Almeida F, Zhang D, Fung AS. An iterative enthalpy method to overcome the limitations in ESP-r's PCM solution algorithm. ASHRAE Trans 2011;117:100–7.

[10] Knoll DA, Keyes DE. Jacobian-free Newton–Krylov methods: a survey of approaches and applications. J Comput Phys 2004;193:357–97.

[11] Idelsohn S, Storti M, Crivelli L. Numerical methods in phase-change problems. Arch Comput Method E 1994;1:49–74.

[12] Hu H, Argyropoulos SA. Mathematical modelling of solidification and melting: a review. Model Simul Mater Sc E 1996;4:371–96.

[13] Poirier D, Salcudean M. On numerical methods used in mathematical modeling of phase change in liquid metals. J Heat Transfer 1988;110:562–70.

[14] Poirier DJ. On numerical methods used in mathematical modelling of phase change in liquid metals. Canada: University of Ottawa; 1986.

[15] Hashemi HT, Sliepcevich CM. A numerical method for solving two-dimensional problems of heat conduction with change of phase. Chem Eng Prog Symp 1967;63:34–41.

[16] Faghri A, Zhang Y. Transport phenomena in multiphase systems. Elsevier Academic Press; 2006.

[17] Zeng X, Faghri A. Temperature-transforming model for binary solid–liquid phase-change problems. Part I: mathematical modeling and numerical methodology. Numer Heat Tr B: Fund 1994;25:467–80.

[18] Zeng X, Faghri A. Temperature-transforming model for binary solid–liquid phase-change problems. Part II: numerical simulation. Numer Heat Tr B: Fund 1994;25:481–500.

[19] Cao Y, Faghri A, Soon CW. A numerical analysis of Stefan problems for generalized multi-dimensional phase-change structures using the enthalpy transforming model. Int J Heat Mass Transfer 1989;32:1289–98.

[20] Cao Y, Faghri AA. Numerical analysis of phase-change problems including natural convection. J Heat Transfer 1990;112:812–6.

[21] Wang S, Faghri A, Bergman TL. A comprehensive numerical model for melting with natural convection. Int J Heat Mass Transfer 2010;53:1986–2000.

[22] Crank J. Free and moving boundary problems. Clarendon Press; 1984.

[23] Eyres NR, Hartree DR, Ingham J, Jackson R, Sarjant RJ, Wagstaff JB. The calculation of variable heat flow in solids. Philos Trans Roy Soc London Ser A, Math Phys Sci 1946;240:1–57.

[24] Voller VR. Implicit finite—difference solutions of the enthalpy formulation of Stefan problems. IMA J Numeri Anal 1985;5:201–14.

[25] Swaminathan CR, Voller VR. Towards a general numerical scheme for solidification systems. Int J Heat Mass Transfer 1997;40:2859–68.

[26] Kaushik SC, Sodha MS, Bhardwaj SC, Kaushik ND. Periodic heat transfer and load levelling of heat flux through a PCCM thermal storage wall/roof in an air-conditioned building. Build Environ 1981;16:99–107.

[27] Chandra S, Kumar R, Kaushik S, Kaul S. Thermal performance of a non-air conditioned building with PCCM thermal storage wall. Energ Convers Manage 1985;25:15–20.

[28] Voelker C, Kornadt O, Ostry M. Temperature reduction due to the application of phase change materials. Energ Buildings 2008;40:937–44.

[29] Zhu N, Wang S, Xu X, Ma Z. A simplified dynamic model of building structures integrated with shaped-stabilized phase change materials. Int J Therm Sci 2010;49:1722–31.

[30] Zhu N, Wang S, Ma Z, Sun Y. Energy performance and optimal control of air conditioned buildings with envelopes enhanced by phase change materials. Energ Convers Manage 2011;52:3197–205.

[31] Drake JB. A study of the optimal transition temperature of PCM (phase change material) wallboard for solar energy storage; 1987.

[32] Huang MJ, Eames PC, Norton B. Thermal regulation of building-integrated photovoltaics using phase change materials. Int J Heat Mass Transfer 2004;47:2715–33.

[33] Huang MJ, Eames PC, Hewitt NJ. The application of a validated numerical model to predict the energy conservation potential of using phase change materials in the fabric of a building. Sol Energ Mat Sol C 2006;90:1951–60.

[34] Huang MJ, Eames PC, Norton B. Comparison of predictions made using a new 3D phase change material thermal control model with experimental measurements and predictions made using a validated 2D model. Heat Transfer Eng 2007;28:31–7.

[35] Almeida F, Zhang D, Fung AS, Leong WH. Comparison of corrective phase change material algorithm with ESP-r simulation. In: Proceedings of building simulation 2011: 12th conference of international building performance simulation association, Sydney; 2011.

[36] Chen C, Guo H, Liu Y, Yue H, Wang C. A new kind of phase change material (PCM) for energy-storing wallboard. Energ Buildings 2008;40:882–90.

[37] Pasupathy A, Velraj R. Mathematical modeling and experimental study on building ceiling system incorporating phase change material (PCM) for energy conservation. In: ASME conference proceedings; 2006. p. 59–68.

[38] Pasupathy A, Velraj R. Effect of double layer phase change material in building roof for year round thermal management. Energ Buildings 2008;40:193–203.

[39] Pasupathy A, Athanasius L, Velraj R, Seeniraj RV. Experimental investigation and numerical simulation analysis on the thermal performance of a building roof incorporating phase change material (PCM) for thermal management. Appl. Therm. Eng. 2008;28:556–65.

[40] Zhang YP, Lin KP, Yang R, Di HF, Jiang Y. Preparation, thermal performance and application of shape-stabilized PCM in energy efficient buildings. Energ Buildings 2006;38:1262–9.

[41] Lin K, Zhang Y, Xu X, Di H, Yang R, Qin P. Modeling and simulation of underfloor electric heating system with shape-stabilized PCM plates. Build Environ 2004;39:1427–34.

[42] Xu X, Zhang Y, Lin K, Di H, Yang R. Modeling and simulation on the thermal performance of shape-stabilized phase change material floor used in passive solar buildings. Energ Buildings 2005;37:1084–91.

[43] Zhou G, Zhang Y, Wang X, Lin K, Xiao W. An assessment of mixed type PCM-gypsum and shape-stabilized PCM plates in a building for passive solar heating. Sol Energ 2007;81:1351–60.

[44] Weinläder H, Pottler K, Beck A, Fricke J. Angular-dependent measurements of the thermal radiation of the sky. High Temp–High Press 2002;34:185–92.

[45] Weinläder H, Beck A, Fricke J. PCM-facade-panel for daylighting and room heating. Sol Energ 2005;78:177–86.

[46] Kedl RJ. Conventional wallboard with latent heat storage for passive solar applications. Energy conversion engineering conference 1990, In: IECEC-90 Proceedings of the 25th intersociety; 1990. p. 222–5.

[47] Kim JS, Darkwa K. Simulation of an integrated PCM–wallboard system. Int J Energ Res 2003;27:215–23.

[48] Kim JS, Darkwa J. Enhanced performance of laminated PCM wallboard for thermal energy storage in buildings. In: Energy conversion engineering conference; 2002. p. 647–51.

[49] Darkwa K, Kim JS. Dynamics of energy storage in phase change drywall systems. Int J Energ Res 2005;29:335–43.

[50] Darkwa K, O'Callaghan PW. Simulation of phase change drywalls in a passive solar building. Appl Therm Eng 2006;26:853–8.

[51] Athienitis AK, Liu C, Hawes D, Banu D, Feldman D. Investigation of the thermal performance of a passive solar test-room with wall latent heat storage. Build Environ 1997;32:405–10.

[52] Ait Hammou Z, Lacroix M. A new PCM storage system for managing simultaneously solar and electric energy. Energ Buildings 2006;38:258–65.

[53] Arnault A, Mathieu-Potvin F, Gosselin L. Internal surfaces including phase change materials for passive optimal shift of solar heat gain. Int J Therm Sci 2010;49:2148–56.

[54] Joulin A, Younsi Z, Zalewski L, Rousse DR, Lassue S. A numerical study of the melting of phase change material heated from a vertical wall of a rectangular enclosure. Int J Comput Fluid D 2009;23:553–66.

[55] Lamberg P, Lehtiniemi R, Henell A-M. Numerical and experimental investigation of melting and freezing processes in phase change material storage. Int J Therm Sci 2004;43:277–87.

[56] Baghban MH, Hovde PJ, Gustavsen A. Numerical simulation of a building envelope with high performance materials. In: COMSOL conference, Paris; 2010.

[57] Hasse C, Grenet M, Bontemps A, Dendievel R, Salle´e H. Realization, test and modelling of honeycomb wallboards containing a phase change material. Energ Buildings 2011;43:232–8.

[58] Gowreesunker BL, Tassou SA, Kolokotroni M. Improved simulation of phase change processes in applications where conduction is the dominant heat transfer mode. Energ Buildings 2012;47:353–9.

[59] Izquierdo-Barrientos MA, Belmonte JF, Rodrı´guez-Sa´nchez D, Molina AE, Almendros-Iba˜n˜ez JA. A numerical study of external building walls containing phase change materials (PCM). Appl Therm Eng 2012;47:73–85.

[60] Ahmad M, Bontemps A, Salle´e H, Quenard D. Experimental investigation and computer simulation of thermal behaviour of wallboards containing a phase change material. Energ Buildings 2006;38:357–66.

[61] Kosny J, Stovall TK, Shrestha SS, Yarbrough DW. Theoretical and experimental thermal performance analysis of complex thermal storage membrane containing bio-based phase change material (PCM). In: Thermal performance of the exterior envelopes of whole buildings XI international conference. Clearwater Beach, FL; 2010.

[62] Tabares-Velasco PC, Christensen C, Bianchi MVA. Validation methodology to allow simulated peak reduction and energy performance analysis of residential building envelope with phase change materials. In: 2012 ASHRAE annual conference. San Antonio, TX; 2012.

[63] Zhuang C, Deng A, Chen Y, Li S, Zhang H, Fan G. Validation of veracity on simulating the indoor temperature in PCM light weight building by EnergyPlus. In: Li K, Fei M, Jia L, Irwin G, editors. Life system modeling and intelligent computing. Berlin/Heidelberg: Springer; 2010. p. 486–96.

[64] Campbell KR. Phase change material as a thermal storage device for passive houses. Portland State University; 2011.

[65] Chan ALS. Energy and environmental performance of building facades integrated with phase change material in subtropical Hong Kong. Energ Buildings 2011;43:2947–55.

[66] Shrestha S, Miller W, Stovall T, Desjarlais A, Childs K, Porter W, et al. Modeling PCM-enhanced insulation system and benchmarking ENERGY-PLUS against controlled field data. In: Proceedings of building simulation 2011: 12th conference of international building performance simulation association. Sydney; 2011.

[67] Tabares-Velasco PC, Griffith B. Diagnostic test cases for verifying surface heat transfer algorithms and boundary conditions in building energy simulation programs. Journal of Building Performance Simulation 2011;1–18.

[68] Tabares-Velasco PC, Christensen C, Bianchi M. Verification and validation of EnergyPlus phase change material model for opaque wall assemblies. Build Environ 2012;54:186–96.

[69] Ghoneim AA, Klein SA, Duffie JA. Analysis of collector-storage building walls using phase-change materials. Solar Energy 1991;47:237–42.

[70] Ghoneim AA. Efficient collection and storage of solar energy. University of Alexandria; 1989.

[71] Stritih U, Novak P. Solar heat storage wall for building ventilation. Renew Energy 1996;8:268–71.

[72] Jokisalo J, Lamberg P, Siren K. Suitability of building construction materials in short-term energy storage-office room simulations. In: Proceedings of IEA annex 10-PCMs and chemical reactions for thermal energy storage, 3rd workshop. Finland; 1999. p. 11–8.

[73] Ahmad M, Bontemps A, Salle´e H, Quenard D. Thermal testing and numerical simulation of a prototype cell using light wallboards coupling vacuum isolation panels and phase change material. Energ Buildings 2006;38:673–81.

[74] Ibáñez M, Lázaro A, Zalba B, Cabeza LF. An approach to the simulation of PCMs in building applications using TRNSYS. Appl. Therm. Eng. 2005;25:1796–807.

[75] Schranzhofer H, Puschnig P, Heinz A, Streicher W. Validation of a TRNSYS simulation model for PCM energy storage and PCM wall construction element. In: Ecostock conference; 2006.

[76] Kuznik F, Virgone J, Johannes K. Development and validation of a new TRNSYS type for the simulation of external building walls containing PCM. Energ Buildings 2010;42:1004–9.

[77] Bontemps A, Ahmad M, Johannes K, Salle´e H. Experimental and modeling study of twin cells with latent heat storage walls. Energ Buildings 2011;43:2456–61.

[78] Thermal energy system specialists LLC -TESS Libraries—individual component libraries. URL: /http://www.trnsys.com/tess-libraries/individual-components.phpS [accessed: October/23/2012].

[79] Heim D. Two solution methods of heat transfer with phase change within whole building dynamic simulation. In: Proceedings of building simulation 2005: 9th conference of international building performance simulation association. Montre´al, Canada; 2005.

[80] Heim D, Clarke JA. Numerical modelling and thermal simulation of PCM–gypsum composites with ESP-r. Energ Buildings 2004;36:795–805.

[81] Rose J, Lahme A, Christensen NU, Heiselberg P, Hansen M. Numerical method for calculating latent heat storage in constructions containing phase change material building simulation 2009. In: 11th international IBPSA conference. Glasgow, Scotland; 2009.

[82] Corina S, Helmut EF, Frederick CW. Development of a simulation tool to evaluate the performance of radiant cooling ceilings. Lawrence Berkeley Laboratory Report, LBL-37300: University of California, Berkeley; 1995.

[83] Kelly Kissock J, Michael Hannig J, Whitney TI, Drake ML. Testing and simulation of phase change wallboard for thermal storage in buildings. In: Morehouse JM, Hogan RE, editors. International solar energy conference. Albuquerque: American Society of Mechanical Engineers; 1998.

[84] Shamsundar N, Sparrow EM. Analysis of multidimensional conduction phase change via the enthalpy model. J Heat Transfer Trans ASME 1975;333–40.

[85] Shamsundar N, Sparrow EM. Effect of density change on multidimensional conduction phase change. J Heat Transfer Trans ASME 1976;98:550–7.

[86] Hamdan MA, Elwerr FA. Thermal energy storage using a phase change material. Solar Energy 1996;56(2):183–9.

[87] Benard C, Gobin G, Martinez F. Melting in rectangular enclosures: experiments and numerical simulations. J Heat Transfer Trans ASME 1985;107:794–802.

[88] Yueng W. Engineering analysis of heat transfer during melting in vertical rectangular enclosures. Int J Heat Mass Transfer 1989;32:689–96.

[89] Webb B, Viskanta R. Analysis of heat transfer during melting of a pure metal from an isothermal vertical wall. Numer Heat Transfer 1986;9:539–58.

[90] Lacroix M. Computation of heat transfer during melting of a pure substance from an isothermal wall. Numer Heat Tran B 1989;15:191–210.

[91] Lacroix M. Contact melting of a phase change material inside a heated parallelepedic capsule. Energ Convers Manage 2001;42:35–44.

[92] Binet B, Lacroix M. Numerical study of natural-convection dominated melting inside uniformly and discretely heated rectangular cavities. J Numer Heat Transfer Part A 1998;33:207–24.

[93] Binet B, Lacroix M. E´ tude nume´rique de la fusion dans des enceintes rectangulaires chauffe´es uniforme´ment ou discre`tement par les parois late´ralesconductrices. Int J Therm Sci 1998;37: 607–620

[94] Gau C, Viskanta R. Melting and solidification of a pure metal on a vertical wall. J Heat Transfer 1986;108:174–81.

[95] Carslaw HS, Jaeger JC. Conduction of heat in solids. Oxford: Clarendon Press; 1959.

[96] Brousseau P, Lacroix M. Study of the thermal performance of a multi-layer PCM storage unit. Energ Convers Manage 1996;37:599–609.

[97] Brousseau P, Lacroix M. Numerical simulation of a multi-layer latent heat thermal energy storage system. Int J Energy Res 1998;22:1–15.

[98] Lane GA. Solar heat storage: phase change material. Background and scientific principles, vol. 1. Boca Raton, FL: CRC Press; 1986. p. 163.

[99] Costa M, Oliva A, Perez Segarra CD, Alba R. Numerical simulation of solid–liquid phase change phenomena. Comput Methods Appl Mech Eng 1991;91:1123–34.

[100] Gau C. Heat transfer during solid–liquid phase transformation of metals in rectangular cavities. Thesis, Purdue University; 1984.

[101] Wolff F, Viskanta R. Melting of a pure metal from a vertical wall. Exp Heat Transfer 1987;1:17–30.

[102] Ho CJ. Solid liquid phase change heat transfer in enclosures. Thesis, Purdue University; 1982.

[103] Ho CJ, Viskanta R. Heat transfer during melting from an isothermal vertical wall. J Heat Transfer 1984;12:12–9.

[104] Costa M, Buddhi D, Oliva A. Numerical simulation of a latent heat thermal energy storage system with enhanced heat conduction. Energ Convers Manage 1998;39(3–4):319–30.

[105] Voller VR. Numer Heat Tran B 1990;17:155.

[106] Goodrich L. Int J Heat Mass Transfer 1978;21:615.

[107] Vakilaltojjar SM, Saman W. Analysis and modeling of a phase change storage system for air conditioning applications. Appl Therm Eng 2001;21:249–63.

[108] Dolado P, La´zaro A, Zalba B, Marı´n JM. Numerical simulation of material behaviour of an energy storage unit with phase change materials for air conditioning applications between 17 °C and 40 °C. In: ECOSTOCK 10th international conference on thermal energy storage; 2006.

[109] Zalba B. Thermal energy storage with phase change Experimental procedure. Ph.D. thesis, University of Zaragoza; 2002.

[110] Zukowski M. Mathematical modeling and numerical simulation of a short term thermal energy storage system using phase change material for heating applications. Energ Convers Manage 2007;48:155–65.

[111] Silva PD, Goncalves LC, Pires LO. Transient behavior of a latent heat thermal energy store: numerical and experimental studies. Appl Energy 2002;73:83–98.

[112] Vynnycky M, Kimura S. An analytical and numerical study of coupled transient natural convection and solidification in a rectangular enclosure. Int J Heat Mass Transfer 2007;50(25–26):5204–14.

[113] Farid MM. Solar energy storage with phase change. J Solar Energy Res 1986;4:11–29.

[114] Zivkovic B, Fujii I. An Analysis of isothermal phase change of phase change material within rectangular and cylindrical containers. Sol Energy 2001;70(1):51–61.

[115] Humphries WR, Griggs EL. NASA Technical Paper, A design handbook for phase change thermal control and energy-storage devices. 1977, p. 1074.

[116] Voller VR. Fast implicit finite-difference method for the analysis of phase change problems. Numer Heat Tran B 1990;17:155–69.

[117] Trp A. An experimental and numerical investigation of heat transfer during technical grade paraffin melting and solidification in a shell-and-tube latent thermal energy storage unit. Sol Energy 2005;79:648–60.

[118] Trp A, Lenic K, Frankovic B. Analysis of the influence of operating conditions and geometric parameters on heat transfer in water-paraffin shell-and-tube latent thermal energy storage unit. Int J Heat Mass Transfer 2007;50(9–10):1790–804.

[119] Gong ZX, Mujumdar AS. Finite-element analysis of cyclic heat transfer in a shell-and-tube latent heat energy storage exchanger. Appl Therm Eng 1997;17(6):583–91.

[120] El-Dessouky H, Al-Juwayhel F. Effectiveness of a thermal energy storage system using phase-change materials. Energ Convers Manage 1997;38(6):601–17.

[121] Prakash J, Garg HP, Datta G. A solar water heater with a built-in latent heat storage. Energ Convers Manage 1985;25(1):51–6.

[122] Bansa NK, Buddhi D. An analytical study of a latent heat storage system in a cylinder. Energ Convers Manage 1982;33(4):235–42.

[123] Farid MM, Kanzawa A. Thermal performance of a heat storage module using PCM's with different melting temperatures: mathematical modeling. Trans ASME J Solar Energy Eng 1989;111:152–7.

[124] Jones BJ, Sun D, Krishnan S, Garimella SV. Experimental and numerical study of melting in a cylinder. Int J Heat Mass Transfer 2006;49:2724–38.

[125] Esen M, Ayhan T. Development of a model compatible with solar assisted cylindrical energy storage tank and variation of stored energy with time for different phase change materials. Energ Convers Manage 1996;37(12):1775–85.

[126] Esen M. Numerical simulation of cylindrical energy storage tank containing phase change material on the solar assisted heat pump system and comparing with experimental results. Ph.D. thesis, Trabzon, Turkey: Dept. of Mechanical Engineering, Karadeniz Technical University; 1994.

[127] Jian-you L. Numerical and experimental investigation for heat transfer in triplex concentric tube with phase change material for thermal energy storage. Sol Energ 2008;82:977–85.

[128] Roy SK, Sengupta S. Melting of a free solid in a spherical enclosure: effects of subcooling. J Sol Energy Eng 1989;111:32–6.

[129] Mack LR, Hardee MC. Natural convection between concentric spheres at low Rayleigh numbers. Int J Heat Mass Transfer 1968;11:387–96.

[130] Barba A, Spriga M. Discharge mode for encapsulated PCMs in storage tanks. Sol Energ 2003;74:141–8.

[131] Fomin SA, Saitoh TS. Melting of unfixed material in spherical capsule with nonisothermal wall. Int J Heat Mass Transfer 1999;42:4197–205.

[132] Bahrami PA, Wang TG. Analysis of gravity and conduction driven melting in a sphere. ASME J Heat Transfer 1987;19:806–9.

[133] Hoshina H, Saitoh TS. Numerical simulation on combined close-contact and natural convection melting in thermal energy storage spherical capsule. In: Proceedings of the 34th national heat transfer symposium of Japan; 1997. p. 721–2.

[134] Yamada K. Master Thesis. Supervised by T.S. Saitoh, Numerical simulation of the latent heat thermal energy storage tank. Tohoku University; 1997.

[135] Adref KT. A theoretical and experimental study of phase change in encapsulated ice store. Ph.D. thesis, England, UK: Department of Mechanical Engineering, Sheffield University; 1995.

[136] Adref KT, Eames IW. Freezing of water within spherical enclosures: an experimental study. In: Proceedings international conference on energy research, development, vol. 2; 1998. p. 643–59.

[137] Eames IW, Adref KT. Freezing and melting of water in spherical enclosures of the type used in thermal (ice) storage systems. Appl Therm Eng 2002;22:733–45.

[138] Ismail KAR, Henriquez JR, da Silva TM. A parametric study on ice formation inside a spherical capsule. Int J Therm Sci 2003;42:881–7.

[139] Shih YP, Chou TC. Analytical solution for freezing a saturated liquid inside or outside spheres. Chem Eng Sci 1971;26:1787–93.

[140] Pedroso IR, Domoto GA. Perturbation solutions for spherical solidification of saturated liquids. J Heat Transfer 1973;95(1):42–6.

[141] Hill JM, Kucera A. Freezing a saturated liquid inside a sphere. Int J Heat Mass Transfer 1983;26(11):1631–7.

[142] London LA, Seban RA. Rate of ice formation. Trans ASME 1943;65:771–8.

[143] Ismail KAR, Henriquez JR. Solidification of PCM inside a spherical capsule. Energ Convers Manage 2000;41:173–87.

[144] Veerappan M, Kalaiselvam S, Iniyan S, Goic R. Phase change characteristic study of spherical PCMs in solar energy storage. Sol Energ 2009;83:1245–52.

[145] Regin AF, Solanki SC, Saini JS. Experimental and numerical analysis of melting of PCM inside a spherical capsule. In: Proceedings of the 9th AIAA/ASME joint thermophysics and heat transfer conference (CD ROM); 2006. p. 12, Paper AIAA 2006-3618.

[146] Lin S, Jiang Z. An improved quasi-steady analysis for solving freezing problems in a plate, a cylinder and a sphere. ASME J Heat Transfer 2003;125:1123–8.

[147] Poots G. On the application of integral methods to the solution of problems involving the solidification of liquid initially at fusion temperature. Int J Heat Mass Transfer 1962;5:525–31.

[148] Beckett PM. Ph.D. thesis. England: Hull University; 1971.

[149] De DN, Allen G, Severn RT. The application of the relaxation method to the solution of non-elliptic partial differential equations. Q J Mech Appl Math 1962;15:53.

[150] Tao LC. Generalized numerical solutions of freezing a saturated liquid in cylinders and spheres. AIChE J 1967;13(1):165.

[151] Riley DS, Smith FI, Poots G. The inward solidification of spheres and circular cylinders. Int J Heat Mass Transfer 1974;17:1507–16.

[152] Bilir L, Ilken Z. Total solidification time of a liquid phase change material enclosed in cylindrical/spherical containers. Appl Therm Eng 2005;25:1488–502.

[153] Voller VR, Cross M. Estimating the solidification/melting times of cylindrically symmetric regions. Int J Heat Mass Transfer 1981;24(9):1457–62.

[154] Assis E, Katsman L, Ziskind G, Letan R. Numerical and experimental study of melting in a spherical shell. Int J Heat Mass Transfer 2007;50(9–10):1790–804.

[155] Katsman L. Melting and solidification of a phase-change material (PCM), Graduation Project 16–04. Heat Transfer Laboratory, Department of Mechanical Engineering, Ben-Gurion University of the Negev; 2004.

[156] Assis E, Katsman L, Ziskind G, Letan R. Experimental and numerical investigation of phase change in a spherical enclosure. In: Proceedings of the fourth European thermal sciences conference; 2004.

[157] Khodadadi JM, Zhang Y. Effects of buoyancy-driven convection on melting within spherical containers. Int J Heat Mass Transfer 2001;44:1605–18.

[158] Tan FL. Constrained and unconstrained melting inside a sphere. Int Commun Heat Mass Transfer 2008;35:466–75.

[159] Tan FL, Hosseinizadeh SF, Khodadadi JM, Fan L. Experimental and computational study of constrained melting of phase change materials (PCM) inside a spherical capsule. Int J Heat Mass Transfer 2009;52:3464–72.

[160] Sasaguchi K, Yoshida M, Nakashima S. Heat transfer characteristics of a latent heat thermal energy storage unit with a finned tube: effect of fin configuration. Heat Transfer—Jpn Res 1990;19(1):11–27.

[161] Sasaguchi K. Heat-transfer characteristics of a latent heat thermal energy storage unit with a finned tube. Heat Transfer—Jpn Res 1990;19(7):619–37.

[162] Lacroix M. Study of the heat transfer behavior of a latent heat thermal energy storage unit with a finned tube. Int J Heat Mass Transfer 1993;36(8):2083–92.

[163] Lacroix M, Benmadda M. Numer Heat Trans A Appl 1997;31(1):71–86.

[164] Zhang Y, Faghri A. Heat transfer enhancement in latent heat thermal energy storage system by using an external radial finned tube. J Enhance Heat Transf 1996;3(2):119–27.

[165] Zhang Y, Faghri A. Analytical solution of thermal energy storage system with conjugate laminar forced convection. Int J Heat Mass Transfer 1996;39(4):717–24.

[166] Kays WM, Crawford MW. Convective heat and mass transfer. 2nd ed. New York: McGraw-Hill; 1980.

[167] Velraj R, Seeniraj RV, Hafner B, Faber C, Schwarzer K. Experimental analysis and numerical modelling of inward solidification on a finned vertical tube for a latent heat storage unit. Sol Energ 1997;60(5):281–90.

[168] Velraj R, Seeniraj RV. Heat transfer studies during solidification of PCM inside an internally finned tube. J Heat Transfer 1999;121:493–7.

[169] Seeniraj RV, Velraj R, Lakshmi Narasimhan N. Thermal analysis of a finned tube LHTS module for a solar dynamic power system. Heat Mass Transfer 2002;38(4–5):409–17.

[170] Cao Y, Faghi F. Performance characteristics of a thermal energy storage module. Int J Heat Mass Transfer 1991;36:3851–7.

[171] Lamberg P. Mathematical modelling and experimental investigation of melting and solidification in a finned phase change material storage. Dissertation for the degree of Doctor of Science in Technology to be presented with due permission of the Department of Mechanical Engineering, Helsinki University of Technology; 2003.

[172] Lamberg P, Sire´n K. Analytical model for melting in a semi-infinite PCM storage with an internal fin. Heat Mass Transfer 2003;39:167–76.

[173] Lamberg P, Sire´n K. Approximate analytical model for solidification in a finite PCM storage with internal fins. Appl Math Model 2003;27:491–513.

[174] Lamberg P. Approximate analytical model for two-phase solidification problem in a finned phase-change material storage. Appl Energy 2004;77:131–52.

[175] Castell A, Sole´ C, Medrano M, Roca J, Garcı´a D, Cabeza LF. Effect of using external vertical fins in phase change material modules for domestic hot water tanks. In: International conference on renewable energies and power, quality (ICREPQ'06); 2006.

[176] Castell A, et al. Natural convection heat transfer coefficients in phase change material (PCM) modules with external vertical fins. Appl Therm Eng 2008;28:1676–86.

[177] Reddy KS. Thermal modeling of PCM-based solar integrated collector storage water heating system. J Sol Energy Eng 2007;129:458–64.

[178] Ahmet K, Ozmerzi A, Bilgin S. Thermal performance of a water-phase change material solar collector. Renew Energy 2002;26(3):391–9.

[179] Gharebaghi M, Sezai I. Enhancement of heat transfer in latent heat storage modules with internal fins. Numer Heat Tran A 2008;53:749–65.

[180] Ismail KAR, Henrı´quez JR, Moura LFM, Ganzarolli MM. Ice formation around isothermal radial finned tubes. Energ Convers Manage 2000;41:585–605.

[181] Ismail KAR, Alves CLF, Modesto MS. Numerical and experimental study on the solidification of PCM around a vertical axially finned isothermal cylinder. Appl Therm Eng 2001;21:53–77.

[182] Shatikian V, Ziskind G, Letan R. Numerical investigation of a PCM-based heat sink with internal fins. Int J Heat Mass Transfer 2005;48:3689–706.

[183] Shatikian V, Ziskind G, Letan R. Numerical investigation of a PCM-based heat sink with internal fins: constant heat flux. Int J Heat Mass Transfer 2008;51:1488–93.

[184] Kayansayan N, Ali Acar M. Ice formation around a finned-tube heat exchanger for cold thermal energy storage. Int J Therm Sci 2006;45:405–18.

[185] Akhilesh R, Narasimhan A, Balaji C. Method to improve geometry for heat transfer enhancement in PCM composite heat sinks. Int J Heat Mass Transfer 2005;48:2759–70.

[186] Bejan A. Heat transfer. New York: John Wiley & Sons; 1993.

[187] Shatikian V, Ziskind G, Letan R. Heat accumulation in a PCM-based heat sink with internal fins. In: 5th European thermal-sciences conference; 2008.

[188] Behunek I, Fiala P. Phase change materials for thermal management of IC packages. Radioengineering 2007;16(2):50–5.

[189] Zhang Y, Yan Su, Zhu Y, Hu X. A General model for analyzing the thermal performance of the heat charging and discharging process of latent heat thermal energy storage systems. Trans ASME J Sol Energy Eng 2001;123:232–6.

[190] Zhu Y, Zhang Y, Jiang Y, Kang Y. Thermal storage and heat transfer in phase change material outside a circular tube with axial variation of the heat transfer fluid temperature. ASME J Solar Energy Eng 1999;121:145–9.

[191] Kang Y, Zhang Y, Jiang Y, Zhu Y. A general model for analyzing the thermal characteristics of a class of latent heat thermal energy storage systems. ASME J Solar Energy Eng 1999;121(4):185–93.

[192] Kang Y, Zhang Y, Jiang Y. Simplified model for the heat transfer analysis of shell-and-tube with phase change material and its performance simulation. Acta Energy Solaris Sinica 1999;20(1):20–5 [in Chinese].

[193] Kang Y, Zhang Y, Jiang Y, Zhu Y. Analysis and theoretical model for the heat transfer characteristics of latent heat storage spherical packed bed. J Tsinghua Univ 2000;40(2):106–9 [in Chinese].

[194] Jiang Y, Zhang Y, Kang Y. Heat transfer criterion of plate thermal storage systems. J Tsinghua Univ 1999;39(11):86–9 [in Chinese].

[195] Benmansour A, Hamdan MA, Bengeuddach A. Experimental and numerical investigation of solid particles thermal energy storage unit. Appl Therm Eng 2006;26(5–6):513–8.

[196] Dumas JP, Be´de´carrats JP, Strub F, Falcon B. Modelization of a tank filled with spherical nodules containing a phase change material. In: Proceedings of the 10th international heat transfer conference, vol. 7; 1994. p. 239–44.

[197] Be´de´carrats JP, Strub F, Falcon B, Dumas JP. Experimental and numerical analysis of the supercooling in a phase-change energy storage. In: Proceedings of the 19th international congress of refrigeration, vol. IIIa; 1995. p. 46–53.

[198] Be´de´carrats JP, Dumas JP. Study of the crystallization of nodules containing a phase change material for cool thermal storage. Int J Heat Mass Transfer 1997;40(1):149–57 [in French].

[199] Kousksou T, Be´de´carrats JP, Dumas JP, Mimet A. Dynamic modeling of the storage of an encapsulated ice tank. Appl Therm Eng 2005;25:1534–48.

[200] Kousksou T, Bedecarrats JP, Strub F, Castaing-Lasvignottes J. Numerical simulation of fluid flow and heat transfer in a phase change thermal energystorage. Int J Energy Technol Policy 2008;6(1/2):143–58.

[201] Be´de´carrats JP, Castaing-Lasvignottes J, Strub F, Dumas JP. Study of a phase change energy storage using spherical capsules. Part I: Experimental results. Energ Convers Manage 2009;50:2527–36.

[202] Bedecarrats JP, Castaing-Lasvignottes J, Strub F, Dumas JP. Study of a phase change energy storage using spherical capsules. Part II: Numerical modelling. Energ Convers Manage 2009;50:2537–46.

[203] Be´de´carrats JP, Strub F, Falcon B, Dumas JP. Phase change thermal energy storage using spherical capsules: performance of a test plant. Int J Refrig 1996;19:187–96.

[204] Cheralathan M, Velraj R, Renganarayanan S. Effect of porosity and the inlet heat transfer fluid temperature variation on the performance of cool thermal energy storage system. Heat Mass Transfer 2007;43:833–42.

[205] Arkar C, Medved S. Influence of accuracy of thermal property data of a phase change material on the result of a numerical model of a packed bed latent heat storage with spheres. Thermochim Acta 2005;438(1–2):192–201.

[206] Wei J, Kawaguchi Y, Hirano S, Takeuchi H. Study on a PCM heat storage system for rapid heat supply. Appl Therm Eng 2005;25:2903–20.

[207] Seeniraj RV, Lakshmi Narasimhan N. The thermal response of a cold LHTS unit with heat leak through side walls. Int Commun Heat Mass Transfer 2005;32:1375–86.

[208] Chen SL, Yue JS. A simplified analysis for cold storage in porous capsules with solidification. ASME J Energy Resour Technol 1991;113:108–16.

[209] Ismail KAR, Henriquez JR. Numerical and experimental study of spherical capsules packed bed latent heat storage system. Appl Therm Eng 2002;22:1705–16.

[210] Ismail KAR, Stuginsky R. A parametric study on possible fixed bed models for PCM and sensible heat storage. Appl Therm Eng 1999;19:757–88.

[211] Chen SL. One dimensional analysis of energy storage in packed capsules. ASME J Sol Energy Eng 1992;114:127–30.

[212] Goncalves LCC, Probert SD. Thermal energy storage: dynamic performance characteristics of cans each containing a phase change material, assembled as a packed bed. Appl Energy 1993;45:117–55.

[213] Adebiyi GA, Nsofor EC, Steele WG, Jalalzadeh-Azar AA. Parametric study on the operating efficiencies of a packed bed for high-temperature sensible heat storage. J Solar Energy Eng 1998;120(1):2–13.

[214] Yagi J, Akiyama T. Storage of thermal energy for effective use of waste heat from industries. J Mater Process Technol 1995;48:793–804.

[215] Watanabe T, Kanzava A. Second law optimization of a latent heat storage system with PCMs having different melting points. Heat Recov Syst CHP 1995;15(7):641–53.

[216] Watanabe T, Hikuchi H, Kanzawa A. Enhancement of charging and discharging rates in a latent heat storage system by use of PCM with different melting temperature. Heat Recov Syst CHP 1993;13(1):57–66.

[217] MacPhee D, Dincer I. Thermodynamic analysis of freezing and melting processes in a bed of spherical PCM capsules. J Sol Energy Eng 2009;131:031017-1-11.

[218] Ettouney H, El-Dessouky H, Al-Ali A. Heat transfer during phase change of paraffin wax stored in spherical shells. ASME J Solar Energy Eng 2005;127:357–65.

[219] Erk HF, Dudukovic MP. Phase-change heat regenerators: modeling and experimental studies. AIChE J 1996;42(3):791–808.

[220] Mesalhy O, Lafdi K, Elgafy A. Carbon foam impregnated with phase change material (PCM) as a thermal barrier. In: Proceeding of the conference of the American Carbon Society. Providence, RI: Brown University; 2004.

[221] Mesalhy O, Lafdi K, Elgafy A, Bowman K. Numerical study for enhancing the thermal conductivity of phase change material (PCM) storage using high thermal conductivity porous matrix. Energ Convers Manage 2005;46:847–67.

[222] Khillarkar DB, Going ZX, Mujumdar AS. Melting of a phase change material in concentric horizontal annuli of arbitrary cross-section. Appl Therm Sci 2000;20:893–912.

[223] Fukai J, Hamada Y, Morozumi Y, Miyatake O. Effect of carbon-fiber brushes on conductive heat transfer in phase change materials. Int J Heat Mass Transfer 2002;45:4781–92.

[224] Fukai J, Hamada Y, Morozumi Y, Miyatake O. Improvement of thermal characteristics of latent heat thermal energy storage units using carbon-fiber brushes: experiments and modeling. Int J Heat Mass Transfer 2003;46:4513–25.

[225] Hamada Y, Ohtsu W, Fukai J. Thermal response in thermal energy storage material around heat transfer tubes: effect of additives on heat transfer rates. Sol Energ 2003;75:317–28.

[226] Elgafy A, Lafdi K. Effect of carbon nanofiber additives on thermal behaviour of phase change materials. Carbon 2005;43:3067–74.

[227] Lafdi K, Matzek M. Effect of nanofiber surface functionalization in carbon nano-fiber based nanocomposites systems. In: Proceedings of 35th international SAMPE technical conference, Carbon nanomaterials session, vol. 35; 2003, p. 1.

[228] Frusteri F, Leonardi V, Maggio G. Numerical approach to describe the phase change of an inorganic PCM containing carbon fibres. Appl Therm Eng 2006;26:1883–92.

[229] Charunyakorn P, Sengupta S, Roy SK. Forced convective heat transfer in microencapsulated phase change material slurries: flow in circular ducts. Int J Heat Mass Transfer 1991;34(3):819–33.

[230] Ahuja AS. Augmentation of heat transport in laminar flow of polystyrene suspensions: Part I. J Appl Phys 1975;46(8):3408–16.

[231] Zhang Y, Faghri A. Analysis of forced convection heat transfer in microencapsulated phase change material suspensions. J Thermophys Heat Transfer 1995;9(4):727–32.

[232] Goel M, Roy SK, Sengupta S. Laminar forced convection heat transfer in microencapsulated phase change material suspensions. Int J Heat Mass Transfer 1994;37(4):593–604.

[233] Alisetti EL. Ph.D. thesis, Forced convection heat transfer to phase change material slurries in circular ducts. University of Miami, Department of Mechanical Engineering; 1998.

[234] Alisetti EL, Roy SK. Forced convection heat transfer to phase change material slurries in circular ducts. J Thermophys Heat Transfer 2000;14:115–8.

[235] Petukhov BS. Heat transfer and friction in turbulent pipe flow with variable physical properties. In: Hartnett JP, Irvine Jr. TF, editors. Advances in heat transfer, vol. 6. New York: Academic Press; 1970. p. 503–64.

[236] Roy SK, Avanic BL. Turbulent heat transfer with phase change material suspensions. Int J Heat Mass Transfer 2001;44(12):2277–85.

[237] Sparrow EM, Hallman TM, Siegel R. Turbulent Heat transfer in the thermal entrance region of a pie with uniform heat flux. Appl Sci Res Ser A 1957;7:37–52.

[238] Hartnett JP. Experimental determination of thermal entrance length flow of water and of oil in a circular pipes. Trans ASME 1955;77:1120–211.

[239] Choi E. Forced convection heat transfer with water and phase-change material slurries: turbulent flow in a circular tube. Ph.D. thesis. PA: Drexel University; 1993.

[240] Choi E, Cho YI, Lorsch HG. Forced convection heat transfer with phase change material slurries: turbulent flow in a circular tube. Int J Heat Mass Transfer 1994;17:207–15.

[241] Hu X, Zhang Y. Novel insight and numerical analysis of convective heat transfer enhancement with microencapsulated phase change material slurries: laminar flow in a circular tube with constant heat flux. Int J Heat MassTransfer 2002;45:3163–72.

[242] Ho CJ, Lin JF, Chiu SY. Heat transfer of solid–liquid phasechange material suspensions in circular pipes: effects of wall conduction. Numer Heat Tran A 2004;45:171–90.

[243] Inaba H, Dai C, Horibe A. Numerical simulation of Rayleigh–Be´nard convection in non-Newtonian phase change-material slurry. Int J Therm Sci 2003;42:471–80.

[244] Inaba H, Zhang Y, Horibe A, Haruki N. Numerical simulation of natural convection of latent heat phase-change-material microcapsulate slurry packed in a horizontal rectangular enclosure heated from below and cooled from above. Heat Mass Transfer 2007;43:459–70.

[245] de Vahl Davis G, Jones IP. Natural convection in a square cavity: a comparison exercise. Int J Numer Methods Fluids 1983;3:227–48.

[246] Ozoe H, Churchill SW. Hydrodynamic stability and natural convection in Ostwald–de Waele and Ellis fluids: the development of a numerical solution. AIChE J 1972;18(6):1196–206.

[247] Cassidy D. Ph.D. thesis, Numerical and experimental investigations of mixed convection in solid–liquid flow for MicroPCM applications. North Carolina State University; 2008.

[248] Royon L, Guiffant G. Forced convection heat transfer with slurry of phase change material in circular ducts: a phenomenological approach. Energ Convers Manage 2008;49:928–32.

[249] Bird RB, Stewart WE, Lightfoot EN. Transport phenomena. New York: John Wiley & Sons; 1960.

[250] Sabbah R, Farid MM. Said Al-Hallaj. Micro-channel heat sink with slurry of water with micro-encapsulated phase change material: 3D-numerical study. Appl Therm Eng 2008;29:445–54.

[251] Kuravi S, Kota KM, Du J, Chow LC. Numerical investigation of flow and heat transfer performance of nano-encapsulated phase change material slurry in microchannels. J Heat Transfer 2009;131:062901-1-9.

[252] Zhang Y, Hu X, Wang X. Theoretical analysis of convective heat transfer enhancement of microencapsulated phase change material slurries. Heat Mass Transfer 2003;40:59–66.

[253] Zeng R, Wang X, Chen B, Zhang Y, Niu J, Wang X, et al. Heat transfer characteristics of microencapsulated phase change material slurry in laminar flow under constant heat flux. Appl Energy 2009;86:2661–70.

[254] Sun D, Annapragada SR, Garimella SV. Experimental and numerical study of melting of particle-laden materials in a cylinder. Int J Heat Mass Transfer 2009;52(13–14):2966–78.

Assessment of Thermal Energy Storage Systems

12.1 INTRODUCTION

Thermal energy storage (TES), a well-recognized technology, has become progressively more attractive over the years because of its capability to address the energy redistribution requirements between peak load and off-peak load periods. It is not only for this reason that TES systems would establish a good balance between energy supply and energy demand; they also contribute to reduced green house gas (GHG) emissions into the environment.

To accomplish these qualities, TES systems have to be designed so they would exhibit relatively low energy degradation and reduced inherent irreversibilities. In other words, TES systems are expected to be efficient on the basis of exergy and energy contents. It is more appropriate to infer that TES systems subjected to exergy analysis can be tuned for their maximum operational performance, especially during the charging and discharging processes.

12.2 EVALUATION OF THERMAL STORAGE PROPERTIES

In the scenario of global energy production, it has always been a bottleneck task to ensure a proper balance between energy supply and energy demand. This may be due to fact that the availability of energy for meeting demand requirements is surplus at one point in time or it would be less at other times. This gap can be effectively bridged through thermal energy storage integration in which the required quantity of thermal energy can be stored during off-peak load periods for offsetting the demand load during on-peak periods. The basic requirement or more likely the operation of a TES system solely depends on material that has the capability to store and release the required quantity of thermal energy either in the form of heat or cold, depending on the application. Three distinct heat storage material types are available for storing thermal energy, namely, sensible heat, latent heat, or thermochemical-based storage materials. Each category of heat storage materials possesses a variety of characteristics with regard to their thermal energy storage potential.

In many real-world applications in which heating or cooling systems are integrated with TES systems, latent heat storage materials are mostly preferred compared to sensible and thermochemical energy storage materials. To assess thermal energy storage capability including thermal conductivity, latent heat enthalpy, phase transition temperature, thermal stability, thermal reliability, freezing, and melting characteristics of latent heat storage materials, they must possess some vital properties for their integration into TES systems. The most desirable properties of latent heat storage materials are summarized in Table 12.1.

Table 12.1 Desirable Properties Governing the Selection of phase change material (PCM) [1]

Thermophysical Properties

- Phase change temperature suitable for building application
- High latent heat of fusion per unit volume so that smaller size of container can be used
- High thermal conductivity to assist in charging and discharging of PCM within the limited time frame
- High specific heat so that additional energy in the form of sensible heat is available to the thermal energy storage system
- Small volume change during phase transition and small vapor pressure at operating temperature to avoid the containment problem
- PCM should melt completely (i.e., congruent melting) during phase transition so that the solid and liquid phases are homogenous
- Thermally reliable (i.e., cycling stability) so that PCM is stable in terms of phase change temperature and latent heat of fusion and can be used in long run

Kinetic Properties

- High rate of nucleation to avoid supercooling of the PCM in liquid phase
- High rate of crystal growth so that heat recovery from the storage system is optimum

Chemical Properties

- Chemically compatible with construction/encapsulated materials
- No degradation after large number of thermal (freeze/melt) cycles to ensure long operation life
- Nontoxic, nonflammable, and nonexplosive to ensure safety
- Corrosion resistant to construction/encapsulated materials

Economic Properties

- Cost effective and commercially available

Environmental Properties

- Low environmental impact and nonpolluting during service life
- Having recycling potential

Based on the aforementioned properties, the selection of latent heat storage materials for specific TES integrated applications can be made, by which the evaluation of thermal storage properties of such materials can be performed using the requisite characterization techniques. The commonly adopted characterization methods for the evaluation of thermal properties of phase change materials (PCMs) are provided in Table 12.2. In particular, the four major thermal analysis methods that are most often practiced for determining thermal stability, crystallization, and phase change characteristics of PCMs are detailed in Table 12.3.

Thermal conductivity is one of the key parameter for PCMs, which describes the ability of PCMs to conduct heat during their phase transformation processes. Most of the organic PCMs possess relatively lower thermal conductivity than inorganic PCMs. As described in Table 12.1, PCMs selected for TES applications must have high thermal conductivity. Generally, the enhancement of thermal conductivity of PCMs can be accomplished by incorporating high thermal conductivity additives or materials, starting from the macro scale to the nanoscale level.

Table 12.2 Common Thermal Analysis (TA) Techniques with their Related Methods, Abbreviation, and Measured Property [2]

TA Measurement Technique	Method	Abbreviation	Output Property
Differential thermometry	Differential thermal analysis	DT/DTA	Temperature difference
Differential scanning calorimetry	—	DSC	Heat flow difference
Thermogravimetry	Thermogravimetric analysis	TG/TGA	Mass change
Themomechanometry	Thermomechanical analysis	TM/TMA	Deformation
Thermoacoustimetry	Thermoptometric analysis	TO/TOA	Optical properties
Exchanged/evolved gas measurement	Exchanged/evolved gas analysis	EGM/EGA	Gas exchange
Thermoelectrometry	Thermoelectrical analysis	TE/TEA	Electrical properties
Thermoacoustimetry	Thermoacoustimetric analysis	TO/TOA	Acoustic properties
Thermomanometry	Thermomanometric analysis		Pressure
Thermomagnetometry	Thermoagnetic analysis		Magnetic properties

Several techniques have been adopted for measuring thermal conductivity of pure PCMs and PCMs embedded with heat transfer enhancement materials (especially nanofluid PCM), as shown in Fig. 12.1. The relative change in the thermal conductivity or improvement of thermal conductivity of PCMs with respect to pure PCMs indicates their suitability for TES applications.

12.3 ENERGY AND EXERGY CONCEPTS

12.3.1 Distinction between energy and exergy

The terms *energy* and *exergy* are more related to the first and second laws of thermodynamics, wherein the former refers to the improvements and losses of energy and the latter refers to the exergy content of entropy and irreversibilities. Remarkable consideration has been given to exergy analysis for many years in almost all fields of science, engineering, and technology. A variety of definitions are being framed by different research groups to explain the concept of exergy. For ease of understanding, the various definitions of exergy are presented in Table 12.4.

The significance of exergy and its usage in several methods can be identified from the following:

- It is a basic tool that addresses the impact of the utilization of energy resources on the environment.
- It is an efficient method that combines the thermodynamic principles of mass and energy conservation with the second law of thermodynamics to perform the design analysis of various energy systems.

Table 12.3 Comparison Among Four Common TA Methods [2]

	Thermogravimetric Analysis (TGA)	Differential Thermal Analysis (DTA)	Differential Scanning Calorimetry (DSC)	T-History
Sample size (mg)	10–150	10–150	1–50	15,000
Measurement time (min)	100	100	100	40
Maintenance	++	++	++	+
Equipment price	++	++	++	+
Phenomenon	Thermal stability/decomposition, sublimation/evaporation/dehydration	Decomposition, glass transition, melting	Melting, glass transition, subcooling degree, reaction (curing/polymerization)	Melting, visual phase change, subcooling degree
Thermophysical properties	$-\%$sample mass loss $f(T,t)$	$-\Delta T\ f(T,t)$ $-H\ f(T,t)$	$-C_p\ f(T,t)$ $-H\ f(T,t)$ $-T_m$	$-C_p\ f(T,t)$ $-H\ f(T,t)$ $-T_m$ $-k$

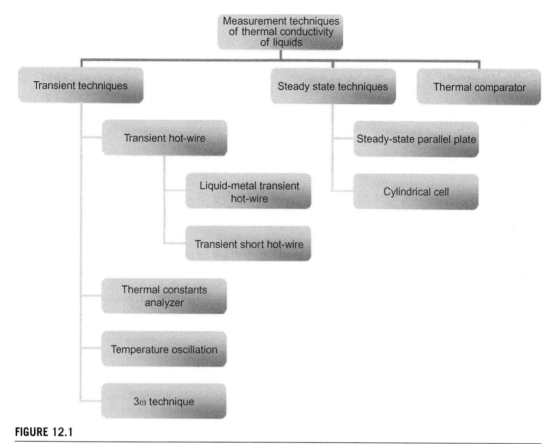

FIGURE 12.1

Different thermal conductivity measurement techniques for nanofluids [3].

- It is the best technique for identifying the energy losses or wastage of energy types or true magnitudes or at locations in the system for enhancing the effectual utilization of energy resources for better yield.
- It is a suitable method that infers the best possible means to design a system more energy efficient through identifying and reducing the inefficiencies of the system appropriately.
- It serves as the key tool for achieving sustainable development.

To have a better understanding of the exergy concepts for designing an energy-efficient system, one must be proficient in identifying the inherent difference existing between exergy and energy. For this purpose, the fundamental distinction between these concepts is given in Table 12.5. The highlighted demarcation between energy and exergy can benefit scientists, engineers, technologists, industrialists, and policy makers to develop energy-efficient systems for maintaining environmental sustainability.

Table 12.4 Various Exergy Definitions [4]

Investigators/ Sources	Exergy Definitions
[7]	Exergy is defined as that part of energy that can be fully converted into any other kind of energy
[8]	Exergy is the shaft work or electrical energy to produce a material in its specified state from materials common in the environment in a reversible way, heat being exchanged only with the environment at temperature T_0
[9,10]	Exergy is a measure of a quality of various kinds of energy and is defined as the amount of work obtainable when some matter is brought to a state of thermodynamic equilibrium with the common components of the natural surroundings by means of reversible processes, involving interaction only with the aforementioned components of nature
[11]	The work equivalent of a given form of energy is a measure of its exergy, which is defined as the maximum work and can be obtained from a given form of energy using the environmental parameters as the reference state
[5]	Exergy is defined as a measure of dispersion potential of energy and matter, and entropy is defined as a measure that indicates the dispersion of energy and matter
[6,12]	Exergy is the minimum theoretical useful work required to form a quantity of matter from substance present in the environment and to bring the matter to a specified state. Exergy is a measure of the departure of the state of the system from that of the environment and is therefore an attribute of the system and environment together
[6,13]	Exergy is the maximum theoretical work that can be extracted from a combined system consisting of the system under study and the environment as the system passes from a given state to equilibrium with the environment—that is, passes to the dead state at which the combined system possesses energy, but no exergy
[6,14]	The property exergy defines the maximum amount of work that may theoretically be performed by bringing a resource into equilibrium with its surroundings through a reversible process
[15]	The maximum fraction of an energy form, which (in a reversible process) can be transformed into work is called exergy. The remaining part is called anergy, and this corresponds to waste heat
[6,16]	Exergy is the concept that quantifies the potential of energy and matter to disperse in the course of their diffusion into their environment, to articulate what is consumed within a system
[17]	Exergy of a thermodynamic system is the maximum theoretical useful work (shaft work or electrical work) obtainable as the system is brought into complete thermodynamic equilibrium with the thermodynamic environment while the system interacts with this environment only
[18,19]	Exergy can be viewed as a measure of the departure of a substance from equilibrium with a specified reference environment, which is often modeled as the actual environment. The exergy of an emission to the environment, therefore, is a measure of the potential of the emission to change or affect the environment. The greater the exergy of an emission, the greater is its departure from equilibrium with the environment, and the greater may be its potential to change or affect the environment
[20]	The exergy of a person in daily life can be viewed as the best job that person can do under the most favorable conditions. The exergy of a person at a given time and place can be viewed as the maximum amount of work he or she can do at that time and place

Table 12.4 Various Exergy Definitions—cont'd	
Investigators/ Sources	**Exergy Definitions**
[21]	Exergy is the maximum amount of work that can be extracted from a physical system by exchanging matter and energy with large reservoirs in a reference state
[22]	In thermodynamics, the exergy of a system is the maximum useful work possible during a process that brings the system into equilibrium with a heat reservoir
[23]	In thermodynamics, exergy is a measure of the actual potential of a system to do work, while in systems energetics, entropy-free energy
[24]	Exergy expresses the quality of an energy source and quantifies the useful work that may be done by a certain quantity of energy
[25]	In thermodynamics, the exergy of a system is the maximum work possible during a process that brings the system into equilibrium with a heat reservoir

Table 12.5 The Main Differences Between Energy and Exergy [26]

Energy	Exergy
Is dependent on the parameters of matter or energy flow only and independent of environment parameters	Is dependent both on the parameters of matter or energy flow and on environment parameters
Has values different from zero (equal to mc^2 in accordance with Einstein's equation)	Is equal to zero (in a dead state by equilibrium with the environment)
Is guided by the first law of thermodynamics for all processes	Is guided by the first law of thermodynamics for reversible processes only (in irreversible processes it is destroyed partly or completely)
Is limited by the second law of thermodynamics for all processes (incl. reversible ones)	Is not limited for reversible processes due to the second law of thermodynamics
Is motion or ability to produce motion	Is work or ability to produce work
Is always conserved in a process, so can neither be destroyed nor produced	Is always conserved in a reversible process, but is always consumed in an irreversible process
Is a measure of quantity	Is a measure of quantity and quality due to entropy

12.3.2 Quality concepts

The concept of exergy plays a pivotal role in describing the energy content of resources through a quality measure. The quantification of energy resources can be performed through the exergy analysis. This makes more sense if the energy content of resources can be multiplied with a suitable quality factor that corresponds to the form of energy. The exergy per unit quantity relates to the physical value of the energy resource with respect to the environment.

The transformation of energy content of an energy resource to the exergy units using the quality factor well describes the primary route for establishing resource budgeting and its integration with conventional energy budgeting. Indeed, exergy can only provide a qualitative analysis of the energy resource

Table 12.6 A List of Some Common Energy Forms [26,27]

Energy Form	Quality Factor
Mechanical energy	1.0
Electrical energy	1.0
Chemical energy	~1.0 (may be more than 1, based on the system definition and states)
Nuclear energy	0.95
Sunlight	0.9
Hot steam (600 °C)	0.6
District heating (90 °C)	0.2–0.3 (depending on the outdoor temperature)
Moderate heating at room temperature (20 °C)	0–0.2 (depending on the outdoor temperature)
Thermal radiation from the earth	0

(physical quality) or the material conversions involved in the process. In that case, if the material is considered an exergy converter, then the quality of interest of the material would decide its efficiency. The common forms of energy with the quality factors are listed in Table 12.6.

12.3.3 Exergy in performance assessment of thermal storage systems

The operational performance of latent thermal energy storage (LTES) systems can be well assessed through applying the concepts of exergy and exergy efficiency. It is well known that based on the first law of thermodynamics, the efficiency of a thermal system can be defined by the ratio of energy output to energy input. In real-time applications, the energy output of thermal systems is much less compared to energy input, which is due to the inherent losses occurring during the thermodynamic process. Thus, by reducing the quantity of energy loss taking place in the thermal system, its energy efficiency can be improved.

However, the thermal systems being designed and assessed on the basis of first law may not be highly energy efficient, because the losses in the system do not actually reveal the degradation of energy. Hence, from the viewpoint of LTES systems, the degradation of energy taking place due to the inherent irreversibilities of the system is much more important, and this degradation has to be given careful attention during the performance assessment.

It is noteworthy that the second law evaluation of the LTES systems yields many more quantifying facts on both the degradation of energy and the energy loss as well [28]. For this reason, the exergy efficiency of the LTES systems is comparatively less than their energy efficiency. Generically, the exergy efficiency can be defined by

$$\Psi = 1 - \left(\frac{\text{Exergy destroyed}}{\text{Exergy input}} \right) \tag{12.1}$$

Thus, the second law of efficiency of the thermal system refers to the ratio of the exergy output (exergy destroyed) to the exergy input. That is, the more the degradation of energy, the more entropy is being generated. In this context, the entropy generation number (N_s) takes the form of the ratio of the exergy destroyed to the exergy input, and Eq. (12.1) can be rewritten as

$$\Psi = 1 - N_s \tag{12.2}$$

From the aforementioned equations, it is obvious that, by reducing the entropy generation number, the operational performance of the thermal system can be enhanced. The factor N_s plays a pivotal role in the optimization of the thermal systems. In the case of LTES systems, the exergy evaluation has to be performed for both charging and discharging processes. This would help to reduce any errors or discrepancies arising from the inherent irreversibilities of the system being investigated.

For the charging process of the LTES system, the exergy efficiency can be obtained by

$$\Psi_{char} = \left(\frac{\text{Exergy stored in the PCM}}{\text{Exergy supplied by the HTF}} \right) \tag{12.3}$$

where, HTF—heat transfer fluid. Because the heat transfer process is time dependent, the exergy efficiency by virtue of exergy rate can be expressed by

$$\Psi_{char} = \left(\frac{\text{Rate of exergy stored in the PCM}}{\text{Rate of exergy supplied by the HTF}} \right) \tag{12.4}$$

By taking into account the pumping power required for circulating HTF in the LTES system, Eq. (12.4) can be rewritten as

$$\Psi_{char} = \left(\frac{\text{Rate of exergy stored in the PCM}}{\text{Rate of exergy supplied by the HTF} + \text{Power input to compressor or pump}} \right) \tag{12.5}$$

The redefinition of Eq. (12.5) in terms of obtaining the maximum possible exergy that is supplied by HTF to the PCM can be estimated by

$$\Psi_{char} = \left(\frac{\text{Rate of exergy stored in the PCM}}{\text{Exergy rate possesed by the HTF before contact with PCM}} \right) \tag{12.6}$$

Similar to the charging process, the exergy efficiency for the discharging process of the PCM in the LTES system can be found by

$$\Psi_{dis} = \left(\frac{\text{Exergy gained by the HTF}}{\text{Initial exergy available with the PCM}} \right) \tag{12.7}$$

By using Eq. (12.7), the total exergy being retrieved by HTF from the maximum exergy available with the PCM during the discharging (melting) process can be easily assessed. In addition, the time dependent discharging process of the PCM leads to the evaluation of exergy efficiency in terms of the exergy rate, which is given by

$$\Psi_{dis} = \left(\frac{\text{Rate of exergy gained by the HTF}}{\text{Rate of exergy released by the PCM}} \right) \tag{12.7}$$

It is pertinent to note that to evaluate the operational performance of the LTES system, the exergy of both the charging and discharging processes have to be included into the exergy analysis. This will ensure the systematic approach of the LTES system's exergy assessment with no errors or discrepancies in the exergy efficiencies obtained for the charging and discharging processes. In this context, the overall exergy efficiency can be evaluated by

$$\Psi_{overall} = \left(\Psi_{char} * \Psi_{dis}\right) \tag{12.8}$$

The overall exergy efficiency of the LTES system as in Eq. (12.8) can also be expressed by

$$\Psi_{overall} = \left(\frac{\text{Exergy extracted from the PCM by the HTF during discharging}}{\text{Exergy input to the PCM during charging}}\right) \tag{12.9}$$

The exergy balance of the LTES system during the charging and the discharging processes is shown in Fig. 12.2. The notations about exergy efficiency of the LTES system for the charging and discharging processes, the comparison between the energy and exergy efficiencies, and the exergy analysis on LTES systems utilizing multiple PCMs as reviewed from some of the major research works are summarized in Tables 12.7–12.9.

12.3.4 Exergy and the environment

A deep interrelationship exists between exergy and the environment, especially applied to TES system applications. The thermal system with high energy efficiency, minimum energy loss, and low degradation of energy can help toward the development of a sustainable environment.

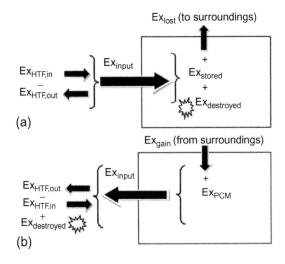

FIGURE 12.2

Exergy balance for a typical LHTS system (a) during charging and (b) during discharging [28].

Table 12.7 Various Forms of Exergy Efficiencies [28]

Efficiency	Expression	Description	References
Charging (Ψ_{char})	1. $\dfrac{Ex_{stored}}{Ex_{HTF}}$	Presents total exergy stored out of supplied	[38]
	2. $\dfrac{\dot{Ex}_{stored}}{\dot{Ex}_{HTF}}$	Presents time-wise variation of exergy efficiency	[39]
	3. $\dfrac{\dot{Ex}_{stored}}{\dot{Ex}_{HTF} + \text{Pump} \leftrightarrows \text{Work}}$	Takes into account the pumping power	[33]
	4. $\dfrac{\dot{Ex}_{stored}}{\dot{Ex}_{HTF,init}}$	Presents maximum possible exergy stored	[40,41]
Discharging (Ψ_{dis})	1. $\dfrac{Ex_{HTF}}{Ex_{PCM,init}}$	Presents total exergy recovered out of supplied	[38]
	2. $\dfrac{\dot{Ex}_{HTF}}{\dot{Ex}_{PCM}}$	Presents maximum possible exergy recovered	[30,40]
Overall ($\Psi_{overall}$)	1. $\dfrac{Ex_{recovered}}{Ex_{supplied}}$	Presents total exergy recovered out of supplied	[30]
	2. $\Psi_{char}\Psi_{dis}$	—	[31,32]
Charging/discharging/ overall	1. $1 - N_s$	Presents the quantity of exergy destroyed	[42]

The increased efficiency of TES systems in turn fulfills the energy redistribution requirements from on-peak to part load periods, thereby enabling the existing thermal system or plant to operate at reduced capacities (either cooling or heating). By this, the need for the establishment of a new or additional plant facility can be reduced, and that results in decreased emissions of GHG.

The incorporation of energy-efficient TES systems into cooling and heating systems can help reduce GHG emissions and contribute to achieving energy security in an environmentally acceptable manner as well. It is interesting to note that the integration of TES systems can help to achieve reduced CO_2 emissions in the range of 14–46%.

The Electric Power Research Institute reports that the utilization of TES systems contributed to a 7% reduction in CO_2 emissions. A recent research study suggests that the integration of ice TES

Table 12.8 List of Works Presenting the Comparison Between Energy and Exergy Efficiencies [28]

References	Latent Heat Thermal Storage (LHTS) Module	Application	Operation Mode	Exergy Efficiency Based on	Exergy Efficiency in Comparison with Energy Efficiency[a]
[29]	Rectangular tank	Cold thermal storage	Cycle	Exergy contents	Less by 80%
[30]	Not available	General	Cycle	Exergy contents	Less by 60%
[31]	Cylindrical tubes	Solar space heating	Cycle	Exergy contents	Less by 37%
[32]	Cylindrical tubes	Solar space heating	Cycle	Exergy contents	Less by 35%
[33]	Cylindrical tank	Solar greenhouse	Charging	Exergy contents	Less by 25%
[34]	Rectangular tank	Solar collector	Charging	Entropy generation	Less by 95%
[35]	Spherical capsules	General	Charging	Entropy generation	Less by 83%
[36]	Shell and tube	General	Discharging	Exergy contents	Less by 47%
[37]	Spherical capsules	Cold thermal storage	Charging and discharging	Entropy generation	Less by 7% (charging); Less by 14% (discharging)

[a]Approximate representative values.

systems with load leveling strategy for office building applications help to reduce the total energy usage by 4% with a significant reduction in emissions ranging from 3000 to 60,000 tons for total system capacities of 352 and 7034 kW. The cost-energy savings potential, energy efficiency, and comparison of TES systems are presented in Tables 12.10–12.14.

In short, TES systems that are subjected to exergy analysis and designed on the basis of exergy efficiency in addition to energy efficiency can achieve enhanced operational performance without sacrificing environmental sustainability [46–55].

Table 12.9 Exergy Based Investigations on LHTS Systems Using Multiple PCMs [28]

References	Number of PCMs	Operation Mode	Approach (Exergy/ Entropy Generation)	Operating/ Design Parameters Considered	Important Findings
[38]	7	Charging	Entropy generation	—	Minimum irreversibility when melting temperatures are linear
[40]	2, 3, and 5	Cycle	Exergy	Heat Transfer Fluid (HTF) inlet temperature, NTU	Exergy efficiency increases by two to three times with multiple PCMs Optimum melting points of PCMs are approximately a geometric progression
[43]	15 or 30	Cycle	Exergy	HTF inlet temperature, Initial temperature of PCM	Faster charging or discharging leads to high exergy efficiency Melting point difference between first and last PCM is key for optimum exergy performance
[44]	2	Cycle	Exergy	Velocity and inlet temperature HTF, NTU	Melting temperatures of first and last PCMs to be close to HTF and atmospheric temperatures, respectively, for optimum exergy efficiency
[45]	3	Cycle	Exergy	Latent heat, NTU	Increase in exergy efficiency due to multiple PCMs is not affected by latent heat of PCM

Table 12.10 Summary of Major Results from Various Case Studies [56]

Project Location	Capacity	Classification of TES			Heat Storage Material/Heat Transfer Medium	Observations	Application
		Type of Storage	Type of System	Type of Operation			
General Motors, Pontiac-MI	17,000 ton-hour	Sensible	Active	Short term	Stratified chilled water	• Peak load shaving capability without the need for additional chiller plant installation • Capital cost savings potential ~$200,000 • Operating cost savings potential ~$80,000/year	Cooling
Washington State University, Pullman, WA, USA	17,750 ton-hour	Sensible	Active	Short term	Stratified chilled water	• Peak load shaving capability for chilled water generating capacity as well as the distribution system capacity • Capital cost savings potential ~$1–$2 million • Operating cost savings potential ~$260,000/year	Cooling

District Cooling Utility, Lisbon, Portugal	39,800 ton-hour	Sensible	Active	Short term	Stratified chilled water	Cooling	• Installed chiller plant capacity reduced by 50% (from 11,373 to 5687 tons) • Provided economically feasible load management to a 8 MW CHP system compared to a 5 MW CHP system without TES • Capital cost savings potential ~$2.5 million • Operating cost savings potential ~$1,160,000/year
University of Alberta Edmonton, Alberta, Canada	60,000 ton-hour	Sensible	Active	Short term	Stratified chilled water	Cooling	• Peak load shaving capability for chilled water generating capacity as well as the distribution system capacity • Immediate capital cost savings potential: 30% • Annual operating cost reduction in campus cooling system: 12%

Continued

Table 12.10 Summary of Major Results from Various Case Studies—cont'd

| Project Location | Capacity | Classification of TES | | | Heat Storage Material/Heat Transfer Medium | Observations | Application |
		Type of Storage	Type of System	Type of Operation			
Chrysler R&D Center, Auburn Hills, MI, USA	68,000 ton-hour	Sensible	Active	Short term	Stratified chilled water	• Installed chiller plant capacity reduced from 17,700 to 11,400 tons) • Capital cost savings potential ~$3.6 million • Annual energy-cost savings potential ~$1 million	Cooling
OUCooling, Orlando, FL, USA	160,000 ton-hour	Sensible	Active	Short term	Stratified chilled water	• Full discharge rate up to 20,000 tons @ 8 h • Demand reduction: 15 MW • Capital cost savings potential ~$5 million • Operating cost savings potential ~$500,000/year	Cooling
Electric Power Utility, Riyadh, Saudi Arabia	193,000 ton-hour	Sensible	Active	Short term	Stratified chilled water	• Cooling capacity sharing to each of 10 large gas turbines: 3000 tons • Increase in the hot weather net power output of the power plant: 30% • Capital cost savings potential ~$10 million	Cooling

Location	Capacity	Storage type	Mode	Duration	Medium	Details	Application
DFW International Airport, Dallas/Ft Worth, TX, USA	90,000 ton-hour	Sensible	Active	Short term	Stratified low temperature fluid	• Full discharge rate up to 29,000 tons • Demand reduction: 17.4 MW • Capital cost savings potential ~$6 million • Operating cost savings potential ~$815,000/year	Cooling
New Jersey Stockton College	800 tons	Sensible	Active	Long term ATES	Groundwater	• COP: 9 • Annual energy savings: 60% • Annual energy savings: 500 MWh/year • Annual CO_2 reduction: 60%	Cooling
Massachusetts Confidential Client	400 tons	Sensible	Active	Long term ATES	Groundwater	• COP: 15 • Annual energy savings: 5610 GJ • Annual energy savings: 61.4% • Annual CO_2 reduction: 263 tons/year • Annual CO_2 reduction: 61.4%	Cooling
Long Island, NY Wyandanch Rising	1050 tons	Sensible	Active	Long term ATES	Groundwater	• COP: 5.2 (cooling), 3.5 (heating)	Cooling/heating

Continued

Table 12.10 Summary of Major Results from Various Case Studies—cont'd

| Project Location | Capacity | Classification of TES | | | Heat Storage Material/Heat Transfer Medium | Observations | Application |
		Type of Storage	Type of System	Type of Operation			
The Netherlands Eindhoven University	5700 tons (20 MW$_t$)	Sensible	Active	Long term ATES	Groundwater	• Annual energy savings: 2600 MWh/year (elec), 37,000 MWh/year (gas) • Annual CO$_2$ reduction: 13,300 tons/year	Cooling/heating
Stockholm, Sweden Arlanda Airport	2900 tons	Sensible	Active	Long term ATES	Groundwater	• COP: 17 • Annual energy savings: 4000 MWh/year (cooling), 10,000 MWh/year (cooling) • Annual CO$_2$ reduction: 7700 tons/year • Annual CO$_2$ reduction: 61.4%	Cooling/heating

Table 12.11 Summary of Major Results from Various Research Studies [57]

PCM Considered	PCMs Integration	Objectives of Study	Key Inferences	Methodology	References
Paraffin Type PCM					
Paraffin-based PCM	The phase-change wallboard containing 20% by paraffin mass	A prototype IEA building located in California climate condition was selected	28% of the peak cooling load was expected to be reduced	Simulation	[58]
Paraffin-based RT 54 Rubitherm GmbH	PCM storage tank coupled with building envelope made of concrete	Building to be incorporated with solar system that is situated in Blacksburg, VA, was chosen for the analysis	Yearly heating energy-cost was reduced by 61.5%	Simulation	[59]
RT25 (12 mm) and S27 (8.6 mm)	PCMs located in transparent plastic containers placed behind a double glazing with an air gap of 10 mm	Investigated on a south facade panel in Würzburg, Germany	25% of energy gains can be reduced in summer; likewise 30% heat losses and 50% solar heat gains can be reduced in winter	Experiment and simulation	[60]
Highly crystalline, n-paraffin-based PCM	PCM infused into the frame walls	A full instrumented test house of 1.83 m×1.83 m×1.22 m in Lawrence, Kansas, USA was investigated	The space cooling load and the average wall peak heat flux were found to be reduced about 8.6% and 15%, respectively	Experiment	[61]
Paraffin-based PCM	PCM is in thin-walled copper pipes and inserted into horizontal slots cut into the polystyrene foam	Fully instrumented test house of 1.83 m×1.83 m×1.22 m was analyzed	The peak heat flux can be reduced by 37% and 62% using a PCMSIP with 10% and 20% PCM concentrations	Experiment	[62]
n-paraffins mixture	PCM impregnated into the ceiling board	Test the performance of the PCM ceiling board in an office building at Tokyo, Japan	Running cost of this LHES system was reduced by 96.6% from the conventional rock-wool ceiling board	Simulation	[63]

Continued

Table 12.11 Summary of Major Results from Various Research Studies—cont'd

PCM Considered	PCMs Integration	Objectives of Study	Key Inferences	Methodology	References
PCM composed of foamed glass beads and paraffin waxes	PCM was embedded directly below OA floor boards in the form of granules	A small experimental system with a floor area of 0.5 m² was investigated	89% daily cooling load can be stored in night using a 30 mm thick packed bed of the granular PCM	Experiment	[64]
Mixture of commercial glycol wax	Incorporated in walls and roofs	An existing building in Campinas, SP, Brazil, was considered	Save 19% and 31% energy for cases using window and central AC units	Simulation and experiment	[65]
Paraffin-based PCM	Building interior and exterior wall structures were finished with PCM wall boards	Thermal performance of PCM wallboards were examined in a residential building (17×13×3 m) located in Boston	Possible cost savings up to $190 was recognized with these PCM wallboards. Economic analysis suggests for a 3-5 years payback period for these PCM wallboards	Simulation using TRNSYS software	[66]
K18 with an average melting temperature of 25.6 °C	Integrated in walls and roofs	Study includes concrete sandwich walls, low-mass steel walls under the typical meteorological weather data in Dayton, Ohio	The peak loads can be reduced by 19%, 30%, and 16% in concrete sandwich walls, steel roofs, and gypsum wallboards, respectively	Simulation	[67]
Graphite added paraffin-based PCM	Incorporated in ceiling	Improve energy storage potential with increase in thermal conductivty	Energy storage capability was enhanced by 12% by using graphite-paraffin-based PCM plates applicable for night ventilation	Experiment	[68]
Paraffin-based PCM	PCM embedded in the ceiling panel	Capture more solar radiation entering through windows onto the PCM for the office building	Heat loss from room can be recovered by 17–36% over the initial gains	Experiment	[69]
Paraffin-based PCM	Two layers of PCM floor structure integrated with other floor construction materials	Investigate the heat storage and release capacities of double-layer PCM floor to offset peak load shaving and acquire energy savings	Energy release rate of PCM-embedded floor structure increased heating and cooling load by 41.1% and 37.9%, respectively	Simulation	[70]

Salt Hydrate Type PCM					
Calcium chloride hexahydrate (CaCl$_2$·6H$_2$O) Salt hydrate	Cylindrical polyvinyl chloride enclosure containing PCM was placed in the storage tank	Study the heat transfer characteristics and energy efficiency of solar-assisted heat pump using this PCM storage technique at the Karadeniz Technical University	Conserved 9390–12,056 kWh of total energy spent on heating in winter season than by using various other heating utilities	Simulation using SOLSIM software and experiment	[71]
Salt type PCM that is held in stasis by a perlite matrix	Between two layers of insulation in a configuration known as resistive, capacitive, resistive	A geometry in which the wall/ceiling structure was assumed as a three-layer plane wall having the PCM in the center layer	The results suggest that 19-57% of maximum reduction in peak load as compared to a purely resistive R-19 wall can be achieved	Simulation	[72]
Calcium chloride	PCM filled in pipe was placed under the floor	Develop solar space heating system for residential buildings in UK	Experimental results suggest that heating energy consumption by heating utilities can be reduced by 18–32%	Experiment	[73]
Fatty Acid Type PCM					
Butyl stearate (25% by weight)	PCM impregnated with gypsum wallboard and acts as lining in interior of the test building room	Test the energy storage capacity of PCM-gypsum wallboard in an outdoor test building room of 2.82×2.22×2.24 m size located in Montreal	Total heating energy savings up to 15% was accomplished using the PCM wallboard	Simulation and experiment	[74]
Fatty acids	PCM combined with gypsum used as wallboard lining in interior of the room	Fully glazed multizone naturally ventilated building was considered for investigation	Substantial reduction of heating energy demand in winter was achieved by a value of 90%	Simulation using ESP-r	[75]
Hexadecane C$_{16}$H$_{34}$	Microencapsulated PCM slurry stored in slurry tank	Demonstrate the energy savings potential of hybrid system that combines chilled ceiling, MPCM slurry storage, and evaporative cooling methodologies for five representative climatic cities in China	Energy savings potential of 80% and 10% were acquired in the northwestern China and southeastern China, respectively	Simulation	[76]

Continued

Table 12.11 Summary of Major Results from Various Research Studies—cont'd

PCM Considered	PCMs Integration	Objectives of Study	Key Inferences	Methodology	References
Capric acid/n-octadecane	Building wall contains PCM spherical capsules of size 64 mm in diameter	Simultaneous management of solar and electrical energy in building was evaluated for a room of 5×5×3 m that has PCM storage wall thickness of 192 mm	Peak electricity utilization was conserved by 30–32% for space heating in building	Simulation	[77,78]
Fatty Acid Ester Type PCM					
Nanomaterials embedded PCMs-ethyl cinnamate, and dimethyl adipate	Nanomaterials embedded PCM spherical encapsulations	Demonstrate the energy redistribution requirements and energy efficiency of hybrid nanocomposite PCM-LTES cooling (HiTES) system suitable for hot and humid climatic condition	Improved thermal properties and heat storage potential. In summer and winter operating conditions, compared to the conventional system, the chiller cooling capacity in HiTES system was reduced by 46.3% and 39.6%, respectively, for part load and on-peak conditions	Simulation and experiment	[79,80]

Table 12.12 Solar Houses in Turkey [81]

Solar House	Location, Year of Construction	Type	Thermal Storage	Heat Transfer Fluid	Solar Energy Covering Heating Load (%)
MTA	Marmaris, 1971	Passive	Trombe wall	Air	30
MTA Chemistry Lab	Marmaris, 1981	Active	Gravel	Air	NA
Cukurova University	Adana, 1981	Passive	Greenhouse+Trombe wall	Air	NA
Ege University	Izmir, 1991	Passive	Greenhouse	Air	85
Ege University—Gama type	Izmir, 1990	Passive	Greenhouse+Trombe wall	Air	85
Hacettepe University	Ankara, 2003	Active+Passive	Gravel	Air	NA
METU	Ankara, 1980	Active+Passive	Water+Greenhouse	Water	22.4
Ankara Municipality	Ankara, 1993	Passive	Air	Air	73
TUBITAK guesthouse	Antalya, 1996	Passive	Greenhouse+Trombe wall	Air	NA
Erciyes University	Kayseri, 1996	Active	Water	Air	84.5
Erciyes University with floor heating	Kayseri, 1998	Active	Water	Water	86
Erciyes University Sports Hall	Kayseri, 2001	Active	Water	Air	73
Denizli PAU	Kayseri, 2007	Passive	Trombe wall+Water	Air+Water	NA
Diyarbakır	Diyarbakır, 2007	Passive	Greenhouse+Trombe wall	Air	NA

Table 12.13 Characteristics and Comparison of the Thermal Energy Storage Systems [82,83]

	Sensible Heat Storage System	Latent Heat Storage System	Thermochemical Storage System
Energy density	Small ~50 kWh m⁻³ of material	Medium ~100 kWh m⁻³ of material	High ~500 kWh m⁻³ of reactant
Volumetric density			
Gravimetric density	Small ~0.02-0.03 kWh kg⁻¹ of material	Medium ~0.05-0.1 kWh kg⁻¹ of material	High ~0.5-1 kWh kg⁻¹ of reactant
Storage temperature	Charging step temperature	Charging step temperature	Ambient temperature
Storage period	Limited (thermal losses)	Limited (thermal losses)	Theoretically unlimited
Transport	Small distance	Small distance	Distance theoretically unlimited [12]
Maturity	Industrial scale	Pilot scale	Laboratory scale
Technology	Simple	Medium	Complex

Table 12.14 Merits and Limitations of High Temperature TES Concepts for Power Generation [84]

System		Advantages	Disadvantages
Active storage	Direct system	• Intermediate heat transfer fluid and steam-generation exchanger is not necessary, improving the efficiency loss in steam generation • Overall plant configuratiis simpler • Lower investment and O&M costs • Allow the solar field to operate at higher temperatures, increasing the power cycle efficiency (reduction of LEC)	• Increase of pipe installation cost (is necessary to work at very high pressures) • Need of auxiliary protective heating systems for start-up, maintenance, and recovery from frozen conditions • Instability of the two phase flow inside the receiver tubes (procedures for filling and draining) • Difficult to control the solar field under solar radiation transients
	Direct steam generation		
	Two tanks	• Cold and hot HTF are stored separately • Low-risk approach • Possibility to raise the solar field output temperature to 450/500 °C (in trough plants), thereby increasing the Rankine cycle efficiency of the power block steam turbine to the 40% range • The HTF temperature rise in the collector field can increase up to a factor of 2.5, reducing the physical size of the thermal storage system	• Very high cost of the material used as HTF and TES • High cost of the heat exchangers and two tanks due to very large tank size requirements • Relatively small temperature difference between the hot and cold fluid in the storage system • Very high risk of solidification of storage fluid, due to its relatively high freeze point (that increases the M&O costs) • The high temperature of both tanks drives to an increase of losses in the solar field • The lowest cost TES design does not correspond to the lowest cost of electricity.

Continued

Table 12.14 Merits and Limitations of High Temperature TES Concepts for Power Generation—cont'd

System		Advantages	Disadvantages
Indirect system	Two tanks	• Cold and hot HTF are stored separately • Low-risk approach • The HTF temperature rise in the collector field can reduce the physical size of the thermal storage system • TES material flows only between hot and cold tanks, not through the parabolic troughs (decrease the risk of solidification of salts)	• Very high cost of the material used as TES • High cost of the heat exchangers and two tanks due to very large tank size requirements • Exchanger between the HTF and TES material is needed • Relatively small temperature difference between the hot and cold fluid in the storage system • The high temperature of both tanks drives to an increase of losses in the solar field • Decrease of the efficiency comparing with two tanks direct system
	Cascaded tanks	• Higher utilization of PCM-storage capacities • More uniform outlet temperature over time	• Heavy increase of cost, due to a higher number of: storage tanks, heat transfer fluid loops, and PCM • Further PCM need to be identified, which also offer sufficient heat of fusion and a satisfying corrosiveness • Not real experiences, only simulation

One tank (thermocline with filler materials)	• Decrease of storage tanks cost, due to this system using only one tank • Low cost of the filler materials (rocks and sand) • In cost comparisons, the thermocline system is about 35% cheaper than the two-tank storage system, due to reduction of storage volume and elimination of one tank.	• Relatively high freeze point of most molten salts formulations (is necessary to maintain a minimum system temperature to avoid freezing and salt dissociation) • More difficult to separate the hot and cold HTF • The high outlet temperature drives to an increase of losses in the solar field • Maintainance of thermal stratification requires a controlled charging and discharging procedure, and appropriate methods or devices to avoid mixing • Design of storage system was complex • Thermodynamically it was an inefficient power plant • This system is riskier with respect to the performance
Passive storage	Concrete/ceramics	• Very low cost of thermal energy storage media, due mainly to the filler cost • High heat transfer rates into and out of the solid medium (due to a good contact between the concrete and piping) • Facility to handling of the material • Low degradation of heat transfer between the heat exchanger and the storage material	• Increase of cost of heat exchanger and of engineering • Long-term instability

Continued

Table 12.14 Merits and Limitations of High Temperature TES Concepts for Power Generation—cont'd

System	Advantages	Disadvantages
PCM-sensible-PCM	• Increasing of capacity storage • Better use of PCM-storage capacities • Reduction of costs, comparing with storage systems with only PCM as storage media • Improvement of storage ratio, comparing with systems with only sensible heat materials • Laboratory-scale experiences	• Necessary to develop technologies to analyze this concept
Chemical storage (with NH_3)	• Unwanted side reactions are not possible, making solar reactors very easy to control • The endothermic reaction takes place at temperatures well suited to solar concentrators • Automatic phase separation of ammonia and hydrogen/nitrogen is provided and can use a common storage tank • There is around 100 years of industrial experience with the Harber-Bosch process	

12.4 CONCISE REMARKS

The assessment of TES systems on the basis of thermal storage properties of heat storage materials and exergy efficiency are considered to be most essential approaches for achieving enhanced operational performance during charging and discharging cycles. The selection of appropriate heat storage material for a specific TES application paves the way for effectual storage and release of thermal energy in demand periods.

TES systems designed on the basis of exergy efficiency (second law) in addition to energy efficiency (first law) can be expected to fulfill energy redistribution requirements in a more effective way. It is noteworthy that the second law evaluation of TES systems yields much more quantifying facts on both the degradation of energy and the energy loss as well.

Furthermore, analyzing the exergy contents of both charging and discharging processes (or the entire cycle of operation) together can minimize any errors or discrepancies associated with exergy efficiencies, thereby enhancing the operational performance of the TES system substantially. The exergetic assessment of TES systems reveal positive attributes on reducing GHG emissions for development of a sustainable future.

References

[1] Memon SA. Phase change materials integrated in building walls: a state of the art review. Renew Sustain Energy Rev 2014;31:870–906.

[2] Solé A, Miró L, Barreneche C, Martorell I, Cabeza LF. Review of the T-history method to determine thermophysical properties of phase change materials (PCM). Renew Sustain Energy Rev 2013;26:425–36.

[3] Paul G, Chopkar M, Manna I, Das PK. Techniques for measuring the thermal conductivity of nanofluids: a review. Renew Sustain Energy Rev 2010;14:1913–24.

[4] Hepbasli A. Low exergy (LowEx) heating and cooling systems for sustainable buildings and societies. Renew Sustain Energy Rev 2012;73–104.

[5] Shukuya M. Energy, entropy, exergy and space heating systems. In: Proceedings of the 3rd international conference: healthy buildings; 1994. p. 369–74.

[6] Sakulpipatsin P. Ph.D. Thesis, Exergy efficient building design. Technical University of Delft; 2008, ISBN 9780-90-6562-175-7.

[7] Rant Z. Exergy and anergy. Wiss Z Tech Univ Dresden 1964;13(4):1145–9.

[8] Rickert L. The efficiency of energy utilization in chemical processes. Chem Eng Sci 1974;29:1613–20.

[9] Szargut J. International progress in second law analysis. Energy 1980;5(8/9):709–18.

[10] Szargut J, Morris DR, Steward FR. Exergy analysis of thermal, chemical and metallurgical processes. New York, NY: Hemisphere Publishing; 1988.

[11] Kotas TJ. The exergy method of thermal plant analysis. London: Butterworth-Heinemann; 1985.

[12] Bejan A. Advanced engineering thermodynamics. New York: John Wiley and Sons; 1997.

[13] Moran MJ, Shapiro HN. Fundamentals of engineering thermodynamics. 3rd ed. New York: John Wiley and Sons; 1998.

[14] Connely L, Koshland CP. Exergy and industrial ecology—part 1: an exergy-based definition of consumption and a thermodynamic interpretation of ecosystem evolution. Int J Energy 2011;1(3):146–65.

[15] Honerkamp J. Statistical physics. Springer; 2002, p. 298.

[16] Ala-Juusela M, editor, Technical editing by Rautakivi A. Heating and cooling with focus on increased energy efficiency and improved comfort, guidebook to IEA ECBCS Annex 37—low exergy systems for heating and cooling of buildings, VTT;

[17] Tsatsaronis G. Definitions and nomenclature in exergy analysis and exergoeconomics. Energy 2007;32:249–53.

[18] Ao Y, Gunnewiek L, Rosen MA. Critical review of exergy-based indicators for the environmental impact of emissions. Int J Green Energy 2008;5:87–104.

[19] Gaudreau K, Fraser RA, Murphy S. The tenuous use of exergy as a measure of resource value or waste impact. Sustainability 2009;1:1444–63.

[20] Cengel YA, Boles MA. Thermodynamics: an engineering approach. 7th ed. McGraw-Hill; 2010.

[21] Wordiq. Exergy-definition, <http://www.wordiq.com/definition/Exergy> [accessed 5.04.11].

[22] Wikipedia. Exergy, <http://en.wikipedia.org/wiki/Exergy> [accessed 5.04.11].

[23] Wiktionary. Exergy,<http://en.wiktionary.org/wiki/exergy> [accessed5.04.11].

[24] Geoseries Glossary. Exergy, <http://www.geoseries.com/glossary.html> [accessed 5.04.11].

[25] Clickstormgroup. Glossary of fuel cell terms, <http://clickstormgroup.us.splinder.com/post/843218/glossary-of-fuel-cell-terms> [accessed 5.04.11].

[26] Dincer I. The role of exergy in energy policy making. Energy Policy 2002;30:137–49.

[27] Wall, G., 1986. ExergyFa useful concept. Ph.D. Thesis, Chalmers University of Technology, S-412 96 Göteborg, Sweden.

[28] Jegadheeswaran S, Pohekar SD, Kousksou T. Exergy based performance evaluation of latent heat thermal storage system: a review. Renew Sustain Energy Rev 2010;14:2580–95.

[29] Rosen MA, Pedinelli N, Dincer I. Energy and exergy analyses of cold thermal storage systems. Int J Energy Res 1999;23:1029–38.

[30] Venkataramayya A, Ramesh KN. Exergy analysis of latent heat storage systems with sensible heating and subcooling of PCM. Int J Energy Res 1998;22:411–26.

[31] Sari A, Kaygusuz K. Energy and exergy calculations of latent heat energy storage systems. Energy Sources 2000;22:117–26.

[32] Sari A, Kaygusuz K. First and second laws analyses of a closed latent heat thermal energy storage system. Chinese J Chem Eng 2004;12:290–3.

[33] Ozturk HH. Experimental evaluation of energy and exergy efficiency of a seasonal latent heat storage system for green house heating. Energy Convers Manage 2005;46:1523–42.

[34] Koca A, Oztop HF, Koyun T, Varol Y. Energy and exergy analysis of a latent heat storage system with phase change material for solar collector. Renew Energy 2008;33:567–74.

[35] Kousksou T, El Rhafiki T, Arid A, Schall E, Zeraouli Y. Power, efficiency, and irreversibility of latent energy systems. J Thermophy Heat Transfer 2008;22:234–9.

[36] Erek A, Dincer I. A New approach to energy and exergy analyses of latent heat storage unit. Heat Transfer Eng 2009;30:506–15.

[37] MacPee D, Duncer I. Thermodynamic analysis of freezing and melting processes in a bed of spherical PCM capsules. ASME J Sol Energy Eng 2009;131:031017/1.

[38] Watanabe T, Kanzawa A. Second law optimization of a latent heat storage system with PCMs having different melting points. Heat Recovery Syst CHP 1995;15:641–53.

[39] Kaygusuz K, Ayhan T. Exergy analysis of solar-assisted heat-pump systems for domestic heating. Energy 1993;18:1077–85.

[40] Gong ZX, Mujumdar AS. Thermodyanamic optimization of the thermal process in energy storage using multiple phase change materials. App Therm Eng 1996;17:1067–83.

[41] Demirel Y, Ozturk HH. Thermoeconomics of seasonal latent heat storage systems. Int J Energy Res 2006;30:1001–12.

[42] El-Dessouky H, Al-Juwayhel F. Effectiveness of a thermal energy storage system using phase-change materials. Energy Convers Manage 1997;38:601–17.

[43] Kousksou T, Strub F, Lasvignottes JS, Jamil A, Bedecarrats JP. Second law analysis of latent thermal storage for solar system. Sol Energy Mater Sol Cells 2007;91:1275–81.

[44] Domanski R, Fellah G. Exergy analysis for the evaluation of a thermal storage system employing PCMs with different melting temperatures. Appl Therm Eng 1996;16:907–19.

[45] Gong ZX, Mujumdar AS. Finite element analysis of a multistage latent heat thermal storage system. Numer Heat Transfer Part A 1996;30:669–84.

[46] Kalaiselvam S, Lawrence MX, Kumaresh GR, Parameshwaran R, Harikrishnan S. Experimental and numerical investigation of phase change materials with finned encapsulation for energy-efficient buildings. J Build Perform Simul 2010;3:245–54.

[47] Kalaiselvam S, Veerappan M, Arul Aaron A, Iniyan S. Experimental and analytical investigation of solidification and melting characteristics of PCMs inside cylindrical encapsulation. Int J Therm Sci 2008;47:858–74.

[48] Madhesh D, Parameshwaran R, Kalaiselvam S. Experimental investigation on convective heat transfer and rheological characteristics of Cu–TiO$_2$ hybrid nanofluids. Exp Therm Fluid Sci 2014;52:104–15.

[49] Veerappan M, Kalaiselvam S, Iniyan S, Goic R. Phase change characteristic study of spherical PCMs in solar energy storage. Sol Energy 2009;83:1245–52.

[50] Kalaiselvam S, Parameshwaran R, Harikrishnan S. Analytical and experimental investigations of nanoparticles embedded phase change materials for cooling application in modern buildings. Renew Energy 2012;39:375–87.

[51] Harikrishnan S, Kalaiselvam S. Preparation and thermal characteristics of CuO–oleic acid nanofluids as a phase change material. Thermochim Acta 2012;533:46–55.

[52] Rismanchi B, Saidur R, BoroumandJazi G, Ahmed S. Energy, exergy and environmental analysis of cold thermal energy storage (CTES) systems. Renew Sustain Energy Rev 2012;16:5741–6.

[53] Beggs CB. Ice thermal storage: impact on United Kingdom carbon dioxide emissions. Building Services Engineering Research and Technology 1994;15:756–63.

[54] Reindl DT. Characterizing the marginal basis source energy emissions associated with comfort cooling systems. USA: Thermal Storage Applications Research Center; 1994.

[55] Rismanchi B, Saidur R, Masjuki HH, Mahlia TMI. Energetic, economic and environmental benefits of utilizing the ice thermal storage systems for office building applications. Energ Build 2012;50:347–54.

[56] Worthington MA. Aquifer thermal energy storage: Feasibility study process and results for district energy systems, 2012. http://www.districtenergy.org/assets/pdfs/2012-Campus-Arlington/Presentations/Tuesday-C/2C1WORTHINGTONATES-FS-for-DES.pdf (accessed: March 2014).

[57] Parameshwaran R, Kalaiselvam S, Harikrishnan S, et al. Sustainable thermal energy storage technologies for buildings: a review. Renew Sust Energy Rev 2012;16:2394–433.

[58] Stetiu C, Feustel HE. Phase-change wallboard and mechanical night ventilation in commercial buildings. Lawrence Berkeley National Laboratory, University of California; 1998 (http://epb.lbl.gov/thermal/docs/pcm2.pdf).

[59] Hassan MM, Beliveau Y. Modeling of an integrated solar system. Build Environ 2008;43:804–10.

[60] Weinlader H, Beck A, Fricke J. PCM-facade-panel for daylighting and room heating. Sol Energ 2005;78:177–86.

[61] Zhang M, Medina MA, King JB. Development of a thermally enhanced frame wall with phase-change materials for on-peak air conditioning demand reduction and energy savings in residential buildings. Int J Energ Res 2005;29:795–809.

[62] Medina MA, King JB, Zhang M. On the heat transfer rate reduction of structural insulated panels (SIPs) outfitted with phase change materials (PCMs). Energy 2008;33:667–78.

[63] Kondo T, Ibamoto T. Research on thermal storage using rock wool phase-change material ceiling board. ASHRAE Transactions 2006;112:526–31.

[64] Nagano K, Takeda S, Mochida T, Shimakura K, Nakamura T. Study of a floor supply air conditioning system using granular phase change material to augment building mass thermal storage – Heat response in small scale experiments. Energ Build 2006;38:436–46.

[65] Ismail KAR, Castro JNC. PCM thermal insulation in buildings. Int J Energ Res 1997;21:1281–96.

[66] Stovall TK, Tomlinson JJ. What are the potential benefits of including latent storage in common wallboard? J Sol Energ Eng Trans, ASME 1995;117:318–25.

[67] Kissock JK, Limas S. Diurnal load reduction through phase-change building components. ASHRAE Trans 2006;112:509–17.

[68] Marin JM, Zalba B, Cabeza LF, Mehling H. Improvement of thermal storage using plates with paraffin–graphite composite. Int. J. Heat Mass Transfer 2005;48:2561–70.

[69] Gutherz JM, Schiler ME. A passive solar heating system for the perimeter zone of office buildings. Energy Sources 1991;13:39–54.

[70] Jin X, Zhang X. Thermal analysis of a double layer phase change material floor. Appl Therm Eng 2011; 10.1016/j.applthermaleng.

[71] Kaygusuz K. Performance of solar-assisted heat-pump systems. Appl Energ 1995;51:93–109.

[72] Halford CK, Boehm RF. Modeling of phase change material peak load shifting. Energ Build 2007;39:298–305.

[73] Kenneth S. Solar thermal storage using phase change material for space heating of residential buildings. University of Brighton, School of the Environment; 2002.

[74] Athienitis AK, Liu C, Hawes D, Banu D, Feldman D. Investigation of the thermal performance of a passive solar test-room with wall latent heat storage. Build Environ 1997;32:405–10.

[75] Heim D, Clarke JA. Numerical modelling and thermal simulation of PCM–gypsum composites with ESP-r. Energ Build 2004;36:795–805.

[76] Wang XC, Niu JL, Van Paassen AHC. Raising evaporative cooling potentials using combined cooled ceiling and MPCM slurry storage. Energ Build 2008;40:1691–8.

[77] Hammou ZA, Lacroix M. A new PCM storage system for managing simultaneously solar and electric energy. Energ Build 2006;38:258–65.

[78] Hammou ZA, Lacroix M. A hybrid thermal energy storage system for managing simultaneously solar and electric energy. Energy Convers Manage 2006;47:273–88.

[79] Parameshwaran R, Kalaiselvam S. Energy conservative air conditioning system using silver nano-basedPCM thermal storage for modern buildings. Energy Build 2014;69:202–12.

[80] Parameshwaran R, Kalaiselvam S. Energy efficient hybrid nanocomposite-based cool thermal storage air conditioning system for sustainable buildings. Energy 2013;59:194–214.

[81] Stritih U, Osterman E, Evliya H, Butala V, Paksoy H. Exploiting solar energy potential through thermal energy storage in Slovenia and Turkey. Renew Sustain Energy Rev 2013;25:442–61.

[82] Pardo P, Deydier A, Anxionnaz-Minvielle Z, Rougé S, Cabassud M, Cognet P. A review on high temperature thermochemical heat energy storage. Renew Sustain Energy Rev 2014;32:591–610.

[83] Gil A, Medrano M, Martorell I, Lazaro A, Dolapo P, Zalba B, et al. State of the art on high temperature thermal energy storage for power generation. Part 1 – concepts, materials and modelisation. Renew Sust Energy Rev 2010;14:31–55.

[84] Medrano M, Gil A, Martorell I, Potau X, Cabeza LF. State of the art on high-temperature thermal energy storage for power generation. Part 2 – case studies. Renew Sust Energy Rev 2010;14:56–72.

Control and Optimization of Thermal Energy Storage Systems

13

13.1 INTRODUCTION

In recent years, the development of thermal storage systems incorporating efficient control techniques has become increasingly attractive. In this sense, the proportional, integral, and derivative (PID) controllers are utilized for regulating the thermal parameters both in building spaces and TES as well. Because of the reasonably fast response delivered by PID controllers toward sudden changes in the controlled parameter, they are preferred for TES systems in buildings compared to conventional On/Off switch controllers. From the viewpoint of energy efficiency, advanced intelligent control methods including fuzzy logic, artificial neural network (ANN), or a combination of them can also be preferred.

The charging and discharging characteristics of TES play a significant role in achieving thermal load shifting during peak demand periods in buildings. The controllers intended herein actually respond to the changing demand conditions by regulating the charging and discharging capacities of the TES. For instance, if the TES is overstuffed (or excess cold energy being stored), it could result in higher energy consumption at the chiller (or cooling plant). This directly reflects on the operating costs of the system.

Similarly, if the discharging rate is too fast, the required peak load shifting could not be achieved in building spaces. This could lead to the rebouncing of peak load giving rise to exergy loss and uneven energy redistribution. Hence, the most important criteria in buildings integrated with TES systems is that a proper control and optimization of the energy system as a whole has to be performed to accomplish reduction in the operating cost without compromising energy efficiency. The major aspects related to the control and optimization of TES systems in buildings are discussed in forthcoming sections.

13.2 CONTROL SYSTEMS AND METHODOLOGIES

The terminology *control* means for the regulation of the functioning of the system through starting and stopping sequences on the basis of load fluctuations. The key steps involved in controlling a system include the following:

- Measuring the variable(s) and collection of data or information
- Processing the collected data or information with other/previously available data or information
- Performing the required control action based on the data or information processing sequence

Generically, these steps are functionally achieved through the incorporation of sensors, control device, or controlling unit amalgamated with the system to be controlled. In most of the control systems operation, a proper control over the functioning of the system can be accomplished by using the following constituents, which are

- Sensor—measures the variable or parameter to be controlled
- Controller—receives the input from the sensor, processes the sensed information, and outputs the required control action for controlling the system
- Energy source—electrical or pneumatic for supplying power to the controller and the sensor for energizing their operation

The major task for a controller is to maintain systems operation well within the set limits or set points of design requirements. This can ensure the system to meet the fluctuating demand load without compromising energy efficiency. Two broad classification of controls are recognized in engineering systems, especially applied to thermal energy systems, and they are as follows:

- Open-loop control
    - Also referred to as the feed forward control with no feedback. Limitations exist to monitor whether the control system is fully operational with respect to the changing demand conditions of the system. This is due to the fact that the output signal is not processed again by the controller to ensure whether the effective control action has been performed.
    - The control action can be made through the sequence or open loop condition, whereby the sensed information executes the cut-in and cut-off action on the power source of the controlled system. One example is the heating control provided in a hot air oven. The resulting temperature in the oven may not be a constant and varies about a set point, by which the steady state operation of the system could not be ensured.
- Closed-loop control
    - This type of control is usually referred to as feedback control. The control action can be made by the controller through the comparison of output information with the desired value to be maintained by the system.
    - The signal (output) corresponding to the variable to be controlled can be fed back to the controller for processing with the set value for achieving the required control over the system performance toward modulating load conditions.
    - In general, control action can be expressed in terms of the error of the controlled variable given by

$$\text{Error} = \text{Set point value} - \text{actual value} \tag{13.1}$$

    For instance, in an air-conditioning application, if the set point value is 24 °C, and the actual value being sensed is 26 °C, then the error in control can be established as −2 °C. With this comparison, the controller activates the compressor of the chiller plant; thereby, the temperature of the space is reduced to the set point value. On the other hand, if the temperature is dropped below the set point and, say, the error is +2 °C, the compressor then is switched Off by means of the controller.
    - The schematic representation of the basic feedback control scheme is depicted in Fig. 13.1. By achieving improved sensitivity and feedback control on the controlling parameter or variable, the system can be made energy efficient in the long run.
    - In reality, most of the conventional thermal energy systems utilize the feedback control scheme for their efficient functioning relative to the erratic changes occurring in the thermal loads apart from their design considerations.

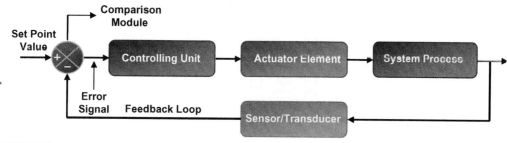

FIGURE 13.1

Schematic representation of the feedback control scheme.

13.2.1 Types of control methodologies

The process of controlling the variable(s) or parameter(s) by virtue of signal processing and conditioning for establishing control over the system's response to fluctuating load conditions is usually termed
control methodology. In the spectrum of control methodologies being adopted in real-time applications, it is appropriate to describe their significance, especially dedicated to thermal energy systems
applications. The generalized classification of a variety of control system methodologies being incorporated for heating, ventilation, and air conditioning systems in buildings is shown in Fig. 13.2. For
brevity, the significance of some major control methodologies is described in the following sections.

The *classical control method* typically makes use of the On/Off, proportional (P), proportional integral (PI), and PID scheme for controlling the process variable or parameter. Using an On/Off controller,
the process variable or parameter can be modulated between the set point values for maintaining the
system under control with respect to design conditions.

The major constraint encountered in an On/Off control method is that the parametric control swings
being established normally deviate relative to the set point value. Thus, the On/Off control method is
less pronounced for modulating load situations in thermal systems, where the process dynamics with
respect to the time delays are uncontrollable in nature. However, the On/Off control scheme is sponta-
neous and easy to handle compared to the conventional P, PI, and PID control methods.

In case of the P, PI, and PID controllers, feedback functionalities are utilized in which the error dynamics of the controlled parameter can be compared with the set point value and actual value obtained
as illustrated in Fig. 13.1. This ensures better accuracy of the process control through modulating the
controlled parameter within the set point value at the cost of reduced time delays.

The control action provided by these controllers can be instantaneous under the changing load conditions
the system would experience. The tuning and/or self-tuning mechanism of the controlled parameters in these
controllers is hard to achieve intrinsically. This can result in the erratic behavior of the system under control,
in case of a mismatch persisting between the tuning conditions and the operating conditions as well.

The *hard control method* is another way of controlling the process dynamics of a system in which
the controller is devised based on the theory of control including gain scheduling control, nonlinear
control, robust control, optimal control, and model-based predictive control (MPC).

In the case of gain scheduling control, the system with nonlinear operation can be segmented into
several piecewise regions, which are linear in nature. The segmented linear regions are then associated
with an individual PI or PID controller assigned with a distinct set of gain values. By scheduling the desired gain values, the required control action can be achieved over the system for its effective operation.

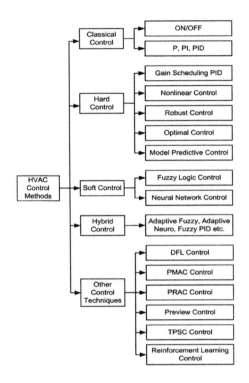

FIGURE 13.2

Classification of control methods in HVAC systems [1].

The nonlinear control design is fundamentally based on Lyapunov's stability theory, feedback linearization, and adaptive control techniques. Control law is devised in such a way that nonlinear operation of the system is tuned toward the stable operating state without compromising control objectives. This control design finds its application in the air distribution system, heat exchanger control, greenhouse emission control, and so on [1].

A robust controller design is adopted in situations where there time-dependent fluctuations or parametric variations can occur in the system under control. This can include the control provided for supply air temperature, supply air flow rate, and thermal zone temperature. The optimal control works on the basis of objective functions being set for the system to accomplish the desired control action over a period of time bound operation.

Usually, the objective functions can be framed for minimization of energy consumption and control effort as well as maximization of thermal comfort conditions to occupants. For instance, consider the control optimization proposed for thermal energy storage (TES) systems including active and passive systems, HVAC systems optimization, variable air volume system optimization, and so on.

The hard control method has been recognized over the years in the field of control systems for controlling operational strategies of thermal energy systems. However, there are some pitfalls in using the gain scheduling control, nonlinear control, robust control, and optimal control, which includes the following:

- Complex mathematical formulations
- Recognition of linear regions and stable states

- Switching logic between recognized regions
- Difficult manual tuning of multiple PID controllers for the regions identified
- Specification of additional parameters that makes the control process cumbersome

Indeed, the MPC technique can be considered advantageous compared to these control theories in the sense that MPC has the capability to combine disturbance rejection, provide effective control over slow-moving dynamics, handle process constraints, and pave the way for incorporation of energy saving strategies into the controller.

In the *soft control method* system control can be effectively achieved through incorporation of advanced intelligent fuzzy logic or ANN-based control strategies into the controller unit. Fuzzy logic control (FLC) can be considered an intelligent way to control process parameters through the formulation of IF and THEN, ELSE-based rule functions related to process control.

Framing different sets of rule statements (or fuzzy rules) in line with the design of membership functions (triangle, trapezoidal, etc.) can help to simplify process control intricacies, provided knowledge on overall system functioning has been thoroughly understood. The FLC can be implemented into conventional PID controllers in the form of either supervisory control or local control.

In the former, the level of control can be based on optimizing the system's response on a global scale. This means that the FLC can help to prioritize the controllers/control options dedicated for the system to achieve energy consumption reduction with good comfort conditions being met. In certain circumstances, the PID acts as the local scope of control, and the integrated FLC can serve as an arbiter or supervisor in resolving the optimization conflicts and responses of the system under control. A deep understanding of the system's operation and its different states of response to the actual conditions limits the utilization of FLC on a large-scale industrial application.

ANN, on the other hand, is also an intelligent way of controlling the system in which the controller is trained and tested using the set of performance data of the system. The training and testing processes are the most essential part of an ANN control design. A large number of measurement data sets can be used for training the ANN controller, a part of which can be used for testing the same controller with actual operating conditions of the system.

Fitting a nonlinear model to the measurement data and the evaluation of weighing factors or gains at each testing stage of the controller as well can help achieve best performance of the system in the long run. The black box approach of formulating the algorithm of an ANN control system makes itself less pronounced among process control and industrial systems.

The *hybrid control method*, as the name indicates, is the combination of both hard and soft control techniques. The interesting part of this control method is that the attributes of hard and soft controls can be blended to accomplish the desired control task, which can otherwise look cumbersome using hard and soft controls individually. By amalgamating soft control on the higher level and hard control on the lower level of the control hierarchy, energy efficiency of the system can be achieved.

The hybrid control technique does have some limitation in the control operation, which can be considered to emanate from individual control methods. The other control techniques represented in Fig. 13.2 have also been proposed for thermal energy systems, especially HVAC control systems. The most essential technical comparison of various control systems proposed for thermal energy systems are summarized in Table 13.1. The tick mark ($\sqrt{}$) shown under the energy consumption heading refers to the energy savings potential attributes of the controller.

Table 13.1 Comparison of Control Systems [2]

Control Systems	Thermal Comfort Control (PMV)	IAQ Control (CO_2)	Visual Comfort Control (Illumination)	Energy Consumption	Global Control Strategies	Priority to Passive Techniques
On/Off	–	–	–	–	–	–
PID	–	–	–		–	–
Fuzzy P control	√	√	√	√	√	√
Fuzzy PID control	√	√	√	√	√	√
Adaptive fuzzy PD	√	√	√	√	√	√
Fuzzy systems	√	√	√	√	√	–
Fuzzy PI control	√	√	√	√	√	√
Adaptive fuzzy PI	√	√	√	√	√	–
Neural network control	√	–	–	√	–	–
Agent-based intelligent control	√	√	√	√	√	√
Predictive control	√	–	–	√		√
Supervisory control	√	√	√	√	√	√
Reinforcement learning control	√	√	√	–	√	√
Ambient intelligent	√	√	√	√	√	–
Self-adaptive control system	√	√	√	√	√	√
Optimal control	–	–	–	√	–	√
Optimal and robust control	–	√	–	√	–	–

13.2.2 **Control methodology of thermal storage systems**

The generalized classification of the techniques being followed for peak demand management in commercial buildings is depicted in Fig. 13.3. It can be clearly seen from the classification that load shedding and load shifting are suitable methods that can be applied to reducing the peak demand in commercial buildings. The method of load shedding is more pronounced in the power supply and demand sides.

In the power supply side, the load shedding method is used to prevent total shutdown of the power supply, when the demand for power exceeds the actual requirement capacity of the network. This is usually performed by a utility company assigned for the said purpose. Load shedding for the power demand side can be performed by using the power control and management facility installed in buildings. This facility can help trigger off any power-consuming equipment that is not required at a specific point in time. The methods related to the hybrid system and on-site generation can also be used for managing the peak load demand in commercial buildings.

Table 13.1 Comparison of Control Systems [2]—cont'd

Control Systems	User Prefer-ences	Learning	Tuning: Fuzzy Systems or GA or Neural Adaptation	Temperature Control	Adaptation	DCV Ventilation Control	Source of Reference
On/Off	–	–	–	–	–	–	[3]
PID	–	–	–	–	–	√	[3]
Fuzzy P control	–	–	–	–	–	√	[4,5]
Fuzzy PID control	–	–	–	–	–	√	[6]
Adaptive fuzzy PD	–	–	–	–	√	√	[6,7]
Fuzzy systems	√	–	–	–	–	–	[8]
Fuzzy PI control	–	–	√	–	–	√	[9]
Adaptive fuzzy PI	–	–	√	–	–	√	[7,10]
Neural network control	–	√	–	–	–	√	[11–14]
Agent-based intelligent control	√	√	√	–	√	√	[15–19]
Predictive control		√		–	–	√	[20,21]
Supervisory control	√		√	–	–	√	[19,22–25]
Reinforcement learning control	√	√	–	–	–	√	[26,27]
Ambient intelligent	√	√	–	–	–	√	[28–30]
Self-adaptive control system	√	–	√	–	–	√	[31–33]
Optimal control	–	–	–	√	√	√	[34]
Optimal and robust control	√	–	–	–	–	√	[35]

Load shifting can be considered to be a widespread and well-established method highly suitable for peak demand management in buildings. This method basically utilizes electricity rate differentials or tariff benefits to establish demand management by shifting the on-peak load to the off-peak/part load periods. The load shifting method adopted in buildings is illustrated in Fig. 13.4, which is self-explanatory.

As discussed in earlier chapters, the concept of offsetting on-peak load demands through the thermal energy (cooling or heating) being stored during part load hours stands as the backbone for achieving peak demand control and management in buildings. Heat energy can be stored in the building thermal mass (BTM), ice thermal storage facility (ITES), phase change material (PCM) cool storage, solar heat storage, and so on. To gain the maximum usefulness of the load shifting control and management in buildings, it is of prime importance to understand the basic three strategies shown in Fig. 13.5.

The first strategy is referred to as the building load prediction, in which the cooling or heating load profile of the building requiring peak load shifting has to be evaluated with greatest accuracy. The prediction

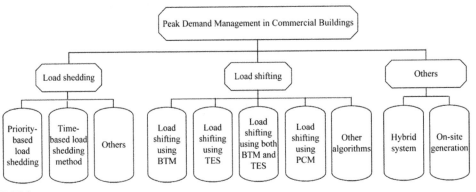

FIGURE 13.3

Classification of peak demand management in commercial buildings [36].

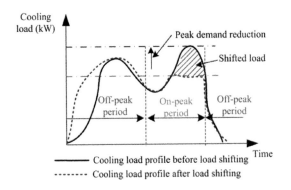

FIGURE 13.4

Schematic diagram of a typical load shifting control [36].

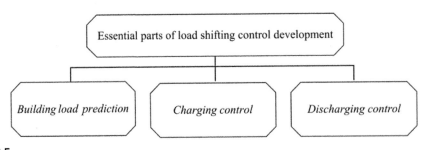

FIGURE 13.5

Essential parts of load shifting control strategy development [36].

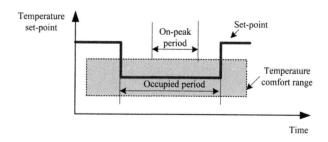

FIGURE 13.6

Conventional night setup control strategy [36].

of the thermal load demand in buildings needs a good understanding of the transient operating conditions while designing the thermal energy systems for meeting the required purpose.

The second strategy is referred to as the charging control, in which the thermal energy to be stored during part load periods has to be determined properly before scheduling for control operations to take part in peak load shifting. Estimating the total thermal energy to be stored during off-peak hours can facilitate having reduced energy consumption at the chiller plant or heating equipment.

Likewise, the third strategy, known as discharging control, also depends on the estimation of the total stored thermal energy to be released during on-peak load hours. In case the discharging time of thermal storage is less, it can result in the mismatch of offsetting the peak load in buildings.

On the other hand, if the thermal energy is released very rapidly, only a part of the peak load can be retrieved from the building, whereas the remaining space cooling/heating demand could still shoot up in building spaces. Thus, implementing proper control over peak load shifting can help contribute toward achieving enhanced energy redistribution and energy management in buildings.

The conventional night setup control strategy shown in Fig. 13.6 infers that the thermal energy being stored in the building fabric components or simply the BTM has not been utilized for reducing the operating costs associated with peak load shifting. It is pertinent to note that the incorporation of BTM concepts in buildings not only helps to shift the peak load to off-peak periods, but can also facilitate the reduction of operational costs, especially peak demand costs. There may be an increase in total energy costs associated with the incorporation of BTM; however, the energy efficiency of the cooling/heating system in buildings can be enhanced.

The method of using BTM for shifting the on-peak hourly load to part load periods has gained momentum in recent years. Factually, load shifting using the BTM can be classified into two major control strategies, namely, charging control and discharging control, as illustrated in Fig. 13.7. Charging control can be subdivided into two modes of objectives, known as peak demand minimization and operating cost minimization.

In peak demand minimization, the main objective is to reduce peaking demand in buildings through the utilization of BTM principles. The thermal energy being captured and stored in the building fabric components during nighttime (or precooling facility) can be retrieved from the storage component during the daytime peak hourly periods.

Similarly, in the operating cost minimization, the key objective is to reduce the operating costs involved in using the BTM method for peak load shifting. This can be effectively accomplished by selecting a proper temperature and time of precooling facility to enable thermal energy to be stored in

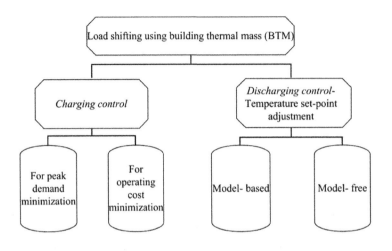

FIGURE 13.7

Charging control and discharging control in load shifting using building thermal mass [36].

the BTM. For getting improved peak load shifting, selection of the precooling storage temperature to be at the lower end, and ensuring the discharging temperature not exceed the maximum temperature in the comfort temperature range are most essential.

Besides, storage time also plays a vital role in acquiring the best control over peak load shifting, which means that the amount of thermal energy to be stored has to be quantified appropriately. Otherwise, the energy spent for storage can increase the operating cost further in addition to the inherent cost factors associated with the thermal system.

In reality, the implementation of charging control strategy is somewhat difficult, which is due to improper selection of the precooling temperature and time estimate for charging the BTM storage. Developing simple and easy ways to identify the actual storage capacity in BTM method in addition to the optimization of the precooling temperature and time can help move an energy efficient approach toward shifting the peak load demand. Based on the research studies performed on BTM storage, it can be seen that BTM cooling storage can help to reduce the peak load demand from 30% to 80% depending on the hourly on-peak periods persisting in buildings.

The other strategy adopted for reducing the peak demand load in buildings is referred to as discharging control using the temperature set point adjustment as shown in Fig. 13.8. In this method, the cold energy discharge during the peak hourly load conditions can be controlled using the model-based and model-free approaches.

The model-based control approach basically requires a model for establishing the control relationship between the room temperature set point and the building load profile. The set point trajectories responsible for minimizing the peak cooling load can be controlled by training the building model with the field measured data. Due to the involvement of model-based control method in terms of training the model using vast measurement data, the implementation is quite cumbersome.

At the same time, there are other possible means of simplifying the control intricacies; for instance, introducing a set point equation and determining the time constants under a trial-and-error approach can enable obtaining reduction in the peak load demand. Likewise, the model-based approaches including semi-analytical (SA), exponential set point equation-based semi-analytical (ESA), and weighted-averaging

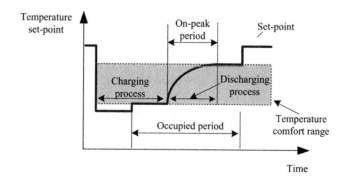

FIGURE 13.8

Temperature set points in charging and discharging processes [36].

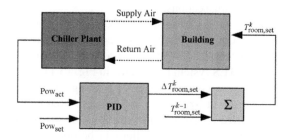

FIGURE 13.9

The schematics of PID demand limiting algorithm [36,38].

(WA) are simple and effective in terms of reducing the peak load demand [37]. However, model-based control strategies are open loop in nature, which may not perform better under time varying or fluctuating load conditions existing in buildings.

As discussed earlier, the utilization of PID controllers for reducing the peak load demand in buildings are much appreciable now because PID controllers work on the principle of feedback control loop or network, as depicted in Fig. 13.9. Thus, they can respond swiftly to sudden changes occurring in the operating conditions of the system, thereby enabling the system to be energy efficient at most operating periods.

Furthermore, the feedback control logic can be beneficial in terms of effectively storing and discharging thermal energy depending on the cooling/heating load persisting in building spaces. This in turn facilitates better peak load shifting compared to the conventional On/Off controllers and cumbersome model-based control approaches.

In the model-free control approach, very simple rules or rule statements can be used in place of expert systems and other learning practices. Three commonly preferred model-free cooling discharging control approaches are being adopted for achieving peak load demand reduction in buildings, which include the following:

- Step change reset strategy
- Linear change reset strategy
- Exponential change reset strategy

In case of the step change reset strategy, the temperature set point can be adjusted from the lower limit to the higher limit in a single step at the commencement of the peak load shifting period. In a linear change reset method, the set point temperature can be increased to the maximum temperature at a constant rate through the entire peak load shifting period.

The exponential change reset strategy involves increase of the set point temperature at a faster rate during the initial period and a slower rate during completion of the peak load shifting period as shown in Fig. 13.8. The relationship between the peak load, storage, and discharge timings and the room temperature, if evaluated and designed properly, can be advantageous for real-time control applications. The summary of some of the major research outcomes related to the load shifting control using the BTM approach is presented in Table 13.2.

A step ahead, the peak load shifting can also be accomplished by using the active TES system, the generic classification of which is shown in Fig. 13.10. It consists of two broad control strategies, which includes heuristic control and optimal control. Each control strategy is significant in terms of acquiring effective peak load shifting in buildings. For brevity, the principles of heuristic and optimal control strategies are discussed in forthcoming sections.

The *heuristic control* strategy is a kind of method to reduce peak load demand in buildings, where it is associated with the TES capacity and priority-based control approaches. In the storage capacity-based approach, control over the peak load demand can be achieved through three distinct operating modes of the chiller plant and the TES system as depicted in Fig. 13.11.

For instance, in the full storage control, the total cooling load during the on-peak demand period can be met fully through the utilization of TES system. That is, the required on-peak cooling energy demand can be satisfied by discharging stored cold energy from the TES system. In a partial storage demand limiting approach, the chiller can be operated below its critical capacity during the on-peak demand period, and through the discharging of TES system, the required peak load shifting can be achieved.

Likewise, in a partial storage load leveling approach, the chiller is operated at its maximum capacity (designed capacity) throughout. In case the cooling load demand exceeds the designed capacity of the chiller, the TES system is put into use by discharging the stored cold energy to offset the peak load excess.

Under the priority-based control strategy, the chiller priority control can be considered the simplest approach for shifting the peak demand load. The functioning of the chiller is continuous, which can be controlled by means of the conventional control strategy being adopted for the demand limiting approach. Moreover, the remaining cooling capacity can be effectively met by the TES system. This control principle can be expressed by [45]

$$u_k = \begin{cases} u_{max} & \text{if } k \text{ is off-peak} \\ 0 & \text{if } k \text{ is on-peak and } Q_k \leq CAP_{ch} \\ -(Q_k - CAP_{ch}) & \text{if } k \text{ is on-peak and } CAP_{ch} \leq Q_k \end{cases} \tag{13.2}$$

where u is the rate of charging or discharging (+ve value represents charging rate and −ve value represents discharging rate), Q is the cooling load, CAP_{ch} is the cooling capacity of the chiller plant, and k is the kth hour.

Similarly, in the constant-proportion control, both the chiller and TES participate equally in meeting the required peak load demand. The cooling load distribution between the chiller and the TES

Table 13.2 Studies On Load Shifting Control Using Building Thermal Mass [35]

Items Studies	References	Objective	Approaches Used		Results
			Cooling Charging	**Cooling Discharging**	
Simulation studies	[39]	Minimized daily peak load	Lowest precooling temperature with longest allowed precooling time	Temperature trajectory determined by detailed building model	Daily peak demand reduction up to 30% in 5 h on-peak period
	[38]	Minimized monthly electricity cost	Optimized precooling temperature and time	Temperature trajectory determined by PID controller	Monthly cost saving about 8.5%
	[37]	Minimized daily peak load	Lowest precooling temperature with longest allowed precooling time	Temperature trajectory determined by three different simple methods (i.e., SA, ESA, and WA)	All three methods work well, and WA is slightly better and easier for application
	[40]	Minimized daily electricity cost	Lowest precooling temperature with fixed precooling time	Temperature trajectory determined by the predictive model	Weekly cost saving is up to 29%
Experimental studies	[41]	Minimized daily peak load	Fixed precooling temperature and time	Temperature trajectory determined by detailed building model	Daily peak demand reduction greater than 30% in 5 h on-peak period
	[42]	Minimized daily peak load	Lowest precooling temperature and longest allowed precooling time	Simple step change reset strategy	Daily peak demand reduction up to 80% in 3 h on-peak period
	[43]	Minimized daily peak load	Fixed precooling temperature and time	Linear and exponential change reset strategies	Linear reset approach cannot achieve the best results, and exponential reset approach only worked well in limited cases
	[44]	Minimized daily peak load	Lowest precooling temperature with fixed precooling time	Dual step change reset strategies	For all test buildings, the peak loads can be reduced by 15-30%

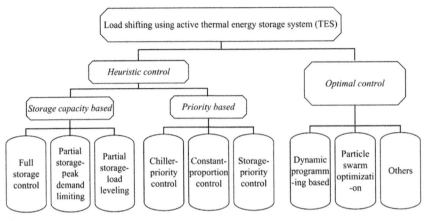

FIGURE 13.10

Classification of load shifting control using thermal energy storage system [36].

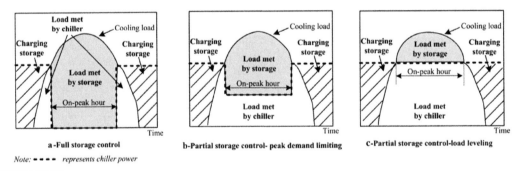

FIGURE 13.11

Storage capacity-based control strategies [36].

decides the control efficiency, where the TES offers a constant fraction f_Q of the cooling load under all conditions. The control principle for this strategy is given by [45]

$$u_k = \begin{cases} u_{max,k} & \text{if } k \text{ is off-peak} \\ \max(-f_Q Q_k, u_{min,k}) & \text{if } k \text{ is on-peak} \end{cases} \tag{13.3}$$

where $u_{max,k}$ and $u_{min,k}$ represent the maximum charging and discharging rates, respectively.

The storage-priority control involves the full melting of the storage medium (e.g., ice) as much as possible to meet the on-peak load demand in buildings. This is usually referred as the full storage control strategy aiming at the maximum extraction of the stored thermal energy for the next available peak load shifting operation. The control principle for this strategy is given by [45]

$$u_k = \begin{cases} u_{\max,k} & \text{if } k \text{ is off-peak} \\ Q_k & \text{if } k \text{ is on-peak and } Q_k \leq -u_{\min,k} \\ u_{\min,k} & \text{if } k \text{ is on-peak and } -u_{\min,k} \leq Q_k \end{cases} \qquad (13.4)$$

In the *optimal control* strategy, the main objective is devised to identify the best possible charging and discharging rates with respect to the different time, thereby achieving minimum operating cost. This control strategy takes the form of a global searching problem with the consideration of the real-time constraints for optimization.

The particle swarm optimization, as a segment of the optimal control strategy, usually refers to the availability of a number of potential solutions to the optimization problem. Each potential solution can be defined as a position of the particle in the optimization. On the other hand, the dynamic programming approach can be considered as the other segment of the optimal control strategy.

The optimization objective can be accomplished by solving the complex problems. The complex problems with overlapping properties can be fragmented into simpler subproblems, which are seen as slightly smaller and with optimal substructure. The principle of the optimal control strategy can be expressed by [46]

$$J = J\,(\bar{u}_1, \bar{u}_2, \ldots, \bar{u}_n) = \min \sum_{i=1}^{n} \left(\alpha_i x E_i + \beta\, x^{\max}_{1 \leq i \leq n} \{PD_i\} \right) \qquad (13.5)$$

where $\bar{u}_1$, $\bar{u}_2$, ..., $\bar{u}_n$ represents the optimal charging and discharging rates in different times.

Other optimal control strategies, including simulated annealing and the evolutionary algorithm, are also the suggestive approaches for achieving optimal peak load shifting control through the integration of the TES system.

In the spectrum of available control strategies, the concept of *MPC* dedicated for shifting the peak load demand in buildings has been attracting special attention recently [1,47]. As pointed out earlier, MPC utilizes a model of the system to predict variation in the thermal load or sudden load disturbances occurring in buildings and the associated changes on the controllable parameter.

The schematic diagram of the conventional and MPC control strategies applied in buildings is shown in Fig. 13.12. It can be clearly seen that, in the conventional building control strategy, either

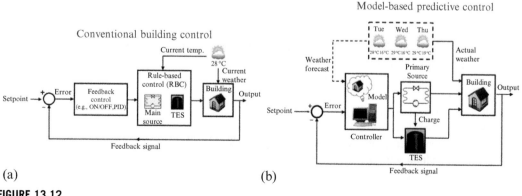

(a)   (b)

FIGURE 13.12

Conventional control compared to MPC [47].

the feedback control or the heuristic control priority approaches are followed. This achieves the corrective measure on the controlled parameter, as well as the swift response of the system to sudden load fluctuations.

If the TES system is integrated, the performance of the conventional controller can be improved, provided the energy distribution aspects between the cooling plant and the TES system are evaluated and designed properly. In the MPC control strategy, when using the model predictions on the thermal load disturbances and controllable inputs, the required control action can be performed for achieving the system's optimized operating cost.

In short, by incporating a stepwise procedure or algorithm in the model structure, the rate of charging and discharging of TES can be obtained relative to on-peak hourly load variations. This would also enable the optimized operation of the TES system as a function of cooling power load and energy cost profiles. For a better understanding of how the MPC control strategy is implemented in a building cooling system integrated with TES, refer to the summarized methodology in Fig. 13.13.

The uncertainty-based control amalgamated into MPC can be considered one of the viable approaches for maintaining the system's operational stability, even under uncertain conditions [48]. Generically, the uncertainty can be referred to as the gap present between the certainty and the present state of information (obtained from decision making) as shown in Fig. 13.14.

It can be inferred from Fig. 13.14 that, although all sorts of information are available with the stakeholder pertaining to the model parameter's specification and distribution type of the model, there can be some specific occasion emanating from the unpredictable part of the model being developed.

This can, in turn, lead to a condition of sporadic uncertainty (or erratic characteristics) in nature and imprecise uncertainty (indefinite characteristics or lack of knowledge) associated with the development of computational models. The uncertainty-based control in MPC can help to sort out the behavior of unpredictable occurrences during the operating periods of the system. By taking into account the uncertain part of the operating strategy, the system can be enabled to perform energy efficiently most of the time.

13.3 OPTIMIZATION OF THERMAL STORAGE SYSTEMS

The term optimization refers to a quantitative approach for identifying methods and obtaining solutions for a problem with stated objectives and constraints. Generically, the optimization process involved in an engineering system can be categorized under three phases, which include the preprocessing phase, optimization phase, and postprocessing phase. For brevity, the optimization program and the essential aspects of the processing phases included in a simulation-based optimization in buildings are shown in Fig. 13.15 and summarized in Table 13.3, respectively.

In general, the optimization problem can be defined by

Minimize $f(x_1, x_2, \ldots, x_n)$

With constraints $C(x_1, x_2, \ldots, x_n) \geq 0$

$x_i \in S_i$

In the aforementioned expressions, the stated problem of having a set of objective functions f has to be minimized along with the set of constraints C, which must be usually equal to or above zero value. Furthermore, each design variable x_i is meant to be constrained to a certain discrete or boundary value, say S_i.

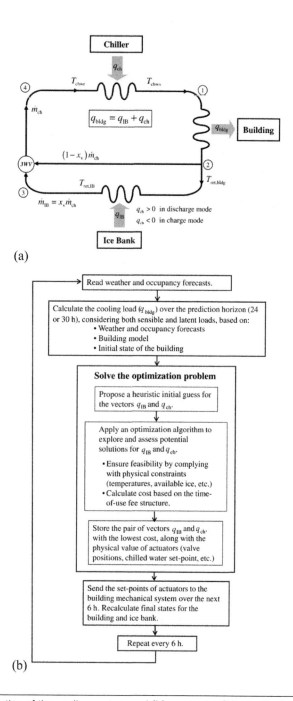

(a)

(b)

FIGURE 13.13

(a) Conceptual representation of the cooling system and (b) summary of the methodology used for MPC strategy [47].

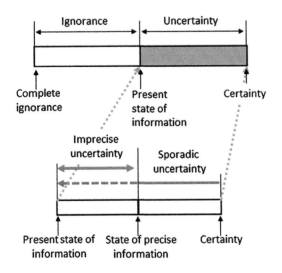

FIGURE 13.14

Characteristics of uncertainty [49].

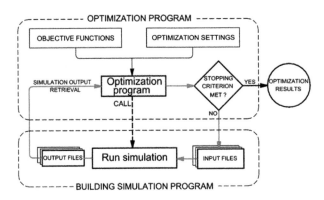

FIGURE 13.15

The coupling loop applied to simulation-based optimization in building performance studies [50].

It is obvious that the objective functions and constraints can be interchanged, depending on the formulation of the problem. The classification of optimization problems and most commonly preferred algorithms dedicated to building performance optimization are summarized in Tables 13.4 and 13.5, respectively (which are self-explanatory). In context to the building cooling/heating systems integrated with TES, optimization applies to the objectives of minimizing the energy consumption and or operating costs involved. At the same time, the optimization can also be performed on TES systems for maximizing the charging and discharging capabilities toward achieving enhanced energy redistribution requirements.

However, in most thermal energy systems, the concept of single-objective optimization rarely occurs because the system is subjected to varying disturbances with time. To counteract the unmatched

Table 13.3 Major Phases in Simulation-based Optimization Studies of Buildings [51–60]

Phase	Major Tasks
Preprocessing	Formulation of the optimization problem: – Computer building model – Setting objective functions and constraints – Selecting and setting independent (design) variables and constraints – Selecting an appropriate optimization algorithm and its settings for the problem in hand – Coupling the optimization algorithm and the building simulation program (Optional) Screening out unimportant variables by using sensitivity analysis to reduce the search space and increase efficiency of the optimization (Optional) Creating a surrogate model (a simplified model of the simulation model) to reduce computational cost of the optimization
Running optimization	Monitoring convergence
Postprocessing	Interpreting optimization results (Optional) Verification and comparing optimization results of surrogate models and "real" models for reliability (Optional) Performing sensitivity analysis on the results Presenting the results

disturbances, the engineering problem can be addressed with multiple objectives under two distinct methods, the weighted-sum approach and multiobjective optimization.

In the weighted-sum approach, the different objectives being framed related to the problem can be collected and combined to produce a single-objective function. This can then be easily optimized in the usual way. On the other hand, using multiobjective optimization, a range of solutions is obtained that gives better insight of the trade-off existing between each objective being framed. This type of optimization is also referred to as the Pareto optimization, wherein the trade-off front or the Pareto front can be realized on the basis of the concept of dominance as depicted in Fig. 13.16.

It is noteworthy from the two objective optimizations in Fig. 13.16 that the highlighted solutions are nondominated, which is due to the fact that no solutions are seen in the shaded area. The Pareto front is made up of the yellow triangles (triangles in print version), which are too nondominated. However, the blue dots (dots in print version) are observed to be dominated, which is due to the fact that better solutions exist in both objectives, and so they are not present on the Pareto front.

In this context, a variety of optimization techniques are being adopted for thermal energy systems in buildings to make them sustainable in the long run. The major attributes of different algorithmic methods of optimization in building systems are represented in Fig. 13.17. The algorithmic methods projected are almost heuristic in nature, which actually do not give optimal solutions; rather, they can result in an efficient method that has a high probability of obtaining solutions that are near to optimum [61].

Table 13.4 Classification of Optimization Problems [61,62]

Classification Schemes Based On	Categories or Classes
Number of design variables	Optimization problems can be classified as one-dimensional or multidimensional optimization, depending on the number of design variables considered in the study
Natures of design variables	Design variables can be independent or mutually dependent
	Optimization problems can be stated as "static"/"dynamic" if design variables are independent/are functions of other parameters (e.g., time)
	Optimization problem can be seen as the deterministic optimization if design variables are subject to small uncertainty or have no uncertainty. In contrast, optimization design variables subject to uncertainty (e.g., building operation, occupant behavior, climate change) define the probabilistic-based design optimization as exemplified in the robust design optimization of [63]
Types of design variables	Design variables can be continuous (accept any real value in a range), discrete (accept only integer values or discrete values), or both. The latter is referred to as mixed-integer programming
Number of objective functions	Optimization problems can be classified as single-objective or multiobjective optimization depending on the number of objective functions. In practice, building optimization studies often use up to two objective functions, but exceptions do exist as exemplified by three-objective function optimization in [64,65]
Nature of objective function	Different optimization techniques can be established depending on whether the objective function is linear or nonlinear, convex or nonconvex, unimodal or multimodal, differentiable or nondifferentiable, continuous or discontinuous, and computationally expensive or inexpensive These result in linear and nonlinear programming, convex and nonconvex optimization, derivative-based and derivative-free optimization methods, heuristic and meta-heuristic optimization methods, simulation-based and surrogate-based optimization
Presence of constraints and constraint natures	Optimization can be classified as constrained or unconstrained problems based on the presence of constraints that define the set of feasible solutions within a larger search space. Dealing with an unconstrained problem is likely to be much easier than a constrained problem, but most of BOPs are constrained
	Two major types of constraints are equality or inequality. A constraints function may have similar attributes to those of objective functions and can be separable or inseparable
Problem domains	Multidisciplinary optimization relates to different physics in the optimization as exemplified in [66]. Such a problem requires much effort and makes the optimization more complex than single-domain optimization

13.3.1 Thermoeconomic optimization

The term *thermoeconomics* has been coined to indicate the significance of a combination of exergetic and economic analysis. The most interesting aspect of thermoeconomics is the inclusion of costing to the exergy content (not the energy) of an energy carrier (exergy costing).

There has been significant usage of the term *thermoeconomic analysis* in recent years by numerous research groups worldwide. This would in fact signify the thermodynamic analysis of a thermal system being performed with economic analysis done separately on the same system.

Table 13.5 Classification of Mostly Used Algorithms Applied to Building Performance Optimization [50]

Family	Strength and Weakness	Typical Algorithms
Direct search family (including generalized pattern search [GPS] methods)	– Derivative-free methods – Can be used even if the cost functions have small discontinuities – Some algorithms cannot give exact minimum point – May be attracted by a local minimum – Coordinate search methods often have problems with nonsmooth functions	Exhaustive search, Hooke-Jeeves algorithms, coordinate search algorithm, mesh adaptive search algorithm, generating set search algorithm, simplex algorithms
Integer programming family	Solving problems that consist of integer or mixed-integer variables	Branch and bound methods, exact algorithm, simulated annealing, tabu search, hill climbing method, CONLIN method
Gradient-based family	– Fast convergence; a stationary point can be guaranteed – Sensitive to discontinuities in the cost function – Sensitive to multimodal function	Bounded BFGS, Levenberg-Marquardt algorithm, discrete Armijo, gradient algorithm, CONLIN method, etc.
Meta-heuristic method Stochastic population-based family	– Not to "get stuck" in local optima – Large number of cost function evaluations – Global minimum cannot be guaranteed	+ Evolutionary optimization family: GA, Genetic programming, Evolutionary programming, Differential evolution, Cultural algorithm + Swarm intelligence: Particle swarm optimization (PSO), Ant colony algorithm, Bee colony algorithm, Intelligent water drop
Trajectory search family	– Easy implementation even for complex problems – Appropriate for discrete optimization problems (continuous variables can also be used), e.g., traveling salesman problems – Only effective in discrete search spaces – Unable to tell whether the obtained solution is optimal or not – Problems of repeated annealing	Simulated annealing, tabu search, hill climbing method
Other		Harmony search algorithm, firefly algorithm, invasive weed optimization algorithm
Hybrid family	Combining the strength and limiting the weakness of the above-mentioned approaches	PSO-HJ, GA-GPS, CMA-ES/HDE, HS-BFGS algorithm

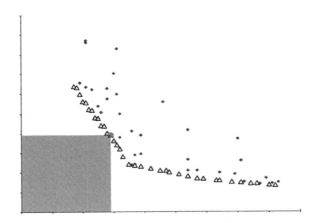

FIGURE 13.16

Pareto front (triangles) and dominated solutions (dots) [67].

However, this terminology is relatively different from that of the term *exergoeconomics*, which clearly points out the interdependency between exergy costing and economics associated with it. Thus, thermodynamic and economic analyses can be amalgamated in exergoeconomics through exergy costing. Besides, exergoeconomics becomes a significant part of thermoeconomics [68].

Four major elements are present in the thermoeconomic analysis, which includes the following:

- Detailed exergy analysis
- Economic analysis performed at the component level of the energy system under investigation
- Exergy costing
- Exergoeconomic evaluation of individual system component

Before dealing with thermoeconomic analysis, it is rather important to be acquainted with the prime objectives being set for the purpose, which includes the following:

- Identification of the location, magnitude, and source of real-time thermodynamic losses in a thermal energy system
- Estimation of exergy losses and exergy destruction related costs
- Assessment of the cost of production for the product or outcome from an energy conversion system containing more than one product
- Provision for enabling optimization and feasibility studies during the design phase of an energy system under development as well as to perform process improvement studies related to the existing system
- Assistance in decision-making processes involving the operation and maintenance of the plant and allocation of research funds accordingly
- Comparison of technical alternatives for overall process optimization

Thermoeconomic analysis is a vast field in which the research studies performed on thermal energy systems are extensive. For a detailed understanding of thermoeconomic optimization cum analysis of different thermal systems, refer to specific literatures. Thus, the scope of this section has been framed

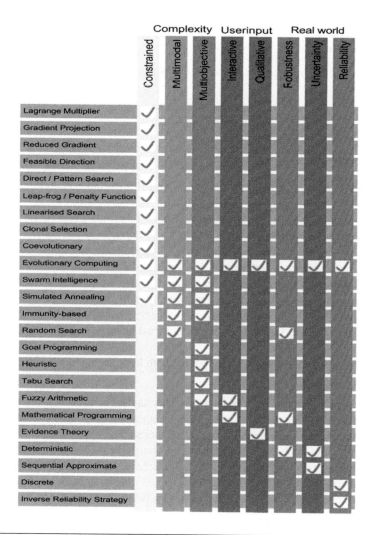

FIGURE 13.17

Attributes of algorithmic methods [61,67].

to present only the essential characteristics pertaining to thermoeconomic analysis of thermal energy systems with an example.

The thermoeconomic evaluation of an energy intensive system or process solely depends on important variables for the kth component as detailed here:

- Exergetic efficiency (ε_k)
- Exergy loss ($E_{L,k}$) and exergy destruction ($E_{D,k}$)
- Exergy ratios ($E_{D,k})/(E_{F,tot})$ and ($E_{L,k})/(E_{F,tot}$)
- Capital costs (Z) associated with the capital investment, operating, and maintenance charges
- Exergy destruction cost, $C_{D,k}$

- Relative cost difference, r_k
- Exergoeconomic factor, f_k

The exergoeconomic factor can be expressed by the relation

$$f_k = \frac{\dot{Z}_k}{\dot{Z}_k + C_{f,k}\left(\dot{E}_{D,k} + \dot{E}_{L,k}\right)} \tag{13.6}$$

To improve the cost effectiveness of the total system, the following rules can be framed for the kth component:

- Particular attention must be given to the kth component if the relative cost difference and cost rates are higher.
- In case of higher values of r_k, using the exergoeconomic factor the cause or the cost source for the rise in r_k must be determined.
- If both r_k and f_k are higher, the cost effectiveness of the system may be achieved to reduce the capital investment for the kth component at the expense of the component efficiency.
- If r_k value is higher with respect to the lower value of f_k, an attempt must be made to enhance the efficiency of the component by increasing the capital investment.
- Eliminate the intermediate steps or subprocesses that may increase the exergy losses or exergy destruction without enabling the reduction of capital cost or investment or any associated fuel costs.
- Increasing the exergetic efficiency of a component with large exergy destruction value can lead to cost effectiveness.

After framing the aforementioned important variables and set of rules, the thermoeconomic evaluation of an energy system can be performed without compromising cost effectiveness. From this perspective, the case study of a thermoeconomic optimization of an ice thermal energy storage (ITES) system for air-conditioning applications in buildings is shortly presented in the forthcoming section, with reference to the research work performed [69]. The schematic diagram of the ITES system being considered for the thermoeconomic optimization is depicted in Fig. 13.18.

The operating cycle of the ITES system is divided into two cycles:

- Charging cycle—components include compressor, condenser, evaporator, cooling tower, pump, and expansion valve.
- Discharging cycle—components include air handling unit (AHU), discharging pump, and ice storage tank.

The working fluid (or refrigerant) used during the charging cycle is R134a, and a water/glycol solution is used as the cooling fluid during the discharging cycle. The operating principle of the ITES system is typically based on the partial storage strategy, which was explained earlier. For the thermoeconomic optimization of the ITES system, two objective functions are devised. The first objective function is given by

$$\text{Obj}(1) = \sum_k \dot{Z}_k + \dot{C}_{elec} + \dot{C}_{env} \tag{13.7}$$

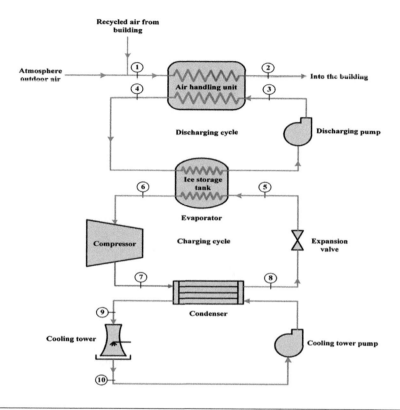

FIGURE 13.18

Schematic diagram of the modeled ITES system [70].

where $\dot{Z}_k$ is the rate of investment and maintenance costs, $\dot{C}_{elec}$ is the operational costs, and $\dot{C}_{env}$ is the rate of penalty cost due to CO_2 emission.

The investment cost of each cycle component based on the cost functions can be expressed as

$$\dot{Z}_k = \frac{Z_k \times CRF \times \Phi}{N \times 3600} \tag{13.8}$$

where N, Φ, and CRF are the operational hours of the ITES system in a year, maintenance factor, and the capital recovery factor, respectively. Again, CRF can be related to the annual interest rate (i) and estimated equipment life time (n) as well. Thus, CRF can be represented by

$$CRF = \frac{i(1+i)^n}{(1+i)^i - 1} \tag{13.9}$$

Likewise, the operating cost of the ITES system can be expressed by

$$\dot{C}_{\text{elec}} = \left[\left(\dot{W}_{\text{Comp}} + \dot{W}_{\text{pump,CT}} + \dot{W}_{\text{fan,CT}}\right) \times \frac{c_{\text{elec,off-peak}}}{3600}\right]$$
$$+ \left[\left(\dot{W}_{\text{pump,dc}} + \dot{W}_{\text{fan,AHU}}\right) \times \frac{c_{\text{elec,on-peak}}}{3600}\right]$$

(13.10)

The rate of penalty cost of CO_2 emission can be given by

$$\dot{C}_{\text{env}} = \frac{\left(m_{CO_2} / 1000\right) \times c_{CO_2}}{N \times 3600}$$

(13.11)

The second objective function associated with the emphasis on thermodynamic inefficiencies and accuracy in the analysis can be defined as

$$\text{Obj}(2) = \sum_k \dot{Z}_k + \dot{C}_{\text{elec}} + \dot{C}_{\text{env}} + \sum_k \dot{C}_{D,k}$$

(13.12)

where the last term of Eq. (13.12) refers to the exergy destruction cost rate pertaining to the kth system component.

By introducing this additional term, the optimal design parameters shift toward more efficient values thermodynamically. The evaluation of the payback period of additional expenses (in years) can be performed by using the expression given by

$$\Delta\left(Z_{\text{op}}\right)\left(\frac{(1+i)^p - 1}{i(1+i)^p}\right) + \Delta Z_{\text{SV}}\left(\frac{i}{(1+i)^p}\right) = \Delta\left(\sum_k Z_k\right)$$

(13.13)

where ΔZ_{SV} is the difference in the salvage values of the ITES and conventional systems. The salvage value can be defined as the estimated value of the asset, which will realize on its sale at the end of its useful lifetime. Thus, the thermoeconomic optimization of the ITES system can be established based on Eq. (13.6) and the set of rules as explained earlier. The thermoeconomic optimization results of the ITES system are presented in Fig. 13.19 and Tables 13.6 and 13.7.

13.3.2 Multiobjective optimization

The multiobjective optimization also referred as to the Pareto optimization has been well recognized in the field of thermal energy systems for a long time. In this optimization method, a set of trade-off optimal solutions or a Pareto set is evaluated. Based on the examination, required solutions are obtained appropriately. The selection of the best solution from the Pareto set is shown in Fig. 13.20.

The Pareto set comprises an infinite number of Pareto points in which the objective is to provide only some elements of the Pareto set rather than the entire set of elements. Depending on the requirements to be satisfied in buildings, the optimum (or best possible) solutions are obtained from the Pareto optimization strategy using the multicriteria decision making tool.

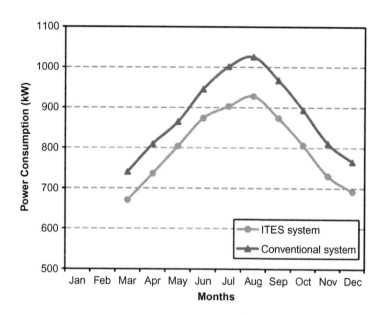

FIGURE 13.19

The comparison of electricity consumption between ITES and conventional systems in a year [70].

Table 13.6 The Optimum Values of Design Parameters (Decision Variables) by Using Objective Functions (1) and (2) [70]

	Optimum Values (Objective Function (1))	Optimum Values (Objective Function (2))
T_3 (°C)	3.86	3.31
T_4 (°C)	12.39	12.80
T_{ST} (°C)	− 2.51	−1.68
T_{EV} (°C)	−5.20	−4.27
T_{Cond} (°C)	38.05	36.11

Furthermore, the integration of multiobjective optimization with the so-called genetic algorithm (GA) seems to be a viable and efficient strategy to access building thermal performance. One such example case study pertaining to the analysis of the ITES incorporating PCM as the partial cold storage for air-conditioning applications is presented next based on the reference work [69].

The schematic diagram of the modeled ITES system incorporating PCM as the partial cold storage is illustrated in Fig. 13.21. For brevity, the objective functions and optimal results obtained from this study are presented next. Refer to this work for mathematical modeling, associated cost functions, and other design parameters of the system further.

The multiobjective optimization performed is comprised of two objective functions including the maximization of the exergy efficiency and the minimization of the total cost rate of the whole system. The objective functions can be expressed as

Table 13.7 The Investment Cost, Exergy Destruction Cost, Exergy Destruction, and Exergoeconomic Factor for Various Components of the ITES System Optimized at Various Ambient Temperatures 30, 35, 40, 45, and 50 °C [70]

Ambient Temperature		AHU	ST	EV	Comp	EX	Cond+CT
30	Investment cost (MUS$)	0.2238	0.0963	0.1093	0.2285	0.0007	0.1883
	Exergy destruction cost (MUS$)	0.2501	0.1343	0.1905	0.1677	0.0062	0.5536
	Exergy destruction (kW)	79.54	97.25	83.08	176.53	18.68	278.06
	Exergoeconomic factor (%)	22.4	20.2	52.1	56.3	14.8	25.3
35	Investment cost (MUS$)	0.2249	0.0964	0.1095	0.2481	0.0007	0.2246
	Exergy destruction cost (MUS$)	0.2575	0.1392	0.1996	0.1779	0.0068	0.5715
	Exergy destruction (kW)	97.16	98.71	83.88	182.01	22.18	285.31
	Exergoeconomic factor (%)	20.8	19.2	50.3	55.6	14.2	24.7
40	Investment cost (MUS$)	0.2262	0.0965	0.1096	0.2741	0.0008	0.2586
	Exergy destruction cost (MUS$)	0.2652	0.1434	0.2091	0.1885	0.0075	0.5886
	Exergy destruction (kW)	119.21	100.19	84.75	187.06	26.17	293.95
	Exergoeconomic factor (%)	19.9	18.4	49.4	54.9	13.8	24.1
45	Investment cost (MUS$)	0.2273	0.0967	0.1098	0.2909	0.0008	0.2906
	Exergy destruction cost (MUS$)	0.2624	0.1476	0.2194	0.2076	0.0083	0.6063
	Exergy destruction (kW)	142.92	102.19	85.49	192.49	30.87	302.77
	Exergoeconomic factor (%)	19.3	17.9	48.8	54.2	12.9	23.7
50	Investment cost (MUS$)	0.2285	0.0969	0.1099	0.3093	0.0009	0.3285
	Exergy destruction cost (MUS$)	0.2799	0.1520	0.2305	0.2199	0.0092	0.6245
	Exergy destruction (kW)	171.51	104.24	86.48	198.06	36.42	311.86
	Exergoeconomic factor (%)	18.4	17.3	48.1	53.6	11.9	23.1

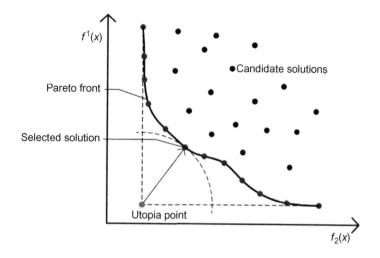

FIGURE 13.20

Selection of the best solution from the Pareto set (closest to the utopia point) [50].

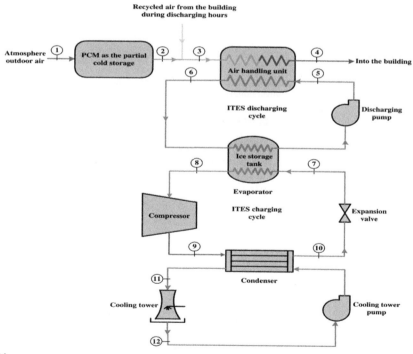

FIGURE 13.21

Schematic diagram of the modeled ITES system incorporating PCM as partial cold storage [69].

For exergy efficiency (objective function 1)

$$\Psi_{tot} = 1 - \frac{\dot{E}_{D,tot}}{\dot{E}_{in}}$$
(13.14)

where $\dot{E}_{D,tot}$ is the sum of the exergy destruction rate of the system components and $\dot{E}_{in}$ is the electricity consumption of the whole system, which includes the compressor, fan, and pumps.

For the total cost rate (objective function 2)

$$\dot{C}_{tot} = \sum_k \dot{Z}_k + \dot{C}_{elec} + \dot{C}_{env}$$
(13.15)

where the designation to the aforementioned terms corresponds to the one defined in the thermoeconomic optimization described earlier.

The estimation of the payback period of additional expenses (in years) can be performed by using the expression given by

$$\Delta(Z_{op})\left(\frac{(1+i)^p - 1}{i(1+i)^p}\right) + \Delta Z_{SV}\left(\frac{i}{(1+i)^p}\right) = \Delta\left(\sum_k Z_k\right)$$
(13.16)

Hence, the multiobjective optimization of the ITES system being performed based on Eqs. (13.14) and (13.15) using the GA has yielded optimization results that are presented in Fig. 13.22 and Tables 13.8–13.10.

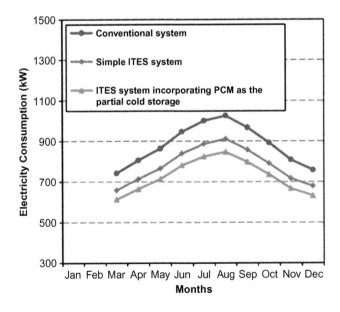

FIGURE 13.22

The comparison of power consumption between ITES system incorporating PCM, simple ITES (without PCM as the partial cold storage), and conventional systems over a year [69].

Table 13.8 The Optimum Values of Hybrid System Design Parameters With Three Methods of Optimization [69]

Design Parameters	Single-Objective Objective Function 1	Single-Objective Objective Function 2	Multiobjective Objective Function 1 & 2
T_5 (°C)	3.36	4.53	4.19
T_6 (°C)	12.45	11.21	11.53
T_{ST} (°C)	−3.93	−1.34	−2.05
T_{EV} (°C)	−5.38	−3.70	−4.37
T_{Cond} (°C)	36.29	38.69	37.93
α	0.1	0.011	0.063

Table 13.9 The Results of Energy and Exergy Analyses of the Hybrid System With Three Methods of Optimization [69]

	Single-Objective Objective Function 1	Single-Objective Objective Function 2	Multiobjective Objective Function 1 & 2
Required PCM (kg)	1.8258×10^4	2.0493×10^3	1.1737×10^4
Annual CO_2 emission (kg)	1.62×10^6	1.87×10^6	1.73×10^6
AHU exergy destruction (kW)	101.55	125.84	113.71
ST exergy destruction (kW)	81.94	96.98	90.93
EV exergy destruction (kW)	68.87	83.88	75.83
Comp exergy destruction (kW)	153.75	177.21	164.23
EX exergy destruction (kW)	17.64	28.24	23.32
Cond + CT exergy destruction (kW)	247.59	278.46	258.49
PCM storage exergy destruction (kW)	55.72	25.67	38.55
Total exergy efficiency (%)	42.17	31.68	38.47

The case study results infer that the power consumption of the hybrid ITES system is comparatively less than the simple ITES and the conventional system. On average, the electricity consumption of the hybrid ITES system decreases by 6.7% and 17.1% as compared to the simple ITES and conventional system, respectively. This can be attributed to the fact that the PCM acts as partial cold storage and meets part of the building cooling load during on-peak hours.

Table 13.10 The Results of Economic Analysis of the Hybrid System With Three Methods of Optimization [69]

	Single-Objective Objective Function 1	Single-Objective Objective Function 2	Multiobjective Objective Function 1 & 2
PCM cost (MUS$)	0.54773	0.06148	0.25211
AHU cost (MUS$)	0.2191	0.2304	0.2217
Discharging pump cost (MUS$)	0.0273	0.0319	0.0309
ST cost (MUS$)	0.0921	0.1005	0.0961
EV cost (MUS$)	0.1428	0.1471	0.1449
Comp cost (MUS$)	0.2792	0.2863	0.2824
EX cost (MUS$)	0.0007532	0.0007681	0.0007548
(Cond þ CT) cost (MUS$)	0.2579	0.2701	0.2612
CT pump cost (MUS$)	0.0324	0.0368	0.0353
CO_2 penalty cost (MUS$ $year^{-1}$)	0.1458	0.1683	0.1557
Total annual cost (MUS$)	1.9769	1.6392	1.7938

Besides, the multiobjective optimization helps to provide optimal solutions for identifying the trade-off between the objective functions. However, the additional cost incurred for the payback period of the hybrid system compared to the simple ITES and the conventional systems is higher. This can be compensated within 4 years because the hybrid system is energy efficient as it consumes less power than the simple ITES and conventional systems.

13.4 CONCISE REMARKS

Control and optimization can be considered two important tools in the field of thermal storage in building applications. The incorporation of control systems starting from the conventional On/Off, PID to the advanced intelligent logic-based hybrid controllers into the thermal energy systems show improved performance of the system under control. The swift response of the control system in actuating the system toward the sudden fluctuations occurring in the load demand can benefit in acquiring energy savings potential.

In this context, the shifting of the peak load demand to off-peak periods through the incorporation of TES systems can be effectively accomplished using different control strategies. Depending on the building thermal load condition, use of specific control strategies to monitor the rate of charging and discharging of TES can help offset peak load demand considerably.

Having this on one side, the energy efficiency of the thermal energy system can be enhanced through performing optimization analysis that takes into consideration real-time parametric disturbances. The evaluation of exergy losses and exergy destruction by means of single-objective or multiobjective optimization strategies can yield the best possible solutions (optimum state) for the problem identified in the system. By analyzing the energy and exergy interactions of the thermal energy system, its energy efficiency can be maximized as well as the operating costs minimized substantially. Thus, the implementation of control and optimization strategies can contribute to achieving better thermal performance of the building integrated TES systems.

References

[1] Afram A, Janabi-Sharifi F. Theory and applications of HVAC control systems—a review of model predictive control (MPC). Build Environ 2014;72:343–55.

[2] Dounis AI, Caraiscos C. Advanced control systems engineering for energy and comfort management in a building environment—a review. Renew Sustain Energy Rev 2009;13:1246–61.

[3] Levermore GJ. Building energy management systems: an application to heating and control. London: E & FN SPON; 1992.

[4] Dounis AI, Manolakis DE. Design of a fuzzy system for living space thermal-comfort regulation. Appl Energy 2001;69:119–44.

[5] Kolokotsa D. Comparison of the performance of fuzzy controllers for the management of the indoor environment. Build Environ 2003;38:1439–50.

[6] Kolokotsa D, Tsiavos D, Stavrakakis G, Kalaitzakis K, Antonidakis E. Advanced fuzzy logic controllers design and evaluation for buildings' occupants thermal–visual comfort and indoor air quality satisfaction. Energy Build 2001;33(6):531–43.

[7] Calvino F, Gennusca ML, Rizzo G, Scaccianoce G. The control of indoor thermal comfort conditions: introducing a fuzzy adaptive controller. Energy Build 2004;36:97–102.

[8] Hamdi M, Lachiever G. A fuzzy control system based on the human sensation of thermal comfort. In: Fuzzy systems proceedings, IEEE world congress on computational intelligence. The 1998 IEEE international conference, vol. 11998. p. 487–92.

[9] Dounis AI, Caraiscos C. Intelligent coordinator of fuzzy controller—agents for indoor environment control in buildings using 3-d fuzzy comfort set. In: IEEE international conference on fuzzy systems; 2007.

[10] Egilegor B, Uribe JP, Arregi G, Pradilla E, Susperregi L. A fuzzy control adapted by a neural network to maintain a dwelling within thermal comfort. In: 5th International 97; 1997.

[11] Kanarachos A, Geramanis K. Multivariable control of single zone hydronic heating systems with neural networks. Energy Convers Manage 1998;13(13):1317–36.

[12] Liang J, Du R. Thermal comfort control based on neural network for HVAC application. In: Control applications 2005, CCA 2005, IEEE conference; 2005. p. 819–24.

[13] Argiriou A, Bellas-Velidis I, Kummert M, Andre P. A neural network controller for hydronic heating systems of solar buildings. Neural Netw 2004;17:427–40.

[14] Argiriou A, Bellas-Velidis I, Balaras CA. Development of a neural network heating controller for solar buildings. Neural Netw 2000;13:811–20.

[15] Hagras H, Callaghan V, Colley M, Clarke G. A hierarchical fuzzy-genetic multi-agent architecture for intelligent buildings online learning, adaptation and control. Inf Sci 2003;150:33–57.

[16] Mo Z, Mahdani A. An agent-based simulation-assisted approach to bi-lateral building systems control. In: Eighth international IBPSA conference; 2003. p. 11–4.

[17] Hagras H, Callaghan V, Colley M, Clarke G, Pounds-Cornish A, Duman H. Creating an ambient-intelligence environment using embedded agents. IEEE Intell Syst 2004;19:12–20.

[18] Qiao B, Liu K, Guy C. A multi-agent for building control. In: Proceedings of the IEEE/WIC/ACM international conference on intelligent agent technology; 2006. p. 653–9.

[19] Dounis AI, Caraiscos C. Fuzzy comfort and its use in the design of an intelligent coordinator of fuzzy controller–agents for environmental conditions control in buildings. J Uncertain Syst 2008;2(2):101–12.

[20] Morel N, Bauer M, El-Khoury M, Krauss J. Neurobat, a predictive and adaptive heating control system using artificial neural networks. Int J Sol Energy 2000;21:161–201.

[21] Yamada F, Yonezawa K, Sugarawa S, Nishimura N. Development of air-conditioning control algorithm for building energy-saving. In: IEEE international conference on control applications; 1999.

[22] Wang S, Xu X. A robust control strategy for combining DCV control with economizer control. Energy Convers Manage 2002;43:2569–88.

[23] Wang S, Xu X. Optimal and robust control of outdoor ventilation airflow rate for improving energy efficiency and IAQ. Build Environ 2004;39:763–73.

[24] Shepherd AB, Batty WJ. Fuzzy control strategies to provide cost and energy efficient high quality indoor environments in buildings with high occupant densities. Build Serv Eng Res Technol 2003;24(1):35–45.

[25] Kolokotsa D, Stavrakakis GS, Kalaitzakis K, Agoris D. Genetic algorithms optimized fuzzy controller for the indoor environmental management in buildings implemented using PLC and local operating networks. Eng Appl Artif Intell 2002;15:417–28.

[26] Dalamagkidis K, Kolokotsa D, Kalaitzakis K, Stavrakakis GS. Reinforcement learning for energy conservation and comfort in buildings. Build Environ 2007;42(7):2686–98.

[27] Anderson CW, Hittle D, Kretchmar M, Young P. Robust reinforcement learning for heating, ventilation and air conditioning control of buildings. In: Si J, Barto AG, Powell WB, Wunsch DII, editors. Handbook of learning and approximate dynamic programming. Hoboken, NJ, USA: IEEE Press/Willey Interscience; 2004. p. 517–34.

[28] Rutishauser U, Joller J, Douglas R. Control and learning of ambience by an intelligent building. IEEE Trans Syst Man Cybern A Syst Hum 2005;35(1):121–32.

[29] Doctor F, Hagras H, Callaghan V. An intelligent fuzzy agent approach for realising ambient intelligence in intelligent inhabited environments. IEEES MC A Syst Hum 2005;35(1):55–65.

[30] Hagras H, Callaghan V, Colley M, Clarke G, Pounds-Cornish A, Duman H. Creating an ambient-intelligence environment using embebedded agents. IEEE Intell Syst 2004;19:12–20 [November/December].

[31] Guillemin A. Using genetic algorithms to take into account user wishes in an advanced building control system [PhD]. E´cole Polytechnique Fe´de´rale De Lausanne; 2003.

[32] Guillemin A, Morel N. An innovative lighting controller integrated in a selfadaptive building control system. Energy Build 2001;33(5):477–87.

[33] Guillemin A, Molteni S. An energy-efficient controller for shading devices self-adapting to the user wishes. Build Environ 2002;37:1091–7.

[34] Kummert M, Andre P, Nicolas J. Optimal heating control in a passive solar commercial building. Sol Energy 2001;69(1–6):103–16.

[35] Wang S, Jin X. Model-based optimal control of VAV air-conditioning system using genetic algorithms. Build Environ 2000;35:471–87.

[36] Sun Y, Wang S, Xiao Fu, Gao D. Peak load shifting control using different cold thermal energy storage facilities in commercial buildings: a review. Energy Convers Manag 2013;71:101–14.

[37] Lee KH, Braun JE. Development of methods for determining demand-limiting setpoint trajectories in buildings using short-term measurements. Build Environ 2008;43:1755–68.

[38] Sun YJ, Wang SW, Huang GS. A demand limiting strategy for maximizing monthly cost savings of commercial buildings. Energy Build 2010;42:2219–30.

[39] Lee K-H, Braun JE. Model-based demand-limiting control of building thermal mass. Build Environ 2008;43:1633–46.

[40] Ma J, Qin J, Salsbury T, Xu P. Demand reduction in building energy systems based on economic model predictive control. Chem Eng Sci 2012;67(1):92–100.

[41] Braun JE, Lee K-H. An experimental evaluation of demand-limiting using building thermal mass in a small commercial building. ASHRAE Trans 2006;112(1):547–58.

[42] Xu P, Haves P, Braun JE, Hope LT. Peak demand reduction from pre-cooling with zone temperature reset in an office building. In: Proceedings of 2004 ACEEE summer study of energy efficiency in buildings, Pacific Grove, CA; 2004.

[43] Xu P, Haves P, Zagreus L, Piette M. Demand shifting with thermal mass in large commercial buildings (field tests, simulation and results). Lawrence Berkeley National Laboratory; 2006, CEC-500-2006-009.

[44] Yin RX, Xu P, Piette MA, Kiliccote S. Study on auto-DR and pre-cooling of commercial buildings with thermal mass in California. Energy Build 2010;42:967–75.

[45] Henze GP, Krarti M, Brandemuehl MJ. A simulation environment for the analysis of ice storage controls. HVAC&R Res 1997;3(2):128–48.

[46] Henze GP, Dodier RH, Krarti M. Development of a predictive optimal controller for thermal energy storage systems. HVAC&R Res 1997;3(3):233–64.

[47] Candanedo JA, Dehkordi VR, Stylianou M. Model-based predictive control of an ice storage device in a building cooling system. Appl Energy 2013;111:1032–45.

[48] Kim SH. An evaluation of robust controls for passive building thermal mass and mechanical thermal energy storage under uncertainty. Appl Energy 2013;111:602–23.

[49] Aughenbaugh JM. Managing uncertainty in engineering design using imprecise probabilities and principles of information economics [Ph.D. Thesis]. Georgia Institute of Technology; 2006.

[50] Nguyen A-T, Reiter S, Rigo P. A review on simulation-based optimization methods applied to building performance analysis. Appl Energy 2014;113:1043–58.

[51] Evins R, Pointer P, Vaidyanathan R, Burgess S. A case study exploring regulated energy use in domestic buildings using design-of-experiments and multi-objective optimisation. Build Environ 2012;54:126–36.

[52] Eisenhower B, O'Neill Z, Narayanan S, Fonoberov VA. Mezic´ I. A methodology for meta-model based optimization in building energy models. Energy Build 2012;47:292–301.

[53] Nguyen AT. Sustainable housing in Vietnam: climate responsive design strategies to optimize thermal comfort [PhD thesis]. Université de Liège; 2013.

[54] Kalogirou SA. Optimization of solar systems using artificial neural-networks and genetic algorithms. Appl Energy 2004;77(4):383–405.

[55] Chen L, Fang QS, Zhang ZY. Research on the identification of temperature in intelligent building based on feed forward neural network and particle swarm optimization algorithm. In: Proceedings of 2010 sixth international conference on natural computation (ICNC); Yantai: Institute of Electrical and Electronics Engineers (IEEE); 2010. p. 1816–20.

[56] Magnier L, Haghighat F. Multiobjective optimization of building design using TRNSYS simulations, genetic algorithm, and artificial neural network. Build Environ 2010;45(3):739–46.

[57] Tresidder E, Zhang Y, Forrester AIJ. Optimisation of low-energy building design using surrogate models. In: Proceedings of the 12th conference of international building performance simulation association. Sydney: IBPSA; 2011. p. 1012–6.

[58] Boithias F, El Mankibi M, Michel P. Genetic algorithms based optimization of artificial neural network architecture for buildings' indoor discomfort and energy consumption prediction. Build Simul 2012;5(2):95–106.

[59] Hasan A, Vuolle M, Sirén K. Minimisation of life cycle cost of a detached house using combined simulation and optimisation. Build Environ 2008;43(12):2022–34.

[60] Tuhus-Dubrow D, Krarti M. Genetic-algorithm based approach to optimize building envelope design for residential buildings. Build Environ 2010;45:1574–81.

[61] Roy R, Hinduja S, Teti R. Recent advances in engineering design optimisation: challenges and future trends. CIRP Ann Manuf Technol 2008;57(2):697–715.

[62] Sahab MG, Toropov VV, Gandomi AH. A review on traditional and modern structural optimization: problems and techniques. In: Gandomi AH, Yang XS, Talatahari S, Alavi AH, editors. Metaheuristic applications in structures and infrastructures. Oxford: Elsevier; 2013. p. 25–47, ISBN: 9780123983640.

[63] Hopfe CJ, Emmerich MT, Marijt R, Hensen J. Robust multi-criteria design optimisation in building design. In: Wright J, Cook M, editors. Proceedings of the 2012 building simulation and optimization conference. Loughborough, Leicestershire: Loughborough University; 2012. p. 118–25, ISBN: 978-1-897911-42-6.

[64] Jin Q, Overend M. Facade renovation for a public building based on a whole-life value approach. In: Wright J, Cook M, editors. Proceedings of the 2012 building simulation and optimization conference. Loughborough, Leicestershire: Loughborough University; 2012. p. 378–85.

[65] Chantrelle FP, Lahmidi H, Keilholz W, Mankibi ME, Michel P. Development of a multicriteria tool for optimizing the renovation of buildings. Appl Energy 2011;88(4):1386–94.

[66] Hamdy M, Palonen M, Hasan A. Implementation of pareto-archive NSGA-II algorithms to a nearly-zero-energy building optimisation problem. In: Wright J, Cook M, editors. Proceedings of the 2012 building simulation and optimization conference. Loughborough, Leicestershire: Loughborough University; 2012. p. 181–8, ISBN: 978-1-897911-42-6.

[67] Evins R. A review of computational optimisation methods applied to sustainable building design. Renew Sustain Energy Rev 2013;22:230–45.

[68] Tsatsaronis G. Thermoeconomic analysis and optimization of energy systems. Prog Energy Combust Sci 1993;19:227–57.

[69] Navidbakhsh M, Shirazi A, Sanaye S. Four E analysis and multi-objective optimization of an ice storage system incorporating PCM as the partial cold storage for air-conditioning applications. Appl Therm Eng 2013;58:30–41.

[70] Sanaye S, Shirazi A. Thermo-economic optimization of an ice thermal energy storage system for air-conditioning applications. Energy Build 2013;60:100–9.

Economic and Societal Prospects of Thermal Energy Storage Technologies

14.1 INTRODUCTION

The exploitation of fossil fuel–based energy sources and the increasing demand for energy have shown clear indications for the development and implementation of energy-efficient technologies. The electricity utility sector is continuously striving hard to meet the peak load demand and to do so with optimum generation costs. In this context, TES technologies have been indentified the most promising toward improving energy economy.

Apart from the technical advantages that the TES system would offer, the cost factors involved in their successful implementation and operation in the long run are equally important and have to be considered starting from the commencement to the completion of the project. Establishing proper documentation of the energetic activities or instances occurring during the project phases can help to minimize the overall cost of the TES system being integrated with the cooling/heating system available in the facility. Furthermore, the assessment of the exergy life cycle cost factors in addition to associated energy costs can be beneficial for making an energy-efficient design of the TES systems for future utilization at the facility when required.

14.2 COMMISSIONING OF THERMAL ENERGY STORAGE (TES) SYSTEMS
14.2.1 Procedure for installation of thermal storage systems

The operational performance of TES systems mainly depends on the installation procedure, which requires more skill and specialized expertise. Numerous factors are involved in the successful installation of TES systems, some of which include [1,2]:

- Clarity in the objectives of setting up TES systems for long-term benefits
- Cooling/heating load profiles of the facility for which TES system is required
- Technical expertise for operating TES systems
- Analysis of utility ownership and TES requirements
- Estimation of electricity cost savings in the long term
- Establishment of cost rate contract with the electricity utility
- Assessment of heat recovery options using TES systems
- Selection of appropriate control strategies to minimize electricity and operational costs

- Determination of the required cooling capacity redistribution between the cooling/heating systems and TES systems
- Review of design considerations and technical aspects for construction of TES systems
- Establishment of project management coordination between the facility owner, engineer, and contractor
- Verification of the personnel operational expertise and other safety aspects associated in commissioning of TES systems
- Ensuring full operational performance of the TES system through conducting technical checks and required tests prior to acceptance of the installation by the facility owner
- Documentation of activities related to the commissioning and installation of the TES system starting from scheme inception to complete construction of the project

Generically, the TES systems can be considered to provide better energy redistribution ability with low life cycle cost to the facility owner. For instance, the chilled water-based TES system performance largely depends on peak load shifting capacity, operating temperature differential, type of thermal storage strategy considered, and so on. In particular, the design operating temperature differential between the supply and the return chilled water network plays a vital role in determining the storage capacity of the stratified TES system. By maintaining a constant as well as a higher temperature gradient of the chilled water TES system, the stratification occurring in the storage tank can be effective, provided proper control strategies are followed during the operation of the system.

The design of storage tank for a TES system is another crucial task for accomplishing better operational performance of the system in context with the installation procedures. The responsible engineering team has to be a well-trained group of professionals dealing with the design principles and experienced in the installation and commissioning activities of the TES systems. This is due to the fact that an oversized storage tank may lead to higher storage time, cost of construction, and cost of operation. In turn, an undersized tank may affect the desired storage capacity of the TES system.

From this perspective, the facility owner, design engineer, and contractor can refer to the procedures set forth by the International Standards and Codes to gain maximum benefits from installation and operation of the TES systems on a long-term basis. The well-known ASHRAE Standard 150 can be a good reference being designated for describing the testing procedures for establishing the cooling capacities and operational efficiencies of the cool thermal energy storage (CTES) systems.

The ASHRAE Standard 150 refers to the "method of testing the performance of cool storage systems" and includes a functional performance test of a TES system at the completion of its installation in accordance to the project requirements of the facility owner. Overall, the ASHRAE Standard 150 covers testing, instrumentation, test methods and procedures, design data and calculations, and the test report of the CTES systems. To make the CTES systems more functional, they are subjected to performance tests in accordance with the ASHRAE Standard 150, which includes

- Charging and discharging test of the cool storage
- Capacity testing of the cool storage
- Efficiency testing of the cool storage

In short, the commissioning and installation of TES systems are integral processes that evolve from the commencement of the conceptual design to the complete construction phase of a project. The stepwise and sequential documentation of the activities related to the cooling/heating energy redistribution between the utility system and the TES system is most vital for the successful installation and operation of the integrated systems on a long-term basis. Furthermore, the parametric testing of TES in accordance to the International Standards and Codes of Practice along with the project requirements of the owner/client can help sustain energy efficiency at low cost economy of the system.

14.3 COST ANALYSIS AND ECONOMIC FEASIBILITY

The installation of TES systems in applications requiring peak load shifting or energy redistribution can be appreciable from the technical viewpoint. However, the preference for TES technologies for real-time cooling/heating applications depends on two important factors, namely, the cost and the economics involved. In this context, the cost analysis and the economic feasibility of *latent thermal energy storage* (LTES) and seasonal TES systems are discussed briefly in the following section.

14.3.1 LTES system

The economic evaluation and the cost savings analysis of the ice thermal energy storage (ITES) system being retrofitted with a conventional A/C system has been studied for an office building located in Malaysia [3]. The long-term benefits of the ITES system over the next 20 years are predicted for three different scenarios. The first scenario is to retrofit for 10% of the conventional A/C system existing in the building, the second and third scenarios being to retrofit 25% and 50% of the existing conventional A/C system, respectively.

Based on the rule of thumb methodology, the capital costs of the retrofit systems are calculated for the three different scenarios. By assuming a steady and constant interest rate of 7% in Malaysia, the utility and the maintenance costs are also predicted for the next 20 years. The prediction results of this study infer that total annual cost of the 50% retrofit ITES system yields 14% cost savings in the best operating condition. Likewise, the other scenarios also result in better cost savings. At the same time, the cost savings potential of full storage is observed to be significantly higher compared to the load leveling strategy.

From the comparison between the installation, maintenance, and electricity costs of the conventional A/C system, it can be seen that the payback period of the full storage ITES system is about 3-6 years. Likewise, the payback period for the load leveling system is found to be between 1 and 3 years. In short, the cost savings achieved through the different operating scenarios of the ITES system facilitate enhanced money savings over the next 20 years for the facility owner, which can be beneficial to society in terms of accomplishing reduced peak load electricity consumption and the reduction of green house gas (GHG) emissions. The test results of the ITES system obtained from this study are presented in Table 14.1.

Other research has studied the operational performance and the cost analysis of LTES system incorporating PCMs with different melting temperature for heating application. Herein, the equivalent money for exergy and the hidden costs associated with the environment have been considered. For achieving a more energy-efficient design of the LTES system, the clearance effect pertaining to exergy analyses on the environment and its equalization with money has been investigated. The schematic representation of the system being investigated is shown in Fig. 14.1, and the thermophysical properties of the selected PCMs are listed in Table 14.2.

Table 14.1 Summary of the Test Results of the ITES System [3]

Scenario	Installation Cost (M$)			Total Costs (M$ Over 20 Years)			Total Costs Savings (M$ Over 20 Years)	
	Conventional System	Full Storage	Load Leveling	Conventional System	Full Storage	Load Leveling	Full Storage	Load Leveling
50%	661	1755	723	38,941	31,716	37,743	7226	1199
25%	331	877	361	19,471	5858	18,871	3613	599
10%	132	351	145	7788	6343	7549	1445	240

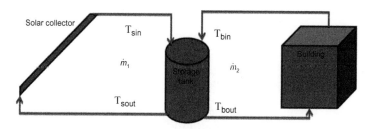

FIGURE 14.1

Schematic of the investigated system [4].

Table 14.2 Properties of Selected PCMs [4]

PCM Type	T_m °C(K)	C_p (kJ/kg K)	L (kJ/kg)	Density (kg/m³)
A32	32 (305)	2.20	130	845
A39	39 (312)	2.22	105	900
A42	42 (315)	2.22	105	905
A53	53 (326)	2.22	130	910
A55	55 (328)	2.22	135	905
A58	58 (331)	2.22	132	910

Table 14.3 Total Economic Analyses of the System for 40 Years [4]

PCM Type	PCM Weight (kg)	PCM Price ($)	Total Price of Lost Energy ($)	Total Price of Lost Exergy ($)	Total Life Cycle Cost ($)
A32	336	2186	1,123,439	4,447,622	5,573,247
A39	416	2707	1,123,439	3,561,181	4,687,627
A42	416	2707	1,123,439	3,193,553	4,319,699
A53	336	2186	1,123,439	1,904,470	3,030,095
A55	324	2105	1,123,439	1,679,538	2,805,082
A58	331	2153	1,123,439	1,347,324	2,472,916

In this LTES system, the total life-cycle-associated economy has been estimated by computing the total energy and exergy lost. Table 14.3 summarizes the results pertaining to the total life cycle cost of this system with each PCM configuration. From the results obtained, it can be observed that the total life cycle cost associated with the exergy lost price decreases with the PCM having higher melting temperature. By virtue of the higher melting temperature and exergy output, inlet and outlet temperatures of the storage are also elevated. Because the exergy input remains constant due to the increase of exergy output, the total exergy lost reduces at a higher melting temperature.

Table 14.3 shows that the PCM A58, which exhibits higher melting point, yields the lowest total life cycle cost. On the other hand, the contribution of exergy in the total life cycle cost of the system is

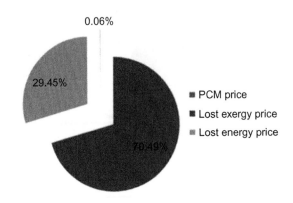

FIGURE 14.2

Lost exergy, lost energy, and PCM cost share in total life cycle costs [4].

higher, as can be observed from Fig. 14.2. Hence, the exergy as a major contributor for determination of the total life cycle cost cannot be neglected, and it has to be considered for the energy-efficient design of the LTES system.

14.3.2 Seasonal TES system

The energy and the cost savings of an aquifer TES system in a Belgian hospital facility has been investigated in recent years [5]. The long-term aquifer seasonal thermal storage system is integrated with the reversible heat pumps in order to provide heating and cooling requirements for a newly constructed hospital in near Antwerp. This monitoring study investigates operational performance as well as the economical aspects of the low temperature aquifer SeTES system, as compared to the conventional heating, ventilation and air conditioning (HVAC) installation over the monitoring period of three years. The existing HVAC system comprises a gas boiler and compression cooling equipment to cater the cooling and heating demand in the hospital building.

The cooling and heating load calculations are performed by taking into account the seasonal performance factor of 3.5 for the cooling machine and the gas-fired boiler efficiency of 85%. The pump energy requirements for the HVAC system in the hospital facility are neglected, due to the fact that they remain same while comparison is made between the aquifer thermal energy storage (ATES) and the conventional systems. The design specifications are summarized in Table 14.4, and the schematic representation of this system can be referred from [5]. The effect of this ATES system integration on CO_2 emissions over the monitoring period of three years are estimated based on the reference values recommended according to the European Standard EN 15603:2008 [6] as presented in Table 14.5. The comparative results presented in Table 14.6 clearly reveal that the installation of the ATES system has eliminated entirely gas consumption for conditioning the ventilation air and is replaced by the source and heat pumps electricity consumption.

In addition, the combined system reduces CO_2 emission by more than 1280 tons over the monitoring period of three years. Moreover, the total primary energy savings potential achieved by this combined system for the climatization of the ventilation air is about 71%. This value corresponds to a reduction of 73% of CO_2 emission, as compared to the reference installation. On the other hand, the

Table 14.4 Design Specifications of the ATES System [5]

Parameter	Value
Maximum flow per well	100 m3/h
Maximum cooling power	1.2 MW
Diameter drilling	0.8 m
Depth wells	65 m
Length filters	36-40 m
Thickness aquifer	30-40 m
Number of cold wells	1
Number of warm wells	1
Distance between cold and warm well	100 m
Undistributed groundwater temperature	11.7 °C
Injection temperature warm well	18 °C
Injection temperature cold well	8 °C

Table 14.5 Primary Energy Factors and CO_2 Emission According EN 15603:2008 [6]

	Total Primary Energy Factor	CO2 Emission Factor (kg/MWh)
European electricity mix (UCPTE[a] countries)	3.31	617
Gas	1.36	277

[a]UCPTE, Union for the Coordination of the Transmission of Electricity.

ATES system provides about 46% of total heating energy demand of the hospital facility. Even if the gas-fired boilers are considered, the reduction of CO_2 emissions can be still achieved up to 38% with this ATES-heat pump combined system.

Based on the results presented in Table 14.7, it can be observed that the installation cost of the ATES system is comparatively higher and a major cost as compared to conventional/reference installation. This is due to the integral elements of the ATES system including wells, pumps, ducts, heat exchanger, and shutoff valves. Also, the supplemental costs including the geological study and other engineering factors are to be considered in the ATES system installation.

Indeed, the energy cost associated with the ATES system installation is comparatively less than the reference installation. That is, about 85% of the energy cost reduction is achievable with the ATES system than with the conventional cooling system installation in the hospital facility considered. The reason for the higher energy cost savings potential of the ATES system is due to the utilization of the natural cooling strategy in the ATES system for sharing the cooling load in the building.

On the other hand, about 55% of the heating energy cost savings is realizable using the ATES system compared to the reference gas-fired boiler installation. Furthermore, even without availing any subsidies, the ATES system is economically feasible with a payback period of 8.4 years. By considering the subsidies, the payback period can be reduced to even less than 4 years.

Table 14.6 Annual Primary Energy Savings and CO_2 Emission Reduction [5]

	Unit	ATES System+Heat Pumps (Installation Klina)				Gas-Fired Boilers+Cooling Machines (Reference-Installation)			
		2003	2004	2005	Total	2003	2004	2005	Total
Electricity consumption	MWh_e	343	242	198	782	313	193	242	748
Gas consumption	GJ	0	0	0	0	6340	5739	4884	16,964
Primary energy consumption	GJ_p	4082	2878	2362	9323	12,350	10,104	9527	31,981
Reduction primary energy consumption	GJ_p	−8268	−7225	−7164	−22,658	−	−	−	−
	%	−67	−72	−75	−71				
CO_2 emissions	ton CO_{2eq}	211	149	122	483	681	561	525	1767
CO_2 emissions reduction	ton CO_{2eq}	469	412	403	1284	−	−	−	−
	%	−69	−73	−77	−73				

Table 14.7 Economical Analysis of the ATES System Compared to the Reference Installation [5]

		ATES System+Heat Pumps (Installation Klina)	Gas-Fired Boilers+Cooling Machines (Reference-Installation)
Investment costs	Underground installation (k€)	299	–
	Overground installation (k€)	266	241
	Heat exchangers	35	
	Pumps, ducts, appendages	46	
	Control equipment	116	
	Larger cooling coils	13	
	Frost protection coils	32	
	Total installation costs (k€)	565	241
	Extra study and engineering costs for ATES installation (k€)	130	–
	Total, subsidies excluded (k€)	695	241
	Subsidies (k€)	244	–
	Total, subsidies included (k€)	451	241
Fuel costs	Cooling supply (MWh)	872	872
	Electricity consumption for cooling (MWh)	33	249
	Total cooling costs (k€)	3.7	27.4
	Heating supply (MWh)	1335	1335
	Gas consumption (MWh)	–	1571
	Gas costs for heating (k€)	–	55.0
	Electricity consumption for heating (MWh)	227	–
	Electricity costs (k€)	25.0	–
	Total heating costs (k€)	25.0	55.0
	Total fuel costs (k€/year)	28.7	82.4
Simple payback time	Subsidies excluded (years)	8.4	–
	Subsidies included (years)	3.9	–

In this context, the costs of investment per water equivalent of 13 real-time and planned projects on the SeTES systems in Germany are shown in Fig. 14.3. It is obvious from Fig. 14.3 that as the storage volume in water equivalent (storage scale) increases, the investment cost per water equivalent decreases. Based on the survey, it is observed that projects requiring hot water storage exhibit higher costs, which is due to the construction of large storage water tanks.

At the same time, the ATES and the borehole TES systems are comparatively cheaper, but geological suitability is a major factor to be considered for their successful installation and efficient operation on a long-term basis. In addition, the inherent heat loss effects of the seasonal storage systems involve supplemental costs for the installation of heat pump units to upgrade of the extracted heat energy to be directly utilized in the building spaces.

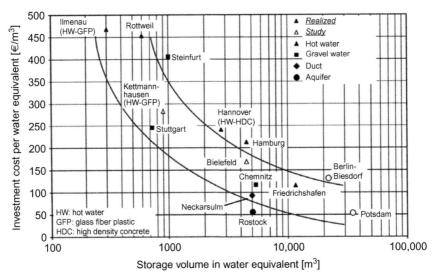

FIGURE 14.3

Investment costs of seasonal heat stores (including design, without VAT) [7,8].

14.4 SOCIETAL IMPLICATIONS OF TES SYSTEMS

The utilization of the sensible heat, latent heat, thermochemical, and long-term seasonal TES technologies for the benefit of the society in terms of energy, cost, and economical aspects have been recognized over years. The promising aspects of the TES systems for fulfilling the energy requirements of the society can be realized from the following points:

- Reduction in the extensive usage of fossil fuels–based energy sources
- Reduced greenhouse gas emissions
- Bridges gap between the energy supply and energy demand
- Peak load redistribution to off-peak load periods; thereby less power consumption at the cooling/heating equipment
- Reduced electricity and energy costs
- Increased external financing options and higher property values
- Economic feasibility

On the other hand, there are some major limitations in the successful installation and the operational aspects of the TES systems from the societal point of view, which include

- Issues related to the storage and release of thermal energy at a constant temperature
- Stratification problems in realizing STES systems implementation
- Inherent thermal losses from the storage system by means of conduction and residual stored energy loss
- Space requirements for the installation of the system

- Disposal of the chemically derived PCMs and thermochemically reactive constituent materials into the environment without proper treatment as per the Standards and Regulations
- Operation and maintenance issues
- Economic risk factors between the utility and the facility operations in terms of electricity pricing policy
- Environmental impacts due to the changes in the hydrological, biological, and geotechnical aspects of the underground
- Possibility for the leakages of the heat transfer medium in the underground TES systems

14.5 **CONCISE REMARKS**

The cost analysis of different TES systems installation discussed indicates possible growth in the economic status of society by means of achieving reduced energy and electricity costs. Next to this, the integration of a TES system with an existing cooling/heating system installation can facilitate for attaining effective energy redistribution needs during peak demand load periods, reduction in the dependency of fossil fuel-based energy source,s as well as greenhouse gas emissions reduction.

To experience the aforementioned promising aspects of the TES systems, their installation procedures starting from the scheme inception to the commissioning stage of the project has to be followed vigilantly. The proper documentation at every step of the project phase involving the economic design aspects are most essential for successful installation and operation of TES systems on a long-term basis. Even though there are certain restrictions in the development of the TES systems, their usefulness has already been realized for socioeconomic development for a sustainable future.

References

[1] Wulfinghoff DR. Cooling thermal storage. Energy efficient manual; 1999.
[2] Hyman LB. Commissioning Chilled Water TES Systems. Engineered Systems Magazine; 2004 (http://gossengineering.com/about/technical-articles-and-books/commissioning-chilled-water-tes-systems/).
[3] Rismanchi B, Saidur R, Masjuki HH, Mahlia TMI. Cost-benefit analysis of using cold thermal energy storage systems in building applications. Energy Proc 2012;14:493–8.
[4] Rezaei M, Anisur MR, Mahfuz MH, Kibria MA, Saidur R. Metselaar. IHSC Performance and cost analysis of phase change materials with different melting temperatures in heating systems Energy 2013;53:173–8.
[5] Vanhoudt D, Desmedt J, Bael JV, Robeyn N, Hoes H. An aquifer thermal storage system in a Belgian hospital: Long-term experimental evaluation of energy and cost savings. Energy Build 2011;43:3657–65.
[6] EN 15603:2008, Energy performance of buildings–overall energy use and definition of energy ratings.
[7] Schmidt T, Mangold D, Müller-Steinhagen H. Seasonal thermal energy storage in Germany. In: ISES Solar World Congress, 14-19 June, Göteborg, Schweden; 2003.
[8] Xu J, Wang RZ, Li Y. A review of available technologies for seasonal thermal energy storage. Sol Energy 2013, http://dx.doi.org/10.1016/j.solener.2013.06.006.

Applications of Thermal Energy Storage Systems

15.1 ACTIVE AND PASSIVE SYSTEMS

The definitions of active and passive thermal energy storage (TES) systems are expected to be very familiar now with regard to their description provided in earlier chapters. In simple terms, an active storage system is one that contains a mechanically assisted component for enabling the thermal energy interactions to take place between the system and the heat source. In a passive storage system, the thermal energy interactions between the system and the heat source occurs by means of natural convection or buoyancy forces (due to density gradient) without the assistance of any external devices. The active and passive TES systems can be considered equally vital in accomplishing energy redistribution requirements in buildings [1]. Depending on the location, ambient conditions, and cooling load profile of the building, the selection of an active or passive TES system can be performed. Based on the peak load demand in buildings, the distribution of the cooling/heating energy between the TES and the building system can be established. Refer to Chapter 5, Fig. 5.7, for a broad classification of active and passive latent thermal energy storage (LTES) systems. The description of some major research works pertaining to the application potential of LTES systems in buildings are gleaned and presented in Table 15.1.

In most building cooling and heating applications, the utilization of LTES system is much preferred to sensible and thermochemical energy storage systems. This is due to the fact that the integration of LTES systems using phase change materials (PCMs) into building components (passive system) or incorporated separately (active system) is very much feasible by their configuration. Also, the heat storage capacity and phase transition characteristics of LTES materials can be adjusted to suit the comfort cooling and heating applications.

Despite having high specific heat and thermal conductivity, the sensible thermal storage system requires more space than the LTES system for storing the same quantity of heat energy. The thermochemical energy storage system has a high energy storage capacity compared to both LTES and sensible thermal storage systems. However, due to the complexity involved in storing and extracting thermal energy by means of reversible chemical reactions, it is somehow cumbersome and requires further research for real-time application in buildings.

15.2 CARBON-FREE THERMAL STORAGE SYSTEMS

The term *carbon-free* literally means for the utilization of the high grade heat energy being derived from the natural and renewable solar radiation for achieving TES requirements in applications ranging from solar heating to solar thermal power systems. The broad categorization of solar TES systems is presented in Fig. 15.1.

Table 15.1 Integration of LTES Systems in Buildings [1]

Application Potential	LTES System Integration in Buildings	LTES System Type	Functional Aspects of LTES System in Buildings	Methodology	Reference
Space heating	Internal wall construction	Passive	Diurnal/short-term heat storage from direct heat gain in a residential building room	Simulation	[2]
	Facade panel	Passive	PCM blended with glazing panels to improve daylighting, space heating and thermal comfort in winter; reduces peak cooling loads in summer	Experimental	[3]
	Floor component	Passive	Cost effective novel form stable PCM containing microencapsulated paraffin regulates indoor temperature	Simulation	[4]
	Internal wall and ceiling construction	Passive	PCM implanted gypsum boards to store heat energy from existing electrical facility in office space; shift on-peak space heating demand and conserve overall electrical energy	Experiment	[5]
	Floor component	Passive	Cost-effective shape stabilized PCM plates stores heat in nighttime using off-peak electricity to compensate on-peak space heating demand in daytime	Experiment	[6]
	Floor component	Passive	Gypsum-concrete PCM mixture as complete and partial carpets included into building flooring regulates floor surface temperature	Simulation and experiment	[7]
Space cooling	Brick construction	Passive	Encapsulated PCM offsets cooling demand from building; to conserve overall electrical energy consumption and reduce CO_2 emission from building	Experiment	[8]
	Ceiling mounted	Passive	PCM stores cool energy from ambient air during nighttime for meeting daytime cooling demand	Experiment	[9]
	Exterior wall	Passive	PCM doped color coatings applied on building exterior fabric minimizes indoor temperature variations, reduces building thermal load; to maintain thermal comfort conditions in indoor space	Experiment	[10]
	Ceiling mounted	Passive	Night ventilation scheme amalgamated with PCM packed bed storage reduces room temperature in day hours and conserves energy spent on cooling and ventilation	Experiment	[11]
	Floor component	Passive	PCM granules made of glass beads and paraffin waxes stores large cooling energy in nighttime to release it on demand during day peak load periods	Experiment	[12]
	Ceiling mounted	Passive	Two separate LTES systems containing sphere encapsulated PCM stores cool energy to reduce heat gain from ventilation air and room return air; save overall building energy, regulate indoor temperature and reduce size of mechanical ventilation system	Experiment	[13]
	Brick construction	Passive	Brick element mixed with PCM to absorb direct heat gain and reduces temperature fluctuations in indoor environment	Simulation and experiment	[14]

Air-conditioning	Ceiling mounted	Passive	PCM integrated with air-heat exchanger to cool indoor air during day peak load conditions with stored cool energy in nighttime	Experiment	[15]
	Ceiling panel	Passive	Ceiling panels embedded with PCM to cool indoor environment	Experiment and simulation	[16]
	Air handling unit	Active	Ice-cool energy storage combined with cold air distribution reduces running cost and save total energy	Experiment	[17]
	Heat pipe	Active	Ice storage combined with helical heat pipe having better solidification and melting characteristics; extracts heat load from indoor space and conserve overall building energy	Experiment	[18]
	Thermal battery	Active	Cold storage integrated with thermal battery improves thermal performance of air conditioning system	Theoretical and experiment	[19]
	Air handling unit	Active	Ice-cool thermal storage system compared with other cool storage systems using fuzzy multicriteria technique for overall performance evaluation	Simulation	[20]
	Fan coil unit	Active	Spherical PCMs packed bed blended with ice-cool thermal storage enhances performance of air conditioning system	Experiment	[21]
	Radiant cooling unit	Active	Ice thermal storage incorporated with air-conditioning system augments the combined effect of radiant cooling and air conditioning in indoor environment	Experiment	[22]
	Ejector unit	Active	Cold storage included with solar assisted ejector unit improves cool thermal storage capability and overall air conditioning system performance	Simulation	[23]
	Air handling unit	Active	Cold storage facility provides optimal control over indoor cooling with increased cost-energy savings and improved thermal capacitance in building	Modeling analysis and simulation	[24]
	Variable air volume (VAV)-air handling unit	Active	PCM-based thermal energy storage (TES) system incorporated with air-conditioning system achieves good thermal comfort and indoor air quality in buildings without sacrificing energy efficiency	Experiment	[25]
	District cooling system	Active	Ice storage system implemented with district cooling facility minimizes overall energy-cost issues related to buildings and conserve energy	Simulation	[26]
	District cooling system	Active	Improved thermophysical properties of PCMs in a cool TES system integrated with district cooling facility reduces chiller size, issues related to peak thermal load-shifts and cost-energy management in buildings	Experiment	[27]
	Vertical ground source heat exchangers	Active	Cool TES integrated with earth heat exchangers and cryogenic cooling system achieve desired space cooling and energy savings	Experiment	[28]

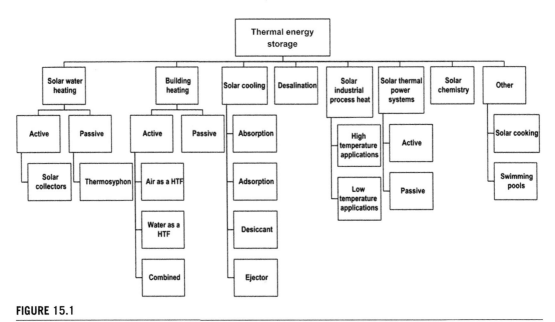

FIGURE 15.1

Classifications of solar TES applications [29].

Solar energy is considered an intermittent source of heat, which has to be stored suitably in order to be reused when demand rises. Two important factors need to be considered for solar TES applications:

- Requirement of larger area for collection of heat energy
- Size of the heat storage facility

The aforementioned factors are most vital in the sense that the maximum quantity of heat energy has to be captured from the incident solar radiation during sun brilliance periods by means of efficient collectors. This can facilitate the provision of more heat energy to the TES system during the storage process. During the discharging process or off-sunshine hours, the stored heat energy can be retrieved for meeting heating demand. Thus, proper sizing of the heat storage facility also contributes to enhancing the energy efficiency of solar TES systems.

In solar thermal power systems, the heat storage medium is forced to circulate through the collector or receiver, which is referred to as the active system. However, in a passive system, the heat transfer medium does not circulate, but rather passes through storage only during the charging and discharging processes. The different storage media utilized in concentrated solar power plants is summarized in Table 15.2.

15.3 LOW ENERGY BUILDING DESIGN

The prime objective of a low energy or low carbon building design is to optimize the orientation, structure, window/glazing location, and size and selection of proper materials pertaining to the building envelope. In addition, amalgamation of energy-efficient heating, ventilation and air conditioning (HVAC) systems

Table 15.2 Different Storage Media Used in Concentrated Solar Power Plants [29,30]

| Storage Medium | Temperature | | Average Density [kg/m³] | Average Heat Conductivity [W/mK] | Average Heat Capacity [W/mK] | Volume Specific Heat Capacity [kWh/m³] | Media Costs per kg [$/kg] | Media Costs per kWh$_t$ [$/kWh$_t$] |
	Cold (°C)	Hot (°C)						
Solid Media								
Sand-rock-mineral oil	200	300	1700	1.0	1.3	60	0.15	4.2
Reinforced concrete	200	400	2200	1.5	0.85	100	0.05	1.0
NaCl (solid)	200	500	2160	7.0	0.85	150	0.15	1.5
Cast iron	200	400	7200	37.0	0.56	160	1.00	32.0
Cast steel	200	700	7800	40.0	0.6	450	5.00	60.0
Silica fire bricks	200	700	1820	1.5	1.00	150	1.00	7.0
Magnesia fire bricks	200	1200	3000	5.0	1.15	600	2.00	6.0
Liquid Media								
Mineral oil	200	300	770	0.12	2.6	55	0.3	4.2
Synthetic oil	250	350	900	0.11	2.3	57	3.00	43.0
Silicone oil	300	400	900	0.10	2.1	52	5.00	80.0
Nitrite salts	250	450	1825	0.57	1.5	152	1.00	12.0
Nitrate salts	265	565	1870	0.52	1.6	250	0.70	5.2
Carbonate salts	450	850	2100	2.0	1.8	430	2.40	11.0
Liquid sodium	270	530	850	71.0	1.3	80	2.00	21.0
Phase Change Materials								
NaNO$_3$	308		2257	0.5	200	125	0.20	3.5
KNO$_3$	333		2110	0.5	267	156	0.30	4.1
KOH	380		2044	0.5	150	85	1.00	24.0
Salt-ceramics	500–850		2600	5.0	420	300	2.00	17.0
NaCl	802		2160	5.0	520	280	0.15	1.2
Na$_2$CO$_3$	854		2533	2.0	276	194	0.20	2.6
K$_2$CO$_3$	897		2290	2.0	236	150	0.60	9.1

with TES is also covered under the low energy building concept. The essential design principles that need to be followed in the design of low carbon/energy buildings are listed here:

- Principle 1: Understanding of the building energy usage
- Principle 2: Utilization of the architecture and fabric components of building
- Principle 3: Minimizing heat losses and gains
- Principle 4: Incorporation of essential building services with low carbon emissions
- Principle 5: Preference for using renewable energy systems
- Principle 6: Implementing energy management within the building

In principle 1, knowing the end-usage energy distribution pattern of the building can not only help to minimize excess energy delivery but can also reduce carbon emissions from the building. The energy breakup profile can be used to identify the component that is consuming higher energy than the designed value.

In principle 2, the architectural form and the fabrics themselves can be used as the source for providing building services including heating, ventilation, cooling, and possible heat storage. Making certain modifications or alterations to the architecture and fabric component in buildings can help reduce the use of the fossil energy-based cooling/heating equipments. This in turn improves the energy efficiency of the buildings.

Principle 3 concerns reducing the unwanted heat losses or heat gains of occupied spaces through providing effective insulation and performing an air tightness checkup regularly based on the appropriate standards recommendation. This criterion is more related to the design phase of the building, in which the estimation of cooling/heating loads must match the actual operational performance of cooling/heating units. In other words, the occupied space must not be overheated or overcooled ever, but must be maintained within comfort conditions as specified in the basis of design.

Principle 4 is closely related to the architects' work, in which more concentration is needed to ensure only energy-efficient building services are being considered starting from the scheme inception to the completion of building construction. Any energy consuming components or devices have to be avoided or their operation can be put under control using energy-efficient control modules.

Principle 5 emphasizes combining renewable energy sources with building services, heat and power systems to gain energy conservative potential and reduce carbon emissions to the environment considerably.

Principle 6 gives clear information to the architects for ensuring whether proper energy monitoring systems and energy management schemes are operative within the building. This criterion also implies that occupants in building spaces should know how to use the building and its services efficiently.

15.4 SCOPE FOR FUTURISTIC DEVELOPMENTS

- In the quest toward developing TES systems for confronting energy challenges and climate change issues, it is evident that a huge potential is available ahead for creating a pathway of research oriented toward a sustainable future. Suggestions and recommendations for future research may include the following:
- The basic ingredient of any TES system is its heat storage material. The operational performance of the TES system purely depends on the characteristics of the heat storage medium or material.

Thus, detailed analyses on thermophysical properties, heat interaction mechanism, and energetic performance of heat storage materials are needed in terms of simulation being validated based on realistic experimentation.

- The inherent discrepancies identified on the thermophysical and thermal storage properties of the energy materials have to be rectified through performing intensive and advanced materials science–based research studies (e.g., heat transfer studies on nanomaterials for TES)
- Performance assessment of various combinations of latent heat storage materials (PCMs) with other sensible heat storage methods to enhance the overall energy efficiency of the cooling/heating systems.
- Development of hybrid cooling/heating storage coupled with renewable energy systems for achieving enhanced energy savings potential with reduced carbon emissions.
- The implementation of energy conservative cold air distribution techniques with a hybrid cooling system would collectively merit toward achieving energy redistribution requirements and energy efficiency for the development of a sustainable future.

References

[1] Parameshwaran R, Kalaiselvam S, Harikrishnan S, et al. Sustainable thermal energy storage technologies for buildings: a review. Renew Sust Energ Rev 2012;16:2394–433.
[2] Peippo K, Kauranen P, Lund P. A multicomponent PCM wall optimized for passive solar heating. Energy Build 1991;17:259–70.
[3] Weinlader H, Beck A, Fricke J. PCM-facade-panel for daylighting and room heating. Sol Energy 2005;78:177–86.
[4] Li J, Xue P, He H, Ding W, Han J. Preparation and application effects of a novel form-stable phase change material as the thermal storage layer of an electric floor heating system. Energy Build 2009;41:871–80.
[5] Qureshi WA, Nair NKC, Farid MM. Impact of energy storage in buildings on electricity demand side management. Energy Convers Manag 2011;52:2110–20.
[6] Lin K, Zhang Y, Xu X, Di H, Yang R, Qin P. Experimental study of under-floor electric heating system with shape-stabilized PCM plates. Energy Build 2005;37:215–20.
[7] Athienities A, Chen Y. The effect of solar radiation on dynamic thermal performance of floor heating systems. Sol Energy 2000;69:229–37.
[8] Castell A, Martorell I, Medrano M, Prez G, Cabeza L. Experimental study of using PCM in brick constructive solutions for passive cooling. Energy Build 2010;42:534–40.
[9] Zalba B, Marn JM, Cabeza LF, Mehling H. Free-cooling of buildings with phase change materials. Int J Refrig 2004;27:839–49.
[10] Karlessi T, Santamouris M, Synnefa A, Assimakopoulos D, Didaskalopoulos P, Apostolakis K. Development and testing of PCM doped cool colored coatings to mitigate urban heat island and cool buildings. Build Environ 2011;46:570–6.
[11] Yanbing K, Yi J, Yinping Z. 'Modeling and experimental study on an innovative passive cooling system – NVP system. Energy Build 2003;35:417–25.
[12] Nagano K, Takeda S, Mochida T, Shimakura K, Nakamura T. Study of a floor supply air conditioning system using granular phase change material to augment building mass thermal storage – Heat response in small scale experiments. Energy Build 2006;38:436–46.
[13] Arkar C, Vidrih B, Medved S. Efficiency of free cooling using latent heat storage integrated into the ventilation system of a low energy building. Int J Refrig 2007;30:134–43.

[14] Alawadhi EM. Thermal analysis of a building brick containing phase change material. Energy Build 2008;40:351–7.

[15] Lazaro A, Dolado P, Marin JM, Zalba B. PCM-air heat exchangers for free-cooling applications in buildings: experimental results of two real scale prototypes. Energy Convers Manag 2009;50:439–43.

[16] Koschenz M, Lehmann B. Development of a thermally activated ceiling panel with PCM for application in lightweight and retrofitted buildings. Energy Build 2004;36:567–78.

[17] Tassou SA, Leung YK. Energy conservation in commercial air conditioning through ice storage and cold air distribution design. Heat Recov Syst CHP 1992;12:419–25.

[18] Fang G, Liu X, Wu S. Experimental investigation on performance of ice storage air-conditioning system with separate heat pipe. Exp Thermal Fluid Sci 2009;33:1149–55.

[19] Chieh JJ, Lin SJ, Chen SL. Thermal performance of cold storage in thermal battery for air conditioning. Int J Refrig 2004;27:120–8.

[20] Jiang WJ, Zhang CF, Jing YY, Zheng GZ. Using the fuzzy multi-criteria model to select the optimal cool storage system for air conditioning. Energy Build 2008;40:2059–66.

[21] Fang G, Wu S, Liu X. Experimental study on cool storage air-conditioning system with spherical capsules packed bed. Energy Build 2010;42:1056–62.

[22] Matsuki N, Nakano Y, Miyanaga T, Yokoo N, Oka T. Performance of radiant cooling system integrated with ice storage. Energy Build 1999;30:177–83.

[23] Diaconu BM, Varga S, Oliveira AC. Numerical simulation of a solar-assisted ejector air conditioning system with cold storage. Energy 2011;36:1280–91.

[24] Kintner-Meyer M, Emery AF. Optimal control of an HVAC system using cold storage and building thermal capacitance. Energy Build 1995;23:19–31.

[25] Parameshwaran R, Harikrishnan S, Kalaiselvam S. Energy efficient PCM-based variable air volume air conditioning system for modern buildings. Energy Build 2010;42:1353–60.

[26] Chan ALS, Chow TT, Fong SKF, Lin JZ. Performance evaluation of district cooling plant with ice storage. Energy 2006;31:2750–62.

[27] He B, Gustafsson EM, Setterwall F. Tetradecane and hexadecane binary mixtures as phase change materials (PCMs) for cool storage in district cooling systems. Energy 1999;24:1015–28.

[28] Hamada Y, Nakamura M, Kubota H. Field measurements and analyses for a hybrid system for snow storage/melting and air conditioning by using renewable energy. Appl Energy 2007;84:117–34.

[29] Stritih U, Osterman E, Evliya H, Butala V, Paksoy H. Exploiting solar energy potential through thermal energy storage in Slovenia and Turkey. Renew Sust Energ Rev 2013;25:442–61.

[30] Pilkington Solar International GmbH. Survey of thermal storage for parabolic trough power plants. Cologne, Germany: National Renewable Energy Laboratory; 2000.

Units and Conversions Factors

Table AI.1 Heat contents

Fuel	Units	Approximate Heat Content
Coal[a]		
Production	Million Btu per short ton	20.136
Consumption	Million Btu per short ton	19.810
Coke plants	Million Btu per short ton	26.304
Industrial	Million Btu per short ton	23.651
Residential and commercial	Million Btu per short ton	20.698
Electric power sector	Million Btu per short ton	19.370
Imports	Million Btu per short ton	25.394
Exports	Million Btu per short ton	25.639
Coal coke	Million Btu per short ton	24.800
Crude Oil		
Production	Million Btu per barrel	5.800
Imports[a]	Million Btu per barrel	5.967
Petroleum Products and Other Liquids		
Consumption[a]	Million Btu per barrel	5.353
Motor or gasoline[a]	Million Btu per barrel	5.048
Jet fuel	Million Btu per barrel	5.670
Distillate fuel oil[a]	Million Btu per barrel	5.762
Diesel fuel[a]	Million Btu per barrel	5.759
Residual fuel oil	Million Btu per barrel	6.287
Liquefied petroleum gases[a]	Million Btu per barrel	3.577
Kerosene	Million Btu per barrel	5.670
Petrochemical feedstocks[a]	Million Btu per barrel	5.114
Unfinished oils	Million Btu per barrel	6.039
Imports[a]	Million Btu per barrel	5.580
Exports[a]	Million Btu per barrel	5.619
Ethanol	Million Btu per barrel	3.560
Biodiesel	Million Btu per barrel	5.359

Continued

Table AI.1 Heat contents—cont'd

Fuel	Units	Approximate Heat Content
Natural Gas Plant Liquids		
Production[a]	Million Btu per barrel	3.566
Natural Gas[a]		
Production, dry	Btu per cubic foot	1.022
Consumption	Btu per cubic foot	1.022
End-use sectors	Btu per cubic foot	1.023
Electric power sector	Btu per cubic foot	1.021
Imports	Btu per cubic foot	1.025
Exports	Btu per cubic foot	1.009
Electricity consumption	Btu per kilowatt-hour	3.412

[a]Conversion factor varies from year to year. The value shown is for 2011.
Btu, British thermal unit.

Table AI.2 Multiples and submultiples of SI units

Prefix	Symbol	Multiplying Factor	
exa	E	10^{18}	1 000 000 000 000 000 000
peta	P	10^{15}	1 000 000 000 000 000
tera	T	10^{12}	1 000 000 000 000
giga	G	10^{9}	1 000 000 000
mega	M	10^{6}	1 000 000
klo	K	10^{3}	1 000
hecto*	H	10^{2}	100
deca*	Da	10	10
deci*	D	10^{-1}	0.1
centi	C	10^{-2}	0.01
milli	M	10^{-3}	0.001
micro	U	10^{-6}	0.000 001
nano	N	10^{-9}	0.000 000 001
pico	P	10^{-12}	0.000 000 000 001
femto	F	10^{-15}	0.000 000 000 000 001
atto	a	10^{-18}	0.000 000 000 000 000 001

*These prefixes are not normally used

Table AI.3 Length units

Millimeters (mm)	Centimeters (cm)	Meters (m)	Kilometers (km)	Inches (in.)	Feet (ft)	Yards (yd)	Miles (mi)
1	0.1	0.001	0.000001	0.03937	0.003281	0.001094	6.21×10^{-7}
10	1	0.01	0.00001	0.393701	0.032808	0.010936	0.000006
1000	100	1	0.001	39.37008	3.28084	1.093613	0.000621
1,000,000	100,000	1000	1	39370.08	3280.84	1093.613	0.621371
25.34	2.54	0.0254	0.000025	1	0.083333	0.027778	0.000016
304.8	30.48	0.3048	0.000305	12	1	0.333333	0.000189
914.4	91.44	0.9144	0.000914	36	3	1	0.000568
1,609,344	160934.4	1609.344	1.609344	63,360	5280	1760	1

Table AI.4 Area units

Millimeter Square (mm²)	Centimeter Square (cm²)	Meter Square (m²)	Inch Square (km²)	Foot Square (in.²)	Yard Square (ft²)
1	0.01	0.0000001	0.00155	0.000011	0.000001
100	1	0.0001	0.155	0.001076	0.00012
1,000,000	10,000	1	1550.003	10.76391	1.19599
645.16	6.4516	0.000645	1	0.006944	0.000772
92,903	929.0304	0.092903	144	1	0.111111
836,127	8361.274	0.836127	1296	9	1

Table AI.5 Volume units

Centimeter Cube (cm³)	Meter Cube (m³)	Liter (l)	Inch Cube (in.³)	Foot Cube (ft³)	US Gallons (US gal)	Imperial Gallons (Imp gal)	US Barrel (Oil) (US brl)
1	0.000001	0.001	0.061024	0.000035	0.000264	0.00022	0.000006
1,000,000	1	1000	61,024	35	264	220	6.29
1000	0.001	1	61	0.035	0.264201	0.22	0.00629
16.4	0.000016	0.01637	1	0.000579	0.004329	0.003605	0.000103
28,317	0.028317	28.31685	1728	1	7.481333	6.229712	0.178127
3785	0.003785	3.79	231	0.13	1	0.832701	0.02381
4545	0.004545	4.55	277	0.16	1.20	1	0.028593
158,970	0.15897	159	9701	6	42	35	1

Table AI.6 Mass units

Grams (g)	Kilograms (kg)	Metric Tons (ton)	Short Ton (shton)	Long Ton (Lton)	Pounds (lb)	Ounces (oz)
1	0.001	0.000001	0.000001	9.84e-07	0.002205	0.035273
1000	1	0.001	0.001102	0.000984	2.204586	35.27337
1,000,000	1000	1	1.102293	0.984252	2204.586	35273.37
907,200	907.2	0.9072	1·	0.892913	2000	32,000
1,016,000	1016	1.016	1.119929	1	2239.859	35837.74
453.6	0.4536	0.000454	0.0005	0.000446	1	16
28	0.02835	0.000028	0.000031	0.000028	0.0625	1

Table AI.7 Density units

Gram/Milliliter (g/ml)	Kilogram/Meter Cube (kg/m³)	Pound/Foot Cube (lb/ft³)	Pound/Inch Cube (lb/in.³)
1	1000	62.42197	0.036127
0.001	1	0.062422	0.000036
0.01602	16.02	1	0.000579
27.68	27,680	1727.84	1

Table AI.8 Volumetric liquid flow units

Liter/ Second (L/S)	Liter/ Minute (L/min)	Meter Cube/Hour (m³/h)	Foot Cube/ Minute (ft³/min)	Foot Cube/ Hour (ft³/h)	US Gallons/ Minute (gal/min)	US Barrels (Oil)/Day (US brl/d)
1	60	3.6	2.119093	127.1197	15.85037	543.4783
0.016666	1	0.06	0.035317	2.118577	0.264162	9.057609
0.277778	16.6667	1	0.588637	35.31102	4.40288	150.9661
0.4719	28.31513	1.69884	1	60	7.479791	256.4674
0.007867	0.472015	0.02832	0.01667	1	0.124689	4.275326
0.06309	3.75551	0.227124	0.133694	8.019983	1	34.28804
0.00184	0.110404	0.00624	0.003899	0.2339	0.029165	1

Table AI.9 Volumetric gas flow units

Normal Meter Cube/Hour (Nm³/h)	Standard Cubic Feet/Hour (scfh)	Standard Cubic Feet/Minute (scfm)
1	35.31073	0.588582
0.02832	1	0.016669
1.699	59.99294	1

Table AI.10 Mass flow units

Kilogram/Hour (kg/h)	Pound/Hour (lb/h)	Kilogram/Second (kg/s)	Ton/Hour (t/h)
1	2.204586	0.000278	0.001
0.4536	1	0.000126	0.000454
3600	7936.508	1	3.6
1000	2204.586	0.277778	1

Table AI.11 High pressure units

Bar (bar)	Pound/ Square Inch (psi)	Kilopascal (kPa)	Megapascal (MPa)	Kilogram Force/ Centimeter Square (kgf/cm²)	Millimeter of Mercury (mmHg)	Atmospheres (atm)
1	14.50326	100	0.1	1.01968	750.0188	0.987167
0.06895	1	6.895	0.006895	0.070307	51.71379	0.068065
0.01	0.1450	1	0.001	0.01020	7.5002	0.00987
10	145.03	1000	1	10.197	7500.2	9.8717
0.9807	14.22335	98.07	0.09807	1	735.5434	0.968115
0.001333	0.019337	0.13333	0.000133	0.00136	1	0.001316
1.013	14.69181	101.3	0.1013	1.032936	759.769	1

Table AI.12 Low pressure units

Meter of Water (mH₂O)	Foot of Water (ftH₂O)	Centimeter of Mercury (cmHg)	Inches of Mercury (inHg)	Inches of Water (inH₂O)	Pascal (Pa)
1	3.280696	7.356339	2.896043	39.36572	9806
0.304813	1	2.242311	0.882753	11.9992	2989
0.135937	0.445969	1	0.39368	5.351265	1333
0.345299	1.13282	2.540135	1	13.59293	3386
0.025403	0.083339	0.186872	0.073568	1	249.1
0.000102	0.000335	0.00075	0.000295	0.004014	1

Table AI.13 Speed units

Meter/Second (m/s)	Meter/Minute (m/min)	Kilometer/ Hour (km/h)	Foot/Second (ft/s)	Foot/Minute (ft/min)	Miles/Hour (mi/h)
1	59.988	3.599712	3.28084	196.8504	2.237136
0.01667	1	0.060007	0.054692	3.281496	0.037293
0.2778	16.66467	1	0.911417	54.68504	0.621477
0.3048	18.28434	1.097192	1	60	0.681879
0.00508	0.304739	0.018287	0.016667	1	0.011365
0.447	26.81464	1.609071	1.466535	87.99213	1

Table AI.14 Torque units

Newton Meter (Nm)	Kilogram Force Meter (kgfm)	Foot Pound (ftlb)	Inch Pound (inlb)
1	0.101972	0.737561	8.850732
9,380,665	1	7.233003	86.79603
1.35582	0.138255	1	12
0.112985	0.011521	0.083333	1

Table AI.15 Dynamic viscosity units

Centipoise[a] (cp)	Poise (poise)	Pound/Foot·second lb/(fts)
1	0.01	0.000672
100	1	0.067197
1488.16	14.8816	1

[a]Centipoises = Centistokes × Specific gravity.

Table AI.16 Kinematic viscosity units

Centistoke[a] (cs)	Stoke (St)	Foot Square/ Second (ft²/s)	Meter Square/ Second (m²/s)
1	0.01	0.000011	0.000001
100	1	0.001076	0.0001
92,903	929.03	1	0.092903
1,000,000	10,000	10.76392	1

[a]Centistokes × specific gravity = centipoise.

Table AI.17 Temperature conversion formulas

Degree Celsius (°C)	$(°F - 32) \times 5/9$
	$(K - 273.15)$
Degree Fahrenheit (°F)	$(°C \times 9/5) + 32$
	$(1.8 \times K) - 459.67$
Kelvin (K)	$(°C + 273.15)$
	$(°F + 459.67)/1.8$

Table AI.18 General information on units, and conversion factors for energy units and currencies

Coal	Mtce	Million tons of coal equivalent
Emissions	ppm	Parts per million (by volume)
	Gt CO_2-eq	Gigatons of carbon-dioxide equivalent (using 100-year global warming potentials [GWP] for different greenhouse gases)
	kg CO_2-eq	Kilograms of carbon-dioxide equivalent
	g CO_2/km	Grams of carbon dioxide per kilometer
	g CO_2/kWh	Grams of carbon dioxide per kilowatt-hour
Energy	Mltoe	Million tons of oil equivalent
	MBtu	Million British thermal units
	Gcal	Gigacalorie (1 calorie $\times 10^9$)
	TJ	Terajoule (1 Joule $\times 10^{12}$)
	kWh	Kilowatt-hour
	MWh	Megawatt-hour
	GWh	Gigawatt-hour
	TWh	Terawatt-hour
Gas	mcm	Million cubic meters
	bcm	Billion cubic meters
	tcm	Trillion cubic meters
Mass	kg	Kilogram (1000 kg = 1 ton)
	kt	Kilotons (1 ton $\times 10^3$)
	Mt	Million tons (1 ton $\times 10^6$)
	Gt	Gigatons (1 ton $\times 10^9$)
Monetary	$ million	1 US dollar $\times 10^6$
	$ billion	1 US dollar $\times 10^9$
	$ trillion	1 US dollar $\times 10^{12}$
Oil	b/d	Barrels per day
	kb/d	Thousand barrels per day
	mb/d	Million barrels per day
	mpg	Miles per gallon
Power	W	Watt (1 Joule per second)
	kW	Kilowatt (1 W $\times$ 109)
	MlW	Megawatt (1 W $\times$ 109)
	GW	Gigawatt (1 W $\times$ 109)
	TW	Terawatt (1 W $\times$ 1012)

Sources: U.S. Energy Information Administration (EIA), Annual Energy Review 2011, DOE/EIA-0384(2011) (Washington, DC, September 2012), and EIA, AEO2013 National Energy Modeling System run REF2013.D102312A. https://www.isa.org/ccst/CCST-Conversions-document.pdf, World Energy Outlook Special Report: Redrawing the Energy-Climate Map © OECD/IEA, 2013, Annex A, pp. 117-118.

Thermal Properties of Various Heat Storage Materials

Table AII.1 List of selected solid and liquid materials for sensible heat storage in building applications

Material	Type	Temperature Range (°C)	ρ (kg/m³)	C_p (kJ/ (kg K))	k (W/(mK)) (at 20 °C)	e (W s$^{1/2}$/ (m² K))
Water	Liquid	0-100	1000	4.19	0.58	49.3
Caloriea HT43	Liquid	12-260	867	2.2	–	–
Ethanol	Liquid	Up to 78	790	2.4	0.171	18.01
Proponal	Liquid	Up to 97	800	2.5	0.161	17.94
Butanol	Liquid	Up to 118	809	2.4	0.167	18.01
Isobutanol	Liquid	Up to 100	808	3	0.133	17.96
Isopentanol	Liquid	Up to 148	831	2.2	0.141	16.06
Octane	Liquid	Up to 126	704	2.4	0.134	15.05
Engine oil	Liquid	Up to 160	888	1.88	–	–
Brick	Solid	20-70	1600	0.84	1.2	1270
Concrete	Solid	20-70	2240	1.13	0.9-1.3	47.73-57.36
Cement sheet	Solid	20-70	700	1.05	0.36	514
Gypsum plastering	Solid	–	1200	0.837	0.42	649
Granite	Solid	20-70	2650	0.9	2.9	2967
Marble	Solid	20-70	2500	0.88	2	2285
Sandstone	Solid	20-70	2200	0.712	1.83	1710
Clay sheet	Solid	–	1900	0.837	0.85	1163
Asphalt sheet	Solid	–	2300	1.7	1.2	2166
Steel slab	Solid	20-70	7800	0.502	50	13,992
Corkboard	Solid	–	160	1.888	0.04	110
Wood	Solid	–	800	2.093	0.16	324
Plastic board	Solid	–	1050	0.837	0.5	663
Rubber board	Solid	–	1600	0.2	0.3	310
PVC board	Solid	–	1379	1.004	0.16	410
Asbestos sheet	Solid	–	2500	1.05	0.16	648
Formaldehyde board	Solid	–	30	1.674	0.03	39

Continued

Table AII.1 List of selected solid and liquid materials for sensible heat storage in building applications—cont'd

Material	Type	Temperature Range (°C)	ρ (kg/m³)	C_p (kJ/ (kg K))	k (W/(mK)) (at 20°C)	e (W s$^{1/2}$/ (m² K))
Thermalite board	Solid	–	753	0.837	0.19	346
Fiberboard	Solid	–	300	1	0.06	134
Siporex board	Solid	–	550	1.004	0.12	257
Polyurethane board	Solid	–	30	0.837	0.03	27
Light plaster	Solid	–	600	1	0.16	712
Dense plaster	Solid	–	1300	1	0.5	806
Aluminum	Solid	Up to 160	2707	0.896	204	703.42
Aluminum oxide	Solid	Up to 160	3900	0.84	30	313.5
Aluminum sulfate	Solid	Up to 160	2710	0.75	–	–
Cast iron	Solid	Up to 160	7900	0.837	29.3	440.16
Pure iron	Solid	Up to 160	7897	0.452	73	510.46
Calcium chloride	Solid	Up to 160	2510	0.67	–	–
Copper	Solid	Up to 160	8954	0.383	385	1149.05
Stone, granite	Solid	Up to 160	2640	0.82	1.7-3.98	61.20-92.82
Stone, sandstone	Solid	Up to 160	2200	0.71	1.83	53.46

(–), not available.

Table AII.2 Commercial PCMs suitable for residential applications

Application	PCM Type	Designation	Melting Temperature (°C)	Heat of Fusion (kJ/kg)	C_p (kJ/kgK)	k (W/(mK))	ρ^a (kg/m³)	Manufacturer[b]
Cooling	Inorganics	S10	10	155	1.9	0.43	1470	A
		ClimSel C10	10.5	126	3.6	0.5–0.7	1420	B
		S13	13	160	1.9	0.43	1515	A
		S15	15	160	1.9	0.43	1510	A
		S17	17	160	1.9	0.43	1525	A
		S19	19	160	1.9	0.43	1520	A
		S21	22	170	2.2	0.54	1530	A
		ClimSel C21	21	144	3.6	0.5–0.7	1380	B
		S23	23	175	2.2	0.54	1530	A
		ClimSelC24	24	126	3.6	0.5–0.7	1380	B
		S25	25	180	2.2	0.54	1530	A
		S27	27	183	2.2	0.54	1530	A
		ClimSel C28	28	162	3.6	0.5–0.7	1420	B
		S30	30	190	1.9	0.48	1304	A
	Organics	A15	15	130	2.26	0.18	790	A
		A17	17	150	2.22	0.18	785	A
		RT21	21	134	–	0.2	880 (s, 15°C) 770 (liq, 25°C)	C
		A22	22	145	2.22	0.18	785	A
		A23	23	145	2.22	0.18	785	A
		A24	24	124	2.22	0.18	790	A
		A25	25	150	2.26	0.18	785	A
		A26	26	150	2.22	0.21	790	A
		RT27	27	184	–	0.2	880 (s, 15°C) 760 (liq, 40°C)	C
		A28	28	155	2.22	0.21	789	A
		RT28	28	245	–	0.2	880 (s, 15°C) 768 (liq, 40°C)	C

Continued

Table AII.2 Commercial PCMs suitable for residential applications—cont'd

Application	PCM Type	Designation	Melting Temperature (°C)	Heat of Fusion (kJ/kg)	C_p (kJ/(kg K))	k (W/(mK))	ρ^a (kg/m³)	Manufacturer[b]
	Eutectics	SP 22 A17	22-24	150	–	0.6	1490 (s, 20°C)	C
							1420 (liq, 40°C)	
		SP 25 A8	25-27	160	–	0.6	1430 (s, 20°C)	C
							1230 (liq, 40°C)	
		SP 26 A9	26-28	170	–	0.6	1460 (s, 20°C)	C
							1440 (liq, 40°C)	
		SP 29 A15	28-30	190	–	0.6	1530 (s, 25°C)	C
							1510 (liq, 45°C)	
Heating	Inorganics	S32	32	200	1.91	0.51	1460	A
		ClimSel 32	32	162	3.6	0.5-0.7	1420	B
		S34	34	115	2.1	0.52	2100	A
		S44	44	100	1.61	0.43	1584	A
		S46	46	210	2.41	0.45	1587	A
		ClimSel 48	48	216	3.6	0.5-0.7	1360	B
		S50	50	100	1.59	0.43	1601	A
		S58	58	145	2.55	0.69	1505	A
		ClimSel 58	58	288	3.6	0.5-0.7	1360	B
		S72	72	127	2.13	0.58	1666	A
		S83	83	141	2.31	0.62	1600	A
		ClimSel 70	70	396	3.6	0.6-0.7	1700	B
	Organics	A32	32	130	2.2	0.21	845	A
		A39	39	105	2.22	0.22	900	A
		A42	42	105	2.22	0.21	905	A
		P116	46.7-50	209	2.89	0.277 (liq), 0.140s	786 (s)	–
		A53	53	130	2.22	0.22	910	A
		A55	55	135	2.22	0.22	905	A
		A58	58	132	2.22	0.22	910	A
		A60	60	145	2.22	0.22	910	A
		A62	62	145	2.2	0.22	910	A
		A70	70	173	2.2	0.23	890	A
		A82	82	155	2.21	0.22	850	A
		A95	95	205	2.2	0.22	900	A

(–), not available.
[a] s, solid; liq, liquid.
[b] A, Environmental Process Limited; B, Climator AB; C, Rubitherm.

Table AII.3 Inorganic substances with potential use as PCM for residential applications

Application	Compound	Type of Melting[a]	Melting Temperature (°C)	Heat of Fusion (kJ/kg)	C_p[b] (kJ/(kg K))	k[b] (W/(mK))	ρ (kg/m³)
Cooling	$LiClO_3 \cdot 3H_2O$	c	8.1	253	1.35 (s)	—	1720
	$ZnCl_2 \cdot 3H_2O$	—	10	—	—	—	—
	$K_2HPO_4 \cdot 6H_2O$	—	13	—	—	—	—
	$NaOH \cdot 7/2CO_2$	—	15	—	—	—	—
	$Na_2CrO_4 \cdot 10H_2O$	—	18	—	1.31 (s)	—	—
	$KF \cdot 4H_2O$	c	18.5	231	1.84 (s)	—	1447 (liq, 20°C); 1455 (s, 18°C)
	$Mn(NO_3)_2 \cdot 6H_2O$	—	25.8	125.9	—	2.34 (s)	1800 (liq, 20°C)
	$CaCl_2 \cdot 6H_2O$	i	29	188.34	1.43 (s); 2.31 (liq)	0.540 (liq, 38.7°C); 0.561 (liq, 61.2°C); 1.09 (s, 23°C)	1562 (liq, 32°C); 1802 (s, 24°C)
Heating	$LiNO_3 \cdot 3H_2O$	—	30	296	—	0.58 (liq); 1.37 (s)	1780 (liq); 2140 (s)
	$Na_2CO_3 \cdot 10H_2O$	i	32–36; 33	246.5; 247	1.79 (s)	—	1442
	$Na_2SO_4 \cdot 10H_2O$	i	32.4	251	1.44 (s)	0.5–0.7; 2194 (s, 24°C)	1420
	$CaBr_2 \cdot 6H_2O$	i	34	115.5	—	—	1956 (liq, 35°C); 2194 (s, 24°C)
	$K(CH_3COO) \cdot 3/2H_2O$	—	42	—	—	—	—
	$K_3PO_4 \cdot 7H_2O$	—	45	—	—	—	—
	$Zn(NO_3)_2 \cdot 6H_2O$	c	36.4	147	1.34 (s); 2.26 (liq)	—	2065 (14°C)

Continued

Table AII.3 Inorganic substances with potential use as PCM for residential applications—cont'd

Application	Compound	Type of Melting[a]	Melting Temperature (°C)	Heat of Fusion (kJ/kg)	C_p[b] (kJ/(kg K))	k[b] (W/(mK))	ρ (kg/m³)
	$Ca(NO_3)_2 \cdot 4H_2O$	i	42.7	–	–	–	–
	$Na_2HPO_4 \cdot 7H_2O$	i	48	281	1.7 (s)	0.514 (32°C)	1520 (s)
					1.95 (liq)	0.476 (49°C)	1422 (liq)
	$Na_2S_2O_3 \cdot 5H_2O$	i	48-49	209.3	3.83 (liq)	–	1666
			48	201-206			
	$Zn(NO_3)_2 \cdot 2H_2O$	c	54	–	–	–	–
	$NaOH \cdot H_2O$	c	58	–	2.18 (s)	–	–
	$Na(CH_3COO) \cdot 3H_2O$		58	226	–	–	1450
				267	2.79-4.57	0.63 (s)	1280
	$Cd(NO_3)_2 \cdot 4H_2O$	c	59.5	–	–	–	–
	$Fe(NO_3)_2 \cdot 6H_2O$	–	60	–	–	–	–
	$Na_2B_4O_7 \cdot 10H_2O$	i	68.1	–	–	–	–
	$Na_3PO_4 \cdot 12H_2O$	i	69	–	–	–	–
	$Ba(OH)_2 \cdot 8H_2O$	i	78	265.7	–	0.653 (liq, 85.7°C)	1937 (liq, 84°C)
						0.678 (liq, 98.2°C)	2070 (s, 24°C)
	$KAl(SO_4)_2 \cdot 12H_2O$	i	80	–	–	–	–
	$Al2(SO_4)_3 \cdot 18H_2O$	–	85.8	–	–	–	–
	$Al(NO_3)_{3} \cdot 8H_2O$	–	88	–	–	–	–

(–), not available.
[a] c, congruent; i, incongruent.
[b] s, solid; liq, liquid.

Table AII.4 Melting point and latent heat of fusion of some metal alloys used in residential heat storage applications

Material	Composition (%)	Melting Temperature (°C)	Heat of Fusion (kJ/kg)	$C_p{}^a$ (kJ/(kg K))	k^a (W/(mK))	ρ (kg/m³)
Gallium	–	30	80.3	0.83 (s/liq)	40.6 (s)	5910
Cerrolow 117	Bismuth 44.7	47.2	68.175	0.15 (s/liq)	16.7 (s)	8860
	Lead 22.6					
	Tin 8.3					
	Cadmium 5.3					
	Indium 19.1					
Cerrolow 136	Bismuth 49	58	90.9	0.13 (s/liq)	16.7 (s)	8570
	Lead 16					
	Tin 12					
	Indium 21					
Cerrolow 158	Bismuth 50	70	159	0.17 (s/liq)	16.7 (s)	9380
	Lead 26.7					
	Tin 13.3					
	Indium 10					
Cerrolow 203	Bismuth 52.5	95	–	–	–	9850
	Lead 32					
	Tin 15.5					

(–), not available; A, approximate.
^{a}s, solid; liq, liquid.

Table AII.5 Organic substances with potential use as PCM for residential applications

Group		Compound	Melting Temperature (°C)	Heat of Fusion (kJ/kg)	C_p (kJ/(kg K))	k^a (W/(mK))	ρ^a (kg/m³)
Paraffin	Cooling	Paraffin C14	4.5	165	–	–	–
		Paraffin C15-C16	8	153	2.2 (s)	–	–
		Polyglycol E400	8	99.6	–	0.187 (liq, 38.6°C) 0.185 (liq, 69.9°C)	1125 (liq, 25°C) 1228 (s, 3°C)
		Dimethyl-sulfoxide (DMS)	16.5	85.7	–	–	1009 (s/liq)
		Paraffin C17-C18	20-22	152	2.2 (s)	–	–
		Polyglycol E600	22	127.2	–	0.189 (liq, 38.6°C) 0.187 (liq, 67.0°C)	1126 (liq, 25°C) 1232 (s, 4°C)
		Paraffin C13-C24	22-24	189	2.1 (s)	0.21 (s)	760 (liq, 70°C) 900 (s, 20°C)
		1-Dodecanol	26	200	–	–	–
		Paraffin C18	28	244	2.2 (s)	0.148 (liq, 40°C) 0.15 (s)	774 (liq, 70°C) 814 (s, 20°C)
			27.5	243.5			
	Heating	Heating Paraffin C20-C33	48-50	189	–	0.21 (s)	769 (liq, 70°C) 912 (s, 20°C)
		Paraffin C22-C45	58-60	189	2.4 (s)	0.21 (s)	795 (liq, 70°C) 920 (s, 20°C)
		Paraffin wax	64	173.6 266	–	0.167 (liq, 63.5°C) 0.346 (s, 33.6°C) 0.339 (s, 45.7°C)	790 (liq, 65°C) 916 (s, 24°C)
		Polyglycol E6000	66	190	–	–	1085 (liq, 70°C) 1212 (s, 25°C)
		Paraffin C21-C50	66-68	189	–	0.21 (s)	830 (liq, 70°C) 930 (s, 20°C)
		1-Tetradecanol	38	205	–	–	–

	Material					
	Paraffin C16-C28	48-50	189	—	0.21 (s)	765 (liq, 70 °C); 910 (s, 20 °C)
	Biphenyl	71	119.2	—	—	991 (liq, 73 °C); 1166 (s, 24 °C)
	Propionamide	79	168.2	—	—	—
	Naphthalene	80	147.7	2.8 (s)	0.132 (liq, 83.8 °C); 0.341 (s, 49.9 °C); 0.310 (s, 66.6 °C)	976 (liq, 84 °C); 1145 (s, 20 °C)
Fatty acids — Cooling	Propyl palmiate	10; 16-19	186	—	—	—
	Caprylic acid	16; 16.3	148.5; 149	—	0.149 (liq, 38.6 °C); 0.145 (liq, 67.7 °C); 0.148 (liq, 20 °C)	901 (liq, 30 °C); 862 (liq, 80 °C); 981 (s, 13 °C); 1033 (s, 10 °C)
	Isopropyl palmiate	11	95-100	—	—	—
	Capric-lauric acid+pentadecane (90-10%)	13.3	142.2	—	—	—
	Isopropyl stearate	14-18	140-142	—	—	—
	Capric-lauric acid (65-35%)	18.0	148	—	—	—
	Butyl stearate	17-21	143	—	—	—
	Capric-lauric acid (45-55%)	19	140	—	—	—
		21	143	—	—	—
	Dimethyl sabacate	21	120-135	—	—	—
	Octadecyl 3-mencapto-propylate 21	143	—	—	—	—
	Myristic acid-capric acid (34-66%)	24	147.7	—	0.164 (liq, 39.1 °C); 0.154 (liq, 61.2 °C)	888 (liq, 25 °C); 1018 (s)
	Octadecylthioglycate	26	90	—	—	—
Heating	Vinyl stearate	27-29	122	—	—	—

Continued

Table AII.5 Organic substances with potential use as PCM for residential applications—cont'd

Group	Compound	Melting Temperature (°C)	Heat of Fusion (kJ/kg)	C_p (kJ/(kg K))	k^a (W/(mK))	ρ^a (kg/m³)
	Lauric acid	42–44	178	2.3 (liq)	0.147 (liq, 50°C)	862 (liq, 60°C)
		44	177.4	1.7 (s)		870 (liq, 50°C)
	Myristic acid	49–51	204.5	2.4 (liq)	–	861 (liq, 55°C)
		54	187			844 (liq, 80°C)
		58	186.6	1.7(s)		990 (s, 24°C)
	Palmitic acid	64	185.4	2.8 (liq)	0.162 (liq, 68.4°C)	850 (liq, 65°C)
		61	203.4	1.9 (s)	0.159 (liq, 80.1°C)	847 (liq, 80°C)
		63	187		0.165 (liq, 80°C)	989 (s, 24°C)
	Stearic acid	69	202.5	2.2 (liq)	0.172 (liq, 70°C)	848 (liq, 70°C)
		60–61	186.5	1.6 (s)		
		70	203			965 (s, 24°C)

(–), not available.
^{a}s, solid; liq, liquid.

Table AII.6 Thermal properties of some eutectics PCMs suitable for residential applications

Group	Appli-cation	Compound	Composition	Melting Temperature (°C)	Heat of Fusion (kJ/kg)	k^a (W/(mK))	ρ^a (kg/m³)
Inorganics	Cooling	$CaCl_2 \cdot 6H_2O + CaBr_2 \cdot 6H_2O$	45+55	14.7	140	–	–
		$CaCl_2 + MgCl_2 \cdot 6H_2O$	66.3+33.3	25	95	–	1590
	Heating	$Ca(NO_3) \cdot 4H_2O + Mg(NO_3)_3 \cdot 6H_2O$	47+53	30	136	–	–
		$Mg(NO_3)_3 \cdot 6H_2O + NH_4NO_3$	61.5+38.5	52	125.5	0.494 (liq, 65.0°C) 0.515 (liq, 88.0°C)	1515 (liq, 65°C) 1596 (s, 200°C)
		$Mg(NO_3)_3 \cdot 6H_2O + MgCl_2 \cdot 6H_2O$	58.7+41.3	59	132.2	0.510 (liq, 65.0°C) 0.565 (liq, 88.0°C)	1550 (liq, 50°C) 1630 (s, 24°C)
		$Mg(NO_3)_3 \cdot 6H_2O + MgCl_2 \cdot 6H_2O$	50+50	59.1	144	–	–
		$Mg(NO_3)_3 \cdot 6H_2O + Al(NO_3)_2 \cdot 9H_2O$	53+47	61	148	–	1850
		$Mg(NO_3)_2 \cdot 6H_2O + MgBr_2 \cdot 6H_2O$	59+41	66	168	–	–
		$LiNO_3 + NH_4NO_3 + NaNO_3$	25+65+10	80.5	113	–	–
		$LiNO_3 + NH_4NO_3 + KNO_3$	26.4+58.7+14.9	81.5	116	–	–
		$LiNO_3 + NH_4NO_3 + NH_4Cl$	27+68+5	81.6	108	–	–
Organics	Cooling	Triethylolethane+water+urea	38.5+31.5+30	13.4	160	–	–
		$C_{14}H_{28}O_2 + C_{10}H_{20}O_2$	34+66	24	147.7	–	–
	Heating	$CH_3CONH_2 + NH_2CONH_2$	50+50	27	163	–	–
		Triethylolethane+urea	62.5+37.5	29.8	218	–	–
		$CH_3COONa \cdot 3H_2O + NH_2CONH_2$	40+60	30	200.5	–	–
		$NH_2CONH_2 + NH_4NO_3$	53+47	46	95	–	–
		$CH_3CONH_2 + C_{17}H_{35}COOH$	50+50	65	218	–	–
		Naphthalene+benzoic acid	67.1+32.9	67	123.4	–	–
		$NH_2CONH_2 + NH_4Br$	66.6+33.4	76	151	–	–

^{a}s, solid; liq, liquid; (–), not available.

Table AII.7 Thermal energy storage properties of pure fatty acids and form-stable composite

Form-Stable Composite PCM (kJ/kg)	Melting Temperature (°C)	Heat of Melting (kJ/kg)	Freezing Temperature (°C)	Heat of Freezing (kJ/kg)
CA	31.35	172.4	25.86	172.67
LA	44.15	178.74	41.48	177.66
MA	54.39	196.04	51.45	198.3
PA	62.73	217.45	58.82	213.98
PnBMA/CA	29.62	67.23	27.65	67.09
PnBMA/LA	41.69	71.12	38.47	71.02
PnBMA/MA	45.97	77.27	41.79	78.35
PnBMA/PA	53.73	86.34	50.46	85.32

Table AII.8 Sorbents materials for ad/absorption phenomena suitable for heat storage in buildings

Phenomena	Class	Type	Characteristics	Water Level Sorption Capacity (g/g^a) and Heat Storage Capacity	Characterization Level
Adsorption	Mesoporous silicates	Silica gels	Pore dimension: from 2 nm for regular density to 15–20 nm for low density ones	0.40: Silica gel type A, 2.2 nm pore size	Material scale
			Pore volume: from 0.3–0.4 to 1.0–1.5 cm^3/g	0.45: Silica gel type RD, 2.2 pore size	
			Surface area: 300–700 m^2/g		
			Charge: >88 °C	HYDES Project: 450 MJ m^{-3} of silica gel	Reactor scale
			Discharge: 32 °C	MODESTORE Project: 180 MJ m^{-3} of silica gel	System scale
			Commercially available and cheap		
		Silica aerogels	Advantages: large surface, pore volume, open system of pores	1.35: 100% SiO$_2$ with CO$_2$ supercritical drying	Material scale
			Drawbacks: instability of the molecular, structure to moisture, low mechanical, stability, low density	1.25: 100% Al$_2$O$_3$ with CO$_2$ supercritical drying	
				1.15: 70% SiO$_2$–30% Al$_2$O$_3$ with CO$_2$ supercritical drying	
		Ordered mesoporous silicates (MCM, SBA, etc.)	Advantages: mono-sized pores, high structure homogeneity, large pore surface and volume	0.83: MCM-48, 2.8 nm pore diameter, 983 m^2/g, surface area	Material scale
			Drawbacks: Cylindrical or hexagonal pores	0.81: KIT-1, 2.9 nm pore diameter, 923 m^2/g surface area	
				0.44: SBA-1, 2.1 nm pore diameter, 940 m^2/g surface area	
				0.84: SBA-15, 5.0 nm pore diameter, 645 m^2/g surface area	

(Continued)

Table AII.8 Sorbents materials for ad/absorption phenomena suitable for heat storage in buildings—cont'd

Phenomena	Class	Type	Characteristics	Water Level Sorption Capacity (g/g[a]) and Heat Storage Capacity	Characterization Level
	Classical zeolites	Alumina-silicates	Largest known class of crystalline porous solids: about 200 types Too high affinity to water at low ratio of Si/Al	0.20–0.45	Material scale
			Charge: 130°C Discharge: 65°C	Heat: 576 MJ/m³ of zeolite 4A	Reactor scale
	Metal-aluminophosphates		Uniform pore size of 0.3–0.8 nm formed by 6-, 10-, 12-, and 18-membered 0.55: SAPO-37 ring channels	0.35: VPI-5, ring size 18, ring size 12 0.14: VAPO-5, ring size 12	Material scale
			Moderate affinity to water	0.32–0.36: GeAPO-5,MnAPO-5, ring size 12 0.46: SAPO-5, ring size 12 0.46: SAPO-40, ring size 12	
	Metal-organic frameworks		Regeneration at 45–50°C Cold generation at 15°C Low affinity to water		—
	Composite sorbents	Silicates+CaCl₂	Charge: 90°C	1.27: CaCl₂/SiO₂ aerogel (29% CaCl₂) 1.17: CaCl₂/SiO₂ xerogel (28.6% CaCl₂) 0.80: CaCl₂/silica gel (with 33.7% CaCl₂) 0.60: CaCl₂/silica gel (with 24% CaCl₂) Heat: 806 MJ/m³	Material scale

		Material	Charge/Discharge	Heat storage	Scale
		Silicates+LiBr		0.80: Libr/SiO$_2$ aerogel (28.6% LiBr), pore size 7.5 0.76: Libr/densified SiO2 aerogel (28.6% LiBr), pore size 12.9	Material scale
		Zeolite 13X+MgSO$_4$	Charge: 150°C Discharge: 30–50°C	597.6MJ/m^3 (15%wt MgSO$_4$) 640.8MJ/m^3 (10%wt MgSO$_4$)	Material scale
		Expanded natural graphite+SrBr2	Charge: 70–80°C Discharge: 35°C (heating), 18°C (cooling) Charge: 62–65°C	Heat: 216MJ/m^3 of reactor Cooling: 144MJ/m^3 of reactor Power: 2.5kW (heating); 4kW (cooling)	Reactor scale
Absorption	Salts	Expanded	Discharge: 33–36°C	0.4 H$_2$O/g of composite material Heat: 450MJ/m^3 of reactor	Reactor scale
		NaOH/H$_2$O	Charge: 100–150°C Discharge: 40–65°C	Heat: 900MJ/m^3 of NaOH (with single stage reactor)	Reactor scale
		LiCl/H$_2$O	Charge: 46–87°C Discharge: 30°C	Heat: 910.8MJ/m^3 of LiCl (with crystallization in the storage tank)	Reactor scale
		CaCl$_2$/H$_2$O	Charge: 70–80°C Discharge: 21°C	Heat storage: 428 MJ/m^3 of solution	Reactor scale
		LiBr/H$_2$O	Charge: 40–90°C Discharge: 30–33°C	Heat storage capacity: 907.2MJ/m^3 of solution	Reactor scale

(−), not available.
[a]Measurements were performed at 25°C and $P/P_0 = 1.0$.

Table AII.9 Material couples suitable for chemical heat storage in residential applications

Material/Couple	Operating Conditions	Performances	Characterization Level
$MgSO_4 \cdot 7H_2O \rightarrow MgSO_4 + 7H_2O$	Charge: 122-150°C Discharge: 122°C	Heat:1512 MJ/m^3 of $MgSO_4$ (theoretical) Hard recovery	Material scale ECN project: Characterization of the material (experimental tests) sample of 10 mg of material
$MgCl_2 \cdot 6H_2O \rightarrow MgCl_2 \cdot 2H_2O + 4H_2O$	Charge: 115-130°C Discharge: 35°C	Heat: 2170.8 MJ/m^3 of $MgCl_2 \cdot 2H_2O$	Material scale IEC Project: material characterization. Sample of 250 mg of material Stabilization with zeolite 4A to be further considered
$MgCl_2 \cdot 6H_2O \rightarrow MgCl_2 \cdot H_2O + 5H_2O$	Charge: 150°C Discharge: 30-50°C	—	Material scale ECN project: material characterization Sample of 300 g of material
$CuSO_4 \cdot 5H_2O \rightarrow CuSO_4 \cdot H_2O + 4H_2O$	Discharge: 40-60°C (heat supply at Z≥0°C to ignite discharge)	Heat: 2066.4 MJ/m^3 of $CuSO_4 \cdot H_2O$ (theoretical)	ITW project: characterization of 100 mg of material
$CaCl_2 \cdot 2.3H_2O \rightarrow CaCl_2 + 2.3H_2O$	Charge: 150°C Discharge (temperature lift); $\Delta T = 62°C$ (reactor and evaporator both at 25°C) $\Delta T = 10°C$ (reactor at 50°C and evaporator at 10°C)	—	ECN project: sample of 40 g of material
Bentonite + $CaCl_2$	Discharge: 35°C	Heat: 667 MJ/m^3 of composite material	ITW Project: Material characterization
$KAl(SO_4)_2 \cdot 12H_2O \rightarrow KAl(SO_4)_2 \cdot 3H_2O + 9H_2O$	Charge: 65°C Discharge: 25 °C	Heat: 864 MJ/m^3 of $KAl(SO_4)_2 \cdot 3H_2O$	Reactor scale PROMES CEA-INES Project: prototype of 25 kg of $KAl(SO_4)_2 \cdot 12H_2O$

Reaction	Temperature	Heat/Cold	Scale	Notes
$Al_2(SO_4)_3 \cdot 18H_2O \rightarrow Al_2(SO_4)_3 \cdot 5H_2O + 13H_2O$	Charge: 150°C Discharge (temperature lift): $\Delta T = 9.8°C$ (reactor and evaporator both at 25°C) $\Delta T \sim 1°C$ (reactor at 50°C and evaporator at 10°C)	–	Reactor scale	ECN Project: Sample of 40g of material
$Na_2S \cdot 5H_2O \rightarrow Na_2S \cdot 1.5H_2O + 4.5H_2O$	Charge: 83°C Discharge: 35°C	Heat: 2808 MJ/m³ of $Na_2S \cdot 5H_2O$ Cold: 1836 MJ/m³ of $Na_2S \cdot 5H_2O$	Reactor scale	ECN project: SWEAT prototype, 3kg of material Short-term heat and cold storage
$SrBr_2 \cdot 6H_2O \rightarrow SrBr_2 \cdot H_2O + 5H_2O$	Charge: 70–80°C Discharge: 35°C	Heat: 216 MJ/m³ of $SrBr_2 \cdot H_2O$ Possible cold recovery at 18°C	Reactor scale	PROMES CEA-INES Project: SOLUX prototype, 170kg of $SrBr_2 \cdot H_2O$
$CaCl_2 \cdot 2H_2O \rightarrow CaCl_2 \cdot H_2O + H_2O$	Charge: 95°C Discharge: 35°C	Heat: 720 MJ/m³ of $CaCl_2$	Reactor scale	BEMS: theoretical study

(–), not available.

Table AII.10 Published data on potential sensible heat storage materials

T_{cold} (°C)	T_{hot} (°C)	Material	Thermal Conductivity (W/mK)	Density (kg/m³)	Average Specific Heat Capacity C_p (kJ/kg K)	Volumetric Specific Heat Capacity (kWh$_{th}$/m³)	Type of Medium
200	300	Sand-rock-oil	1	1700	1.3	60	Solid
200	400	Reinforced concrete	1.5	2200	0.85	100	Solid
200	400	Cast iron	37	7200	0.56	160	Solid
200	500	NaCl	7	2160	0.85	150	Solid
200	700	Cast steel	40	7800	0.6	450	Solid
200	700	Silica fire bricks	1.5	1820	1	150	Solid
200	1200	Magnesia fire bricks	5	3000	1.15	600	Solid
250	350	Synthetic oil	0.11	900	2.3	57	Liquid
250	450	Nitrite salts	0.57	1825	1.5	152	Liquid
270	530	Liquid sodium	71	853	1.3	80	Liquid
300	400	Silicone oil	0.1	900	2.1	52	Liquid
180	1300	Lithium liquid salt	38.1	510	4.19		Liquid
15	400	Dowtherm A	0.1171 at 155°C	867	2.2		Liquid
0	345	Therminol 66		750	2.1		Liquid

Table AII.11 Published data on potential latent heat storage materials

T_{melt} (°C)	Material	Latent Heat of Fusion (J/g)	Thermal Conductivity (W/mK)
307	$NaNO_3$	177	0.5
318	77.2 mol% NaOH-16.2% NaCl-6.6% Na_2CO_3	290	
320	54.2 mol% LiCl-6.4% $BaCl_2$-39.4% KCl	170	
335	KNO_3	88	0.5
340	52 wt% Zn-48% Mg	180	
348	58 mol% LiCl-42% KCl	170	
380	KOH	149.7	0.5
380	45.4 mol% $MgCl_2$-21.6% KCl-33% NaCl	284	
381	96 wt% Zn-4% Al	138	
397	37 wt% Na2CO$_3$-35% K_2CO_3-Li_2CO_3	275	2.04
443	59 wt% Al-35% Mge6% Zn	310	
450	48 wt% NaCl-52% $MgCl_2$	430	0.96
470	36 wt% KCle64% $MgCl_2$	388	0.83
487	56 wt% Na_2CO_3-44% Li_2CO_3	368	2.11
500	33 wt% NaCl-67% $CaCl_2$	281	1.02
550	LiBr	203	
632	46 wt% LiF-44% NaF_2-10% MgF_2	858	1.20
660	Al	398	250
714	$MgCl_2$	452	

Table AII.12 Materials that have been characterized for their thermal storage capability

Material	Measured			Theoretical	
	Latent Heat (kJ/kg)	T_{melt} (°C)	Purity of Material Tested	Latent Heat (kJ/kg)	T_{melt} (°C)
29 mol% Zn/71% Mg eutectic alloy	138	343	Commercially pure	247–464	–
37.5 mol% Mg/62.5% Al eutectic alloy	310	451	Commercially pure	458–477	–
17.5 mol% Cu/82.5% Al eutectic alloy	351	548	Commercially pure	359–380	–
13 mol% Si/87% Al eutectic alloy	515	579	Commercially pure	571	–
17 mol% Cu/16.2% Mg/Al ternary eutectic alloy	360	506	Commercially pure	400–406	–
12.6 mol% Cu/5.1% Mg/Al ternary eutectic alloy	545	560	Commercially pure	449–549	–
59 mol% LiCl/41% KCl	234.6	352	Reagent grade vacuum dried for 6h.	242–272, 170	348
KNO3	98.9	337	Analytical reagent grade, vacuum dried for 16h.	95.2–120.4	337
58.7 mol% LiCl/41.3% KCl	218.32	354.4	99.5% pure KCl/99% pure LiCl	234.9	355
95.5 wt% KNO$_3$–4.5% KCl	82.86	319.68	>99%	74	320
80.69 wt% KNO$_3$–7.44% KCl-11.87% KBr	76.6	N/Aa	>99%	140	342
34.81 wt% NaCl-32.29% KCl-32.9% LiCl	138	352.89	>99%	281	346
60 wt% MgCl$_2$–20.4% KCl-19.6% NaCl	199	387.60	>99%	400	380

Table AII.13 Container materials used for high temperature salts

Salt Used	Container Material	Operating Temperatures
KNO_3; KNO_3/KCl; $NaNO_3$ KNO_3/$NaNO_3$/$Ca(NO_3)_2$ binary and ternary mix	AISI 1015 (PCM) AISI K01200 (Tank) 304 and 316SS and A36 carbon steel	T_{min}−270°C; T_{min}−350°C 570°C for SS; 316°C for C steel
KNO_3/$NaNO_3$/$Ca(NO_3)_2$	316SS	450 and 500°C
$MgCl_2$/KCl	316SS and high nickel alloys	850°C for 100h
Fluoride salt eutectics	Inconel 617	727°C for 20,000-30,000h

Table AII.14 Comparison of various heat storage media (for sensible storage materials, energy is stored in the temperature range 25-75°C)

	Heat Storage Material			
	Sensible Heat Storage		Phase Change Materials	
Property	Rock	Water	Paraffin Wax	$CaCl_2 \cdot 6H_2O$
Latent heat of fusion (kJ/kg)	[a]	[a]	174.4	266
Specific heat capacity (kJ/(kg K))	0.9	4.18	–	–
Density (kg/m³) at 24°C	2240	1000	1802	795
Storage volume for storing 1 GJ (m³)	9.9	4.8	3.2	4.7
Relative volume[b]	3.1	1.5	1.0	1.5

(–) Specific heat capacity is not of interest for latent heat storage.
[a]*Latent heat of fusion is not of interest for sensible heat storage.*
[b]*Equivalent storage volume, reference taken on paraffin.*
Sources: Tatsidjodoung P, Pierrès NL, Luo L. A review of potential materials for thermal energy storage in building applications. Renew Sustain Energy Rev 2013;18:327–49; Kuravi S, Trahan J, Goswami DY, Rahman MM, Stefanakos EK. Thermal energy storage technologies and systems for concentrating solar power plants. Prog Energy Comb Sci 2013;39:285–319.

Rules of Thumb for Thermal Energy Storage Systems Design

Thermal energy storage using solid storage media:

- Space heating—300-500 kg or rock per square meter of collector area
- Size of rock piece—1-5 cm

Thermal mass design:

- Mass surface to glass area ratio—6:1
- Most effective thickness in masonry materials—First 100 mm
- Most effective thickness in wood—First 25 mm

Solar heating sizing:

- Total collector area—10% of floor area
- Single glazed flat plate collector yields 2 times ambient temperature
- Double glazed and vacuum tube collector yields 5 times ambient temperature
- Storage size—60 L per square meter of collector area
- Collector field temperature difference—55-85 °C (absence of sensible storage)—140 °C (presence of sensible storage)
- Collector operating temperature—½ (collector field temperature difference above the saturation steam temperature)

Large seasonal storage:

- Minimum recommended load: 500 MWh
- Water storage—1.5-2.5 m² solar collectors per MWh of load for 40-60% of load
  1-2 m³ per m² of solar collector
  Thermal insulation: minimum 40 cm at 0.04 W/mK
  Solar productivity: 200-300 kWh/m² in mid-Europe climate
- Duct storage—1.5-3 m² solar collectors per MWh of load for 40-60% of load
  2-6 m³ of storage per m² of solar collectors
  minimum 20,000 m³ if insulation only on top 15-50 W/m of borehole (double U-pipe, quartz bentonite filling)
  Storage cost: 30-60 Euro/m of borehole
- Aquifer storage—1.52-6 m³ of water per m² of collectors
  Minimum 50,000 m³ of storage for a no heat pump system
  Minimum depth: 10-20 m
  Should preferably be used for cold storage
  Storage cost: 5-20 Euro/m³ but very dependent on local conditions
  A strong regional groundwater flow (1 m/month) can ruin the store

Ice thermal energy storage:

- For 500 sq. ft per ton—70 sq. ft tank cools about 10,000 sq. ft of floor space
- For full storage—70 sq. ft tank cools about 7000 sq. ft of floor space
- 144 Btu/lb @ 32 °F + 0.48 Btu/lb for each 1 °F below 32 °F
- 3.2 cu. ft/ton h
- Only the latent heat capacity of ice should be accounted for when designing ice storage systems
- Encapsulated ice—17-22 gal/ton h
- Ice on coil—18-26 gal/ton h

Chilled water thermal energy storage:

- For 10 °F ΔT—19.3 cu. ft/ton h
  623.1 Btu/cu. ft; 83.3 Btu/gal
- For 12 °F ΔT—16.1 cu. ft/ton h
  747.7 Btu/cu. ft; 100.0 Btu/gal
- For 16 °F ΔT—12.4 cu. ft/ton h
  996.9 Btu/cu. ft; 133.3 Btu/gal
- For 20 °F ΔT—9.6 cu. ft/ton h
  1246.2 Btu/cu. ft; 166.7 Btu/gal

Indoor space heating storage:

- For one degree reduction in indoor temperature—5% heating cost savings can be obtained

SOURCES

Ataer OE. Energy storage systems—storage of thermal energy. Encyclopedia of Life Support Systems (EOLSS). http://www.powerfromthesun.net/Book/chapter15/chapter15.html.

Harrigan, RW. Handbook for the conceptual design of parabolic trough solar energy systems. Process heal applications, SAND81-0763. Albuquerque: Sandia National Laboratories. July; 1981.

Keenan JH, Keyes FG, Hill PG, Moore JG. Steam tables: thermodynamic properties of water including vapor, liquid and solid phases. New York: John Wiley & Sons; 1978.

Hadorn J-C. Storage solutions for solar thermal energy. Freiburg Solar Academy 2004. www.Calmac.com.

Arthur A, Bell JR. HVAC equations data and rules of thumb 2006.

Doty S. Commercial energy auditing reference handbook. 2nd ed. Fairmont Press; 2010.

Parametric and Cost Comparison of Thermal Storage Technologies

Table AIV.1 Typical Parameters of Thermal Energy Storage Systems

TES System	Capacity (kWh/t)	Power (MW)	Efficiency (%)	Storage Period (h, d, m)	Cost (€/kWh)
Sensible (hot water)	10-50	0.001-10	50-90	d/m	0.1-10
PCM	50-150	0.001-1	75-90	h/m	10-50
Chemical reactions	120-250	0.01-1	75-100	h/d	8-100

Table AIV.2 Economic Viability of TES Systems as a Function of the Number of Storage Cycles Per Year

TES type	Cycles Per Year	5-Year Energy Savings (kWh)	5-Year Economic Savings (€)	Investment Cost (€/kWh)
Seasonal storage	1	500	25	0.25
Daily storage	300	150,000	7500	75
Short-term storage (3 c/day)	900	450,000	22,500	225
Buffer storage (10 c/day)	3000	1,500,000	75,000	750

Table AIV.3 State of Development, Barriers, and Main R&D Topics for Different TES Technologies

Technology	Status (%) Market/R&D	Barriers	Main R&D Topics
Sensible Thermal Energy Storage			
Hot water tanks (buffers)	95/5		Super insulation
Large water tanks (seasonal)	25/75	System integration	Material tank, stratification
UTES	25/75	Regulation, high cost, low capacity	System integration

Continued

Table AIV.3 State of Development, Barriers, and Main R&D Topics for Different TES Technologies—cont'd

Technology	Status (%) Market/R&D	Barriers	Main R&D Topics
High temperature solids	10/90	Cost, low capacity	High temperature materials
High temperature liquids	50/50	Cost, temperature <400°C	Materials
Latent Thermal Energy Storage (PCM)			
Cold storage (ice)	90/10	Low temperature	Ice production
Cold storage (other)	75/25	High cost	Materials (slurries)
Passive cooling (buildings)	75/25	High cost, performance	Materials (encapulation)
High temperature PCM (waste heat)	0/100	High cost, mat. stability	Materials (PCM containers)
Thermochemical Energy Storage			
Adsorption TES Absorption TES Other chemical reactions	5/95	High cost, complexity	Materials and reactor design

Table AIV.4 Estimated Cost of 3600 MWh$_t$ Two-Tank/Thermocline TES System (Approximately 12 h of TES for 100-MWe Plant)

Component	Indirect Storage System Two-Tank	Direct Storage Systems Two-Tank	Thermocline
Solar field HTF, type	Therminol	HitecXL	HitecXL
Outlet temperature (°C)	391 (736°F)	450 (842°F)	450 (842°F)
Storage fluid, type	Solar salt	HitecXL	HitecXL
Fluid cost, (k USD)	51,200	71,200	26,000
Filler material, type	NA	NA	Quartzite
Filler cost, (k USD)	0	0	8700
Tank(s), number	3 Hot, 3 Cold	2 Hot, 2 Cold	2 Thermocline
Tank cost, (k USD)	23,400	18,200	12,100
Salt-to-oil heat exchanger, (k USD)	9000	0	0
Piping/solar field heat tracing	0	10,600	10,600
Total (k USD)	91,900	108,900	62,000
Specific cost (USD/kWh$_t$)	26	33	19
Development status	Early commercial	Prefeasibility study	Prefeasibility study

Sources: Thermal Energy Storage Technology Brief, IEA-ETSAP and IRENA© Technology Brief E17, 2013 (www.etsap. org—www.irena.org); Assessment of thermal energy storage for parabolic trough solar power plants (http://www.nrel.gov/ csp/troughnet/pdfs/kearney_tes_overview.pdf).

Summary of Thermal Energy Storage Systems Installation

Table AV.1 Operational Solar Thermal Facilities with Thermal Energy Storage Systems

Project	Type	Storage Medium	Nominal Temperature (°C)		Storage Concept	Plant Capacity	Storage Capacity
			Cold	Hot			
SSPS-DCS test facility Almeria, Spain	Parabolic trough	Santotherm 55	225	295	1 tank thermocline	1.2 MWth	5 MWht
Nevada Solar One Nevada, USA	Parabolic trough	Dowtherm A	318	393	Oversized field piping	64 MWe	0.5 h
Holaniku at Keahole Point Hawaii, USA	Parabolic trough	Water	n.a.[a]	200	Indirect storage	2 MWth, 500 kWe	2 h
Planta Solar-10 Sevilla, Spain	Central receiver	Pressurized water	240	260	Steam accumulator	11 MWe	50 min/ 20 MWht
Planta Solar-20 Sevilla, Spain	Central receiver	Pressurized water	n.a.	250–300	Steam accumulator	20 MWe	50 min
La Florida Badajoz, Spain	Parabolic trough	Molten solar salt[b]	292	386	2-tank Indirect	50 MWe	7.5 h
Andasol-1 Granada, Spain	Parabolic trough	Molten solar salt	292	386	2-tank Indirect	50 MWe	7.5 h/ 1010 MWht
Andasol-2 Granada, Spain	Parabolic trough	Molten solar salt	292	386	2-tank Indirect	50 MWe	7.5 h/ 1010 MWht
Extresol-1 Badajoz, Spain	Parabolic trough	Molten solar salt	292	386	2-tank Indirect	50 MWe	7.5 h/ 1010 MWht
Manchasol-1 Ciudad Real, Spain	Parabolic trough	Molten solar salt	292	386	2-tank Indirect	50 MWe	7.5 h
Manchasol-2 Ciudad Real, Spain	Parabolic trough	Molten solar salt	292	386	2-tank Indirect	50 MWe	7.5 h
La Dehesa Badajoz, Spain	Parabolic trough	Molten solar salt	292	386	2-tank Indirect	50 MWe	7.5 h

Continued

Table AV.1 Operational Solar Thermal Facilities with Thermal Energy Storage Systems—cont'd

Project	Type	Storage Medium	Nominal Temperature (°C)		Storage Concept	Plant Capacity	Storage Capacity
			Cold	Hot			
Puerto Errado 1 Murcia, Spain	Linear Fresnel	Saturated steam	n.a.	270	Steam accumulator	1.4 MWe	n.a.
Archimede Sicily, Italy	Parabolic trough	Molten solar salt	290	550	2-tank direct	5 MWe	8 h/ 100 MWht
Torresol Gemasolar Seville, Spain	Central receiver	Molten solar salt	292	565	2-tank direct	17 MWe	15 h
Dahan Beijing, China	Central receiver	Saturated steam/oil	290	350	Combined steam accumulator/ concrete	1 MWe	1 MWht

[a]n.a., not available.
[b]Molten solar salt, 60% sodium nitrate/40% potassium nitrate.

Table AV.2 List of Cool Thermal Energy Storage Projects/Installations

Name	Technology	Rated Power (kW)	Location	Status
Redding Electric Utilities— Peak Capacity, Demand Response, HVAC Replacement Program	Ice Thermal Storage	1000	Redding, California, United States	Operational
Glendale Water and Power—Peak Capacity Project	Ice Thermal Storage	1500	Glendale, California, United States	Operational
Southern California Edison—HVAC Optimization Program with energy storage	Ice Thermal Storage	750	Rosemead, California, United States	Operational
St. Kilian Parish and School	Ice Thermal Storage	100	7076 Franklin Road, Cranberry Township, Pennsylvania 16066, United States	Operational
Bethel Park High School	Ice Thermal Storage	375	309 Church Road, Bethel Park, Pennsylvania 15102, United States	Operational
Duquesne University	Ice Thermal Storage	600	600 Forbes Avenue, Pittsburgh, Pennsylvania 15282, United States	Operational

Table AV.2 List of Cool Thermal Energy Storage Projects/Installations—cont'd

Name	Technology	Rated Power (kW)	Location	Status
University of Arizona	Ice Thermal Storage	3000	1339 E. Helen Street, Tucson, Arizona 85717, United States	Operational
El Capitan	Ice Thermal Storage	150	6834 Hollywood Boulevard, Los Angeles, California 90028, United States	Operational
Mission City Office Complex	Ice Thermal Storage	500	2365 Northside Drive, San Diego, California 92108, United States	Operational
Fossil Ridge High School	Ice Thermal Storage	200	5400 Ziegler Road, Fort Collins, Colorado 80528, United States	Operational
O-I World Headquarters	Ice Thermal Storage	250	Perrysburg, Ohio 43551, United States	Operational
Redding Electric Utilities— Peak Capacity, Demand Response, HVAC Replacement Program Phase 2	Ice Thermal Storage	6000	Redding, California, United States	Contracted
Nissan Technical Center North America Inc	Ice Thermal Storage	1425	Farmington Hills, Michigan 48331, United States	Operational
JC Penney Headquarters	Ice Thermal Storage	4425	6501 Legacy Drive, Plano, Texas 75301, United States	Operational
Encinitas Civic Center	Ice Thermal Storage	75	1140 Oakcrest Park Dr., Encinitas, California 92024, United States	Operational
Nordstrom, Inc.	Ice Thermal Storage	1200	1450 Ala Moana Blvd., Honolulu, Hawaii 96814, United States	Operational
Shell Point Retirement Village	Ice Thermal Storage	4800	15071 Shell Point Blvd., Fort Myers, Florida 33908, United States	Operational
CLPCCD—Utility Infrastructure Project	Ice Thermal Storage	890	25555 Hesperian Blvd., Hayward, California 94545, United States	
SCPPA Thermal Energy Storage Program	Ice Thermal Storage	2427	1160 Nicole Court, Glendora, California, United States	Operational

Continued

Table AV.2 List of Cool Thermal Energy Storage Projects/Installations—cont'd

Name	Technology	Rated Power (kW)	Location	Status
1500 Walnut	Ice Thermal Storage	210	1500 Walnut Street,, Philadelphia, Pennsylvania 19102, United States	Operational
The State of North Carolina	Chilled Water Thermal Storage	2590	Salsbury St., Raleigh, North Carolina 27601, United States	Operational
University of Central Florida	Chilled Water Thermal Storage	3000	4000 Central Florida Blvd., Orlando, Florida, United States	Operational
Cache Creek Casino	Chilled Water Thermal Storage	1300	14455 California 16, Brooks, California 95606, United States	Operational
VA Medical Center	Chilled Water Thermal Storage	2300	4500 South Lancaster Road, Dallas, Texas 75216, United States	Operational
Lackland Air Force Base	Chilled Water Thermal Storage	580	1030 Reese, San Antonio, Texas 78299, United States	Operational
Geisinger Health System	Chilled Water Thermal Storage	700	100 North Academy Avenue, Danville, Pennsylvania 17822, United States	Operational
University of Texas Pan-Am	Chilled Water Thermal Storage	875	1201 W. University Dr., Edinburg, Texas 78539, United States	Operational
Texas Instruments Manufacturing Plant	Chilled Water Thermal Storage	6400	Dallas, Texas, United States	Operational
Federal Government Facility Chilled Water TES	Chilled Water Thermal Storage	274	Quantico, Virginia, United States	Operational
American Online Data Center	Chilled Water Thermal Storage	1500	Chantilly, Virginia, United States	Operational
Disney California Adventure	Chilled Water Thermal Storage	2000	1313 Disneyland Dr., Anaheim, California 92802, United States	Operational

Sources: Kuravi S, Trahan J, Goswami DY, Rahman MM, Stefanakos EK. Thermal energy storage technologies and systems for concentrating solar power plants. Prog Energy Comb Sci 2013;39:285-319; http://www.energystorageexchange.org/projects.

Abbreviations

A/C	air conditioning
AD	anaerobic digestion
AFM	atomic force microscopy
AgNP	silver nanoparticles
AHU	air handling unit
ANN	artificial neural network
ASHRAE	American Society of Heating Refrigerating and Air Conditioning Engineers
ATES	aquifer thermal storage
BREEAM	Building Research Establishment's Environmental Assessment Method
BSRIA	Building Services Research and Information Association
BTES	bore-hole thermal storage
BTM	building thermal mass
CAES	compressed air energy storage
CASBEE	Comprehensive Assessment System for Building Environmental Efficiency
CAV	constant air volume
CCGT	combined cycle gas turbine
CCU	centralized controlling and interface unit
CES	cryogenic energy storage
CFB	circulating fluidized bed
CFD	computational fluid dynamics
CHP	combined heat and power
CHW	chilled water
CLTD	cooling load temperature difference
CO_2	carbon dioxide
COP	coefficient of performance
CRF	capital recovery factor
CSP	concentrated solar power
CT	cooling tower
CTES	cool thermal energy storage
CTF	conduction transfer function
CWS	chilled water storage
DALY	disability adjusted life years
DCV	demand controlled ventilation
DD	de-ionized double distilled water
DFIG	doubly fed induction generator
DFL	direct feedback linear control
DSC	differential scanning calorimetry
EACE	ecologically allowable CO_2 emissions
ECBMR	enhanced coal bed methane recovery
ECV	economizer cycle ventilation
EDAX	energy dispersive X-ray spectroscopy

EDLC	electrochemical double layer capacitors
EES	electrical energy storage
EGR	enhanced gas recovery
EIA	energy Information Administration
EM	explicit time stepping marching
EOR	enhanced oil recovery
ERU	energy recovery unit
ESA	exponential set-point equation-based semi-analytical
ETAP	environmental Technologies Action Plan
ETS	energy transfer station
EU-ETS	European Union emission trading scheme
EV	evaporator
EX	expansion valve
FCU	fan coil unit
FDM	finite difference method
FEM	finite element method
FESEM	field emission scanning electron microscope
FG	fixed grid
FHS	floor heating system
FHU	floor heating unit
FLC	fuzzy logic control
FTIR	Fourier transform infrared spectroscopy
FVM	finite volume method
GA	genetic algorithm
GDP	gross domestic product
GHG	greenhouse gas
G-S	Gauss-Seidel iterative method
GTEC	global total energy consumption
GWP	global warming potential
HDPE	high density polyethylene
HEV	hybrid electric vehicles
HEX	heat exchanger
HP	heat pump
HPRSS	high-pressure reactive solvent scrubbing
HRTEM	high resolution transmission electron microscope
HTESS	hybrid thermal energy storage system
HTF	heat transfer fluid
HTM	heat transfer medium
HVAC	heating, ventilation, and air conditioning
HyNC	hybrid nanocomposite
IAQ	indoor air quality
IC	individual contribution to the EACE
IEA	International Energy Agency
IPCC	Intergovernmental Panel on Climate Change

ITES	ice thermal energy storage
LCA	life cycle assessment
LCI	life cycle inventory
LCOE	levelized cost of energy
LCT	low carbon technology
LEED	leadership in energy and environmental design
LHES	latent heat energy storage
LTES	latent thermal energy storage
MD	molecular dynamics
MM	moving mesh
MPC	model-based predictive control
MPCM	microencapsulated phase change material
MWCNT	multiwalled carbon nanotube
NTES	nanoparticles embedded latent thermal energy storage
NZEB	net zero energy buildings
NZEIT	Net Zero Environmental Impact Times
OECD	Organization for Economic Co-operation and Development
ORC	Organic Rankine cycle
PC	pulverized coal
PCM	phase change material
PHS	pumped hydro energy storage
PID	proportional, integral, and derivative
PMAC	pulse modulation adaptive controller
PMSG	Permanent Magnet Synchronous Generator
PMV	predicted mean vote
PRAC	pattern recognition adaptive controller
PSA	particle size analyzer
PV	photovoltaics
RC	resistant and capacitors
ROI	return on investment
SA	semi-analytical
SAED	selective area electron diffraction
SAT	supply air temperature
SeTES	seasonal thermal energy storage
SMES	superconductive magnetic energy storage
ST	storage tank
STES	sensible thermal energy storage
TCES	thermochemical energy storage
TDMA	tri-diagonal matrix algorithm
TEM	transmission electron microscope
TES	thermal energy storage
TG-DTA	thermogravimetric-differential thermal analysis
TPSC	two parameter switching control
TTM	temperature transforming model

UCG	underground coal gasification
USGBC	U.S. Green Building Council
UTES	underground thermal energy storage
VAV	variable air volume
VVM	variable viscosity of the medium
WA	weighted-averaging
XRD	X-ray diffraction
ZEB	zero energy building

Glossary

Absorption In chemistry, absorption is a process of taking up a substance by another substance. Example: water absorption by cotton.

Absorption In physics, absorption is to take up light, heat, or energy by molecules and converting it to heat.

Active storage systems Thermally isolated systems from the environment that otherwise would interact with it.

Adsorption Phenomenon of adhesion of molecules (liquid, gas, or dissolved substance) on to the surface of a material.

Aquifer Underground reservoir of groundwater as a geological formation that can supply with wells and springs.

Base-load The minimum electric power required for a given time at a steady rate on around the clock basis.

Batteries Storage devices for electricity that convert chemical energy to electrical energy by the reaction between the electrodes and the electrolyte in the device.

Building services The infrastructure comprising lighting, ventilation, temperature, and water facilities for a habitable building.

Carbon content Quantity of carbon present in various redox states in the material that is released during combustion in the form of carbon dioxide or carbon monoxide.

Carbon footprint Gaseous emissions from human production or consumption activities leading to climate changes.

Carbon nano-tubes Small graphite structures that have the capability to bind and store hydrogen.

Carbon neutral Any process or product that does not emit carbon dioxide to the atmosphere over its lifetime.

Carbon sink These are the systems, natural or man-made, that can absorb and store carbon dioxide from the atmosphere. Examples: Trees, plants, oceans.

Charging Storing cooling or heating capacity by removing heat from a cool storage device or adding heat to heat storage device, respectively.

Chiller priority Partial storage system using the chiller to meet maximum load by functioning at the design capacity as a control strategy. On load exceeding capacity of chiller, thermal storage supplements the operation of chiller.

Climate change Long-term and significant change in the weather of a region or earth.

Cogeneration Simultaneous generation of electrical energy and any other form of thermal energy from the same source of energy and used for heating, cooling, and other industrial application.

Combined heat and power Generation of electricity and waste heat during energy conversion and that is used to cater to heating demands and distributions on site.

Convection Transfer of heat energy in a fluid through mass transfer due to some portions being buoyant because of their higher temperature.

Cooling Cooling of a space occupied in comparison to other object in a building.

Coolth Stored energy due to the internal energy difference between a comparatively cooler medium and the surroundings.

Decarbonization Elimination of carbon from fossil fuel energy systems either before or after combustion by chemical separation processes, or separation of flue gases, respectively.

Decentralized energy technologies End use devices or production, conversion, and transfer of energy that operate independent of a large-scale energy system.

Demand limiting Limiting the capacity of HVAC equipment during on-peak period using a partial storage strategy.

Distribution system A part of the energy system to provide useful, low voltage energy to end users.

District heating/cooling The process in which a centralized energy system generates and distributes energy to various locations using energy carrier for service delivery.

Emission Release of particles or gaseous substances into the surroundings from a commercial, industrial, residential, or motor vehicle source.

Energy demand Energy consumed by the building's installed system to maintain a habitable indoor environment.

Energy efficiency Consuming less energy to generate the same level of energy service.

Energy intensity It is the output energy per unit of input energy for an energy conversion system and is the inverse of energy efficiency.

Energy management Computerized control of the operation characteristics of energy using equipment and processes leading to maximum energy efficiency.

Energy security Protection of energy sources and infrastructure from natural, economic, and/or political risk factors.

Energy storage Retention of energy at a point of time for utilization in future.

Energy Capacity to do work (in the form of potential energy) or the capability of motion (kinetic energy).

Full storage Cool storage catering to the cooling load requirement during a predefined on-peak demand period with discharge from the thermal storage system.

Geothermal energy Heat energy obtained from underneath the earth's crust.

Global warming Increase in the average temperature of the earth's atmosphere leading to changes in the climatic conditions.

Greenhouse effect Phenomenon by which certain gases called greenhouse gases allow sunlight to pass through, but absorb the heat energy radiated back from the earth's surface.

Heat exchangers Devices that transfer heat energy from region of higher temperature to the region of lower temperature.

Heat storage Retention of thermal energy above the normal temperature of the process or area.

Ice-on-coil (ice-on-pipe) Technology to store ice on the outside of tubes or pipes that is present in an insulated water tank.

Intermittency of renewable energy sources Energy source that is available in a fluctuating manner due to disturbances of natural services of terrestrial or solar systems.

Load profile Instantaneous thermal load compilation over a period of 24 h.

Mass storage Retention of energy as sensible heat that is stored in buildings, interior equipments, and furnishings.

Nanoscience Study of unique properties of matter or materials at the nanoscale level.

Nanofluidics Science or engineering dealing with the flow of fluid (liquid or gas) through spaces that are at the nanoscale level.

Nanocomposite Combination of two or more materials of which at least one has a nanoscale measurement and is dispersed throughout the other solid material.

Nanoscale The size representation of particles or structures in the range of 1-100 nm for the study of many important fundamental phenomena and properties.

Net-zero energy consumption buildings (ZEB or NZEB) Buildings having highly efficient energy performance and very little or no energy that has been produced off site.

Nominal storage capacity Theoretical capacity usually greater than the usable storage capacity in thermal storage devices. This is not comparable with the usable capacities of alternative storage systems.

On-peak demand period Time period when electrical grid demands and power cost are higher with added demand charges by the supplying utility.

Passive storage systems Systems in constant contact and interaction with the environment thermally.

Power quality Electricity that is free from variations in voltage, current, or frequency providing a smooth flow for advantageous ulitization for industrial and other customers.

Power Rate of transfer of energy, which is a measure of capacity.

Primary energy consumption The crude energy utilized directly from the source without any conversion process.

Renewable energy Energy that cannot be depleted permanently and can be replenished by the natural cycles of earth.

Solar energy The energy provided by the sun. Photovoltaic panels or concentrated solar thermal plants convert this energy to electricity.

Solar thermal collectors These are the devices that collect the sun's heat energy and convert it into electricity.

Storage cycle Complete charging and discharging period starting and completing at the same rate in a thermal storage device.

Sustainability Catering to the present requirements without compromising the future generation ability to meet their needs.

Sustainable technologies Reliable and sustained technologies without any damage to the environment for future generations.

System capacity The maximum cooling capacity provided by the chillers, thermal storage, and the entire cooling system.

Thermal comfort The feeling or the expression of satisfactory mind state with the thermal environment.

Thermal mass Temperature regulating material in a building having high specific heat capacity.

Underground thermal energy storage Storage of thermal energy in the ground beneath a building that can be recouped later for further use.

Waste heat Energy lost in the form of heat during energy conversion or transfer and not available for use.

Water walls Water-filled containers for collecting and controlling solar energy using passive heating and cooling technique.

Zoned air-conditioning Systems with temperature control independently available for each area due to the presence of thermostat controls in these parts.

Zoned building Various zones present in the building are conditioned as per their needs and consist of passive and active building models.

List of Specific Websites

www.ashrae.org
www.baseconsultants.com/IEA32
www.berg-group.com
www.carbontrust.co.uk
www.chicago-bridge.com
www.clipsol.com
www.cristopia.com
www.dunham-bush.com
www.epri.com
www.eren.doe.gov/femp
www.hoval.com
www.ines-solaire.com
www.muel.com
www.preload.com
www.rubitherm.com
www.sce.com/pls
www.solarch.ch
www.solarenergy.ch
www.soltop.com
www.solvis.com
www.sunwell.com
www.tampatank.com
www.trane.com
www.waffle-crete.com
http://ec.europa.eu/energy/renewables/targets_en.htm
http://lccdcasestudies.usgbc.org
http://main.hvac.chalmers.se/cshp/default.htm
http://pcmenergy.com
http://www.breeam.org/

http://www.calmac.com
http://www.climator.com
http://www.cristopia.com
http://www.doerken.de
http://www.ecologyandsociety.org/vol16/iss1/art22/
http://www.solarthermalworld.org/content/underground-thermal-energy-storage-2011
http://www.elle-kilde.dk/altener-combi/
http://www.epsltd.co.uk
http://www.icax.co.uk
http://www.iea.org/
http://www.iea-eces.org/energy-storage/storage-techniques/underground-thermal-energy-storage.html
http://www.iea-shc.org/task26/
http://www.ise.fraunhofer.de
http://www.mfc.co.jp
http://www.nrel.gov/
http://www.pcmproducts.net
http://www.rubitherm.de
http://www.solar-district-heating.eu/Portals/0/Factsheets/SDH-WP3_FS-7-2_Storage_version1.pdf
http://www.teappcm.com
http://www.top50-solar.de/
http://www.worldgbc.org/
https://www.bsria.co.uk/

Index

Note: Page numbers followed by *f* indicate figures, and *t* indicate tables.

Printed and bound by CPI Group (UK) Ltd, Croydon, CR0 4YY

08/05/2025

01864891-0003